# ADVANCES IN FUZZY SETS, POSSIBILITY THEORY, AND APPLICATIONS

# ADVANCES IN FUZZY SETS, POSSIBILITY THEORY, AND APPLICATIONS

Edited by

Paul P. Wang

*Duke University*
*Durham, North Carolina*

PLENUM PRESS • NEW YORK AND LONDON

Library of Congress Cataloging in Publication Data

Main entry under title:

Advances in fuzzy sets, possibility theory, and applications.

Includes bibliographical references and index.
1. Fuzzy sets—Addresses, essays, lectures. I. Wang, Paul P.
QA248.A39 1983 511.3'2 83-11077

ISBN-13: 978-1-4613-3756-0 e-ISBN-13: 978-1-4613-3754-6

DOI: 10.1007/978-1-4613-3754-6


A Division of Plenum Publishing Corporation
233 Spring Street, New York, N.Y. 10013

The Changes is a book
From which one may not hold aloof.
Its tao is forever changing—
Alteration, movement without rest,
Flowing through the six empty places,
Rising and sinking without fixed law,
Firm and yielding transform each other.
They cannot be confined within a rule,
It is only change that is at work here.

I Ching (Book of Changes)

## PREFACE

Since its inception by Professor Lotfi Zadeh about 18 years ago, the theory of fuzzy sets has evolved in many directions, and is finding applications in a wide variety of fields in which the phenomena under study are too complex or too ill-defined to be analyzed by conventional techniques. Thus, by providing a basis for a systematic approach to approximate reasoning and inexact inference, the theory of fuzzy sets may well have a substantial impact on scientific methodology in the years ahead, particularly in the realms of psychology, economics, engineering, law, medicine, decision-analysis, information retrieval, and artificial intelligence.

This volume consists of 24 selected papers invited by the editor, Professor Paul P. Wang. These papers cover the theory and applications of fuzzy sets, almost equal in number. We are very fortunate to have Professor A. Kaufmann to contribute an overview paper of the advances in fuzzy sets. One special feature of this volume is the strong participation of Chinese researchers in this area. The fact is that Chinese mathematicians, scientists and engineers have made important contributions to the theory and applications of fuzzy sets through the past decade. However, not until the visit of Professor A. Kaufmann to China in 1974 and again in 1980, did the Western World become fully aware of the important work of Chinese researchers. Now, Professor Paul Wang has initiated the effort to document these important contributions in this volume and to expose them to the western researchers. A list of Chinese researchers in fuzzy set theory and applications is given in the appendix. The publication of this volume will certainly stimulate mutual exchanges of research ideas and results between Chinese and Western scientists and engineers working in the area of fuzzy sets and applications.

K. S. Fu
President, North American Fuzzy
Information Processing Society
Lafayette, Indiana

December 1982

ACKNOWLEDGEMENTS

The recent opening of the People's Republic of China also opened for study the area of Chinese scientific advances. The researchers are grateful for this new opportunity resulting in a tremendous exchange of ideas which benefits both the Chinese and American people in addition to the universal world of science.

This editor would like to acknowledge and to express his appreciation for the financial assistance he has received for international travel (summer 1981) from the Josiah Charles Trent Memorial Foundation, Inc. The funds for this grant were given to Dr. and Mrs. James Semans through a trust provided by Mrs. Sarah P. Duke. For a project of this nature, it was necessary to take a second trip to China (summer 1982) in order to fully understand the scope of the fuzzy set research activities there. It is important to realize that the support of academic leaders; such as Mr. Zhu Jiusi, President of Huazhong University of Science and Technology, and Mr. Wang Pin-Yang, Chief Engineer of the Electric Power Science Research Institute, is extremely crucial for Chinese research activities to flourish. The appreciation of the value of international communication of my late Dean of Engineering, Dr. Aleksandar Vesic should also be recognized.

It is impossible to mention the names of all the researchers who have helped by offering suggestions about the content. Even more important, is their willingness in providing their best manuscripts to be included in this volume. However, I am profoundly grateful to Mr. Sam Earp of the E. E. Department at Duke University for many stimulating discussions in the course of putting these papers together. Dr. Enrigue H. Ruspini's assistance in preparing Chapter I which improves the overall presentation is also gratefully acknowledged. I also owe thanks to Mr. Wang Ju of U.C.L.A. and Academia Sinica for a portion of indexing work, and to Mr. Zhang Jinwen of Academia Sinica for providing much of the necessary information for the Appendix. I also want to thank Jane S. Culver for typing the manuscripts.

Finally, I am indebted to my chairman Dr. H. Craig Casey, Jr. for his support.

Paul P. Wang
April 1983 in Durham, N.C.

CONTENTS

# FUZZY SET THEORY: PAST, PRESENT AND FUTURE

Paul P. Wang[1], Sam Earp[1] and Enrique H. Ruspini[2]

[1]Department of Electrical Engineering
Duke University
Durham, North Carolina 27706

[2]Hewlett Packard Laboratories
1501 Page Mill Road
Palo Alto, California 94304

In this volume twenty four previously unpublished papers have been selected to provide a reference work in the important area of fuzzy set theory and its applications. These contributions provide significant extensions of fuzzy set and possibility theories, as well as some novel applications of their concepts.

This initial chapter briefly reviews the rationale for the intoduction of the concept of fuzzy set and its subsequent theoretical development. Current trends in fuzzy set theory and its applications are then discussed. The discussion focuses particularly on the description of the characteristics of a large family of current technological problems that are well suited to treatment using fuzzy theoretical concepts. In this context certain recent criticisms to the applicability and technological necessity of the concept of fuzzy set are analyzed.

Each of the subsequent chapters of this volume is then briefly summarized. The chapter concludes with a brief assessment of the future role of fuzzy set technology in the solution of complex system analysis problems.

The concept of fuzzy set was introduced by L. A. Zadeh in 1965. Having dedicated his previous efforts to the study of theoretical and practical issues in system analysis and being already recognized as one of the leading contributors to modern control theory and system science, Professor Zadeh turned his attention in the 1960's to issues related to the inability of existing technology to make substantial contributions to the understanding of large complex systems. Observing that human capabilities for cognitive analysis of those systems were not well matched by the available analytical technology, he sought to identify those characteristics of cognitive analysis which were not found in existing formal concepts and techniques.

His analysis led him to two basic observations. First, humans had a capability to understand and analyze imprecise concepts which was not properly understood or emulated by existing analytical methods. Further, the current methodologies showed a concern for precise representation of certain system aspects that were not only irrelevant to the analysis goals but that were an actual impairment in reaching the system understanding objectives. The result of these basic considerations was his proposal to introduce a new concept, that of fuzzy set, as a basic notion in the representation and analysis of complex real world systems. This important proposal was a significant departure from the ideas supported by prevailing schools of epistemological thought that required precision as a sine qua non imposition on properly defined concepts.

The new notion was intended to formalize the idea of a class of real world entities where each such entity has a certain participation in the class capable of being measured along a continuous variable. Unlike probabilities which are primarily intended to represent a degree of knowledge about real entities, the degrees of membership defining the strength of participation of an entity in a class are a representation of the degree by which a proposition is partially true (in a certain specified sense). This basic epistemic difference between probabilistic concepts, derived from considerations about the uncertainty of propositions about the real world, and fuzzy set concepts, closely related to multivalued logic treatments of issues of imprecision in the definition of entities, properties and attributes, lead to the introduction of new formal methodologies that emphasized use of different quantitative relations between the degrees of membership to related classes (i.e., maximum and minimum operations instead of the real addition and product operations which are customary in standard probability calculus). Further, as pointed out by Gaines (1978), the new formalisms preserved a most desirable characteristic of the two-valued logic formulas they extended, that of strong truth-functionality. The increased computational ability gained as a consequence of the simpler relations linking the degrees of membership of related classes have also been a major reason for the appeal of the new methods over traditional probabilistic approaches which were also questionable on simple epistemic grounds.

In spite of the clear philosophical and methodological differences between fuzzy sets and random functions these distinctions remain to this day one of the major objections to the use of the new concept among its uninformed detractors.[1]

Since 1965 Professor Zadeh has continued to argue vigorously, together with an increasing number of researchers worldwide, for the use of fuzzy set concepts in problems related to the understanding and control of large systems.

His research efforts have been centered around two closely related system analysis issues. The first of these issues is that of the development of mechanisms for the representation and manipulation of the meaning of linguistic utterances about the state of the real world. The second is the use of these representations as constraints in the possible states of a real system being analyzed. This use of fuzzy theoretical concepts as a basis for a theory of possibility, was first proposed by Professor Zadeh in 1980 in the context of considerations about the nature of cognitive processes used in approximate reasoning. Again the emphasis was on the essential tradeoff between gains in the understanding of complex systems at the expense of unneeded precision. Now the scope had been enlarged, however, to allow its application of the formalization of important logical deductive processes. The result has been a theory which avoids the counterintuitive contradictions of other approaches and that can be effectively applied to practical problems.

The initial introduction of the concept of fuzzy set in 1965 was followed by numerous theoretical and practical developments in the ensuing years. The continued interest in fuzzy set theory and the vitality of the field are well evidenced by the contributions presented in this volume. These include both theoretical and applied results produced by researchers in the United States, Europe, Japan and the Far East. We are particularly pleased to devote nearly one third of this edited volume to papers from the People's Republic of China. Fuzzy set theory is one of those recently evolved technologies that has generated a great deal of interest among Chinese researchers. Today there are numerous research laboratories throughout China devoted exclusively to research on fuzzy set and possibility theories and their applications. Researchers outside China have been keenly aware of the scope of the research work being performed in the PRC and vice-versa. This awareness has recently been manifested in the desire,

[1]It is only fair to add that this state of confusion has not been helped by some authors which have embraced the use of fuzzy theoretical concepts without in many cases fully understanding the real nature of the conceptual differences between random functions and fuzzy sets.

strongly expressed in both sides, for a continuing exchange of research contributions between Chinese, American and European institutions. We are particularly pleased to provide, through this volume, an opportunity to continue and further this exchange.[2]

The contributions included in this paper also reflect the diversity of current research efforts in fuzzy set theory. Before summarizing the individual contents of each contribution we proceed to discuss general characteristics of the state of the art in fuzzy set technology as evidenced by the present collection of contributions. In the context of this discussion we will also comment on certain recent criticisms to theoretical and applied results involving use of fuzzy set concepts.

Most of the initial work in the development and application of the concept of fuzzy set was theoretical in nature. While the rationale for the introduction of the concept was the practical need to understand and control a class of complex real world systems it was necessary at first to provide a solid formal framework that extended all mathematical constructs used for the modeling and representation of system characteristics. Further it was also required to develop a most needed understanding of the distinctions between the new epistemic approach and other existing approaches (notably those based on the concept of probability) in order to recognize their relative scope of applicability.

It is also natural that, in the past, the mathematical theory of fuzzy sets has seen a greater development than its applications. The concept of fuzzy set is a generalization of one of the most basic notion in Mathematics, that of set. Since, from this simple basis, a complex and rich framework of mathematical theories has been developed, it can be reasonably expected that a nontrivial conceptual extension at the very foundation of this formidable formal apparatus will trigger a series of generalizations of derived notions and entities.

Most of these generalizations are not, as sometimes criticized, trivial exercises in the enhancement of existing notions which are not mandated by any practical need. Multivalued logic researchers, prior to the introduction of the concept of fuzzy set, were well acquainted with the difficulties inherent in extending basic concepts as the result of enhancing the range of possible truth values of formal propositions. These problems are of three types.

---

[2]In order to help this exchange the Appendix to this volume includes compiled information about the names, addresses and areas of research interest of Chinese scientists.

First, multiple extensions are possible, each resulting in different formal properties. Second, in most cases, all such extensions lose some of the desirable properties of the conventional concept being extended thus making more complex, not more trivial, the choice of a generalization approach. The well known problems associated with the production of a fuzzy set theory that preserves all the algebraic properties of conventional set theory are the simplest example of this basic difficulty. Finally, in spite of common misconceptions, these extensions are the consequence of a perceived practical need not appropriately met using conventional entities. The choice of the extension approach that best suits the particular needs of an applied problem poses additional investigative problems which, in almost every case, are not obvious.

Clearly, straightforward extensions produced solely on theoretical feasibility considerations have been made and will undoubtedly continue to be proposed in the context of fuzzy set theory. The ultimate validity of those extensions is essentially an issue in the philosophy of mathematics which is independent of the need for the notion of fuzzy set. In this regard we will not expect even the most ardent detractor of the theory of fuzzy sets to single our this theory as unique among mathematical frameworks in its production of results for the sake of theoretical richness and elegance.

At present, however, the technological trend in fuzzy set technology is one of change and transition. This transition, initiated a few years, ago has seen an increased interest on the applicability of the theoretical notions derived in the past to current important problems, notably in the areas of pattern recognition, decision analysis and approximate deductive reasoning. Broadly speaking, the most successful applications have been in areas where either alternative approaches (such as probability theory) have failed to provide appropriate results, or, more generally, in problems where the complexity and size of the underlying physical system was very large.

Criticisms about the overall applicability of fuzzy set technology, on the other hand, have not been directed towards the advantages to be gained by use of fuzzy set concepts in those applications. Rather, most critics have confined their primary attention to the applicability of fuzzy set theory to those systems currently studied through well defined mathematical parametric models. A common confusion between the actual physical systems being modeled and their mathematical counterparts has helped to promote this misunderstanding. When this dichotomy is ignored by confusion of object system and model, the applicability of a methodology intended to deal with imprecision and ill-definition is necessarily limited. Attempts to apply such a methodology are frequently inappropriate or, at best, result in the rediscovery or confirmation of known results rather than in the advancement of the frontiers of knowledge.

The applicability of fuzzy set theory is better assessed in the context of those problems not adequately addressed by current parametric models. These models commonly presuppose identifiability and measurability of all relevant parameters or characteristics of a physical system. Many systems of current technological interest are characterized by ill-defined and difficult to observe parameters (as is the case, for example, with medical diagnosis) or by the need to control their behavior so as to attain certain imprecise goals (as in the case with the economy of a country). It is for these systems that fuzzy set theory has the greatest applicability and it is their context that such applicability should be judged. Current experience provides strong evidence of the usefulness of fuzzy set approaches in these cases.

This utility is particularly evident in areas of current technological concern where probabilistic models are inadequate. Among the relevant problems of importance, those associated with the automation of approximate reasoning must be singled out for their importance. While the problems associated with the use of conventional probabilistic concepts for deductive inference purposes do not differ substantially from those that have been apparent since the inception of probability theory, these shortcomings are particularly evident in applications involving large, complex systems. These shortcomings are of three types: the conflicts between an objective (frequentist) and a subjective approach, the identification and manipulation of a large number of parameters (e.g., the myriad of conditional and marginal probabilities required to generate certainty factors in approximate reasoning models), and the interpretation of analytical results. These difficulties are fundamental in that physical reality is required to conform to one of several logically conflicting philosophies, problems in the derivation of appropriate parameterizations are not even addressed by the theory, and experimental results are interpreted in questionable ways.

A frequentist approach pretends to an objectivity that is clearly inapplicable when the model parameters, such as expectations, cannot be derived by experimentation. Subjective Bayesian approaches make use of subjective probabilities and are therefore vulnerable to criticism because the fundamental tenet of the scientific method, that data should determine conclusions, is implicitly ignored. It is also the case that statistical methods are often predicted on a parameterization that is arbitrary or unwieldy. The approach is axiomatic in that an adequate parameterization for a problem must be available ab initio. This is not true in general for complex systems, as evidenced by the interest in nonparametric methods.

Nonparametric methods are, however, heuristic and flawed in that they suffer from the same interpretative defects as parametric statistics.

The theory of fuzzy sets allows a holistic approach to problems of the type discussed above. Experimental results may be interpreted in a meaningful and useful manner due to the close connections between fuzzy set theory and natural languages.

In fuzzy set approaches, subjectivity does not necessarily determine or even influence experimental outcomes as in subjective Bayesian approaches. Rather, subjectivity is treated in a logical manner as only those qualities of data that are relevant to the solution of a problem are abstracted and used in assigning degrees of membership.

In our first presentation in this volume, Kaufmann describes the increasing impact of fuzzy set theory using a comprehensive survey of papers and books published in 1980. Many theoretical and application-oriented details are included. A concise overview of progress in the theory and application of fuzzy sets is also given in Chapter II.

As we have previously emphasized introduction of the concept of fuzzy set as a formal extension of one of the most basic notions in Mathematics has led to the development of fuzzy counterparts of established mathematical disciplines. One important branch, fuzzy topology, has interested researchers since 1968. Pu and Liu have been invited to report on current research in this particular area in the People's Republic of China. In Chapter III Pu and Liu provide a detailed account of "who is doing what and how". Furthermore, relationships between their research activities and those of fuzzy topology researchers outside China have also been described in detail.

Liu devotes half of the discussions of Chapter IV to discussions of fuzzy convex sets. At the very outset of the development of Fuzzy set theory, Zadeh paid special attention to the investigation of this topic. Two important theorems, the separation theorem and the theorem on the shadows of fuzzy convex sets have been since extended and refined by Lowen and Zadeh, respectively.

Under additional assumptions, Liu is able to provide additional insights on the nature of shadows of fuzzy convex sets. Also in this chapter readers will find new results obtained considering fuzzy convex sets in Euclidean spaces which are extendable to linear spaces over real or complex fields.

In Chapter V, Cerruti characterizes compactness and ultracompactness of fuzzy topological spaces by means of a nearness relation between points of X and of an admissible Y of X. Cerruti proposes admissible extensions as a useful tool to extend classical topology concepts into fuzzy topology.

In the next chapter Huang and Tong show some invariant properties of fuzzy events under five operators, proceeding then to the development of fuzzy probability and possibility spaces. Also presented in Chapter VI are some sufficient and necessary conditions for fuzzy independence and noninteraction between fuzzy events.

A novel algebraic system, designated as SLOPE, is proposed by Cao. He is able to show in Chapter VII that a module over a SLOPE is a generalization of the fuzzy subsets. Several important theorems related to the structure of a slope module and the slope matrices can also be found in that chapter.

Wang discusses in Chapter VIII some problems concerning fuzzy statistics and random subsets. The rain space, a framework treating the fuzzy subsets as random subsets, is introduced by Wang. In this space, the measurability of random subsets is equivalent to strong measurability. Applying the graph of random subsets, he gives a correspondence theorem which combines <u>falling</u> subsets and measurable fuzzy subsets.

An extension of previous theoretical results on the important subject of fuzzy relationships is presented by Bouchon and Cohen. In Chapter IX, they study fuzzy binary relations R defined on a finite set E and associated with a distance relation d = 1-R. They also give metrical properties dealing with spheres and cliques in E. These properties are then used to construct partitions of E. In addition, Bouchon and Cohen derive combinatorial results for the number of fuzzy relations and partitions generalizing classical enumeration coefficients. All these results were derived assuming the membership functions take values in a finite set.

In Chapter X, Zhang presents the proofs of 23 lemmas, 3 basic lemmas and 7 theorems concerning fuzzy set structures with strong implications. Fuzzy set structures with strong implications have some very interesting logical characteristics - it is weaker than the normal fuzzy set structures, but a little stronger than some fuzzy set structures.

On a different note, Turksen discusses inference regions for fuzzy propositions in Chapter XI. Inference regions defined by upper and lower bounds are determined for fuzzy relational propositions. He shows that certain current fuzzy reasoning models produce results within these bounds.

Two types of grammars, fuzzy tree grammar and fuzzy forest grammar, are studied by Chu in the next chapter. The basic concept of a fuzzy tree and its properties are explained via graph theory. The fuzzy string grammar, defined by Lee and Zadeh, is then extended to the fuzzy tree grammar. Fuzzy tree grammar is also a generalization of the ordinary tree grammar as proposed by Brainerd. The fuzzy forest grammar, on the other hand, is the result of a further extension of the above development. Chu concludes that there are two interesting connections between a fuzzy forest and a binary tree: a fuzzy forest language and an n-fold fuzzy language proposed by Mizumoto.

Chapter XIII discusses the use of fuzzy production rules for decision making. The advantages of this formalism with respect to classical decision theoretic methodologies are presented by Lesmo, Saitta and Torasso. They also point out the uniformity of representation which allows the designer to easily implement an interface with the external user. An aspect of the formalism, which is analyzed in detail, concerns its suitability for developing methods for automatic learning of fuzzy production rules. An efficient algorithm is also outlined and an example of its use is also reported in this chapter by the authors.

Whalen and Schott employ the methodologies of fuzzy logic, production systems and quasi-natural language to construct a prototype decision support system. The system combines the expert's rules with the user's knowledge of the current environment to deduce suggested decisions for user evaluation. These suggestions are presented in natural-language form. Utilization of fuzzy concepts is essential for the development of their system.

To obtain computer understanding of a scene, imprecise information from several sources must be combined in reasonable ways. In their paper, Jain and Haynes discuss how elements of fuzzy set theory can be used to represent imprecise information and to combine this information. Their paper closes with a discussion of a system under design which exploits the inherent fuzziness of vision processes and the dynamism of the data at every level.

The paper by Henri Prade indicates how the field of information processing may be extended and enriched by incorporating fuzzy set methodologies into a natural language processor. The information itself may be descriptive or qualitative, with a natural interpretation in terms of membership grades. The types of processing that are considered include fuzzy decision procedures and fuzzy guidance procedures. The latter utilizes imprecise directions and distances, with landmarks as checkpoints. First, the notion of a fuzzy instruction, where the operands, operator, or results are fuzzy entities, is developed. The semantic pattern matching is considered as an information retrieval device, and finally there is a preliminary discussion of the idea of fuzzy data types.

Following this chapter, Dubois outlines in Chapter XVII a new methodology for transportation network synthesis, based on man-machine interaction, decomposition of global choice into partial decisions, and the use of fuzzy set theory. Specifically, a procedure is described for dealing with the so-called optimal network problem. Dubois' work consists of two parts: one which builds a candidate set of network alterations, and another where choices are made.

In Chapter XVIII, Chen et al propose a new concept of utility, called fuzzy utility. Some properties of the fuzzy utility function are investigated and fuzzy optimum consumption equilibrium is established. It is claimed that, with ordinary utility theory as a special case, fuzzy utility theory may be identified as not only a more general method for economic analysis, but also as a new method to gain insight into consumer behavior in microeconomics.

Two papers concerning the better design of a "fuzzy controller" are presented in the volume. Mamdani et al investigate the use of fuzzy logic for implementing a rule-based approach, and Sugeno and Takagi propose a new approach based on a fuzzy model of a system. Rule-based methods have been investigated for process control applications since the mid-1970's. Emphasis has been placed on the use of fuzzy logic for implementing linguistic control rules. It is pleasing to see the method is now being used commercially for the control of cement kilns. Mamdani et al present here a simpler controller design in greater detail. A procedure for designing a self-organizing controller is also outlined. Sugeno and Takagi, on the other hand, present a new fuzzy control rule derived from the fuzzy behavior of a system represented by a fuzzy implication. This new idea is illustrated by numerical examples.

The next chapter deals with methods for the improvement of hazard detection in circuits. Mukaidono's approach makes use of fuzzy switching functions which can be employed to describe both the steady state and some transients of binary switching functions. The detection and classification of various kinds of hazards in combinatorial switching circuits is illustrated by methods derived by consideration of the canonical forms of fuzzy switching functions realized by the circuit.

Bruce and Kandel explore in Chapter XXII the application of fuzzy set theory to the modeling of protection structures for computer systems. This fuzzy version of risk analysis models the amount of protection provided by a mechanism by assignment of a grade of membership of that mechanism in the fuzzy set of "mechanisms" which provide total protection against the threat. An example of the modeling procedure for a specific scenario of three possible threats to system operation is given.

The development and evaluation of descriptive models of human problem solving behavior with emphasis on fault diagnosis tasks are reported in the chapter by Rouse. Approaches that rely in pattern recognition concepts, use of problem structure, and rule-based formalisms are considered and contrasted. The behavioral assumptions underlying each approach are discussed and issue associated with measurement and evaluation are considered. We are pleased to note that Rouse places particular stress on the practical issues associated with the application of fuzzy set theory to the modeling of human problem solving.

In the final chapter, Yee Leung develops the analytical framework necessary for the problem of multicriteria conflict resolution in a fuzzy environment. The conventional concept of a 'perfect solution', or ideal, has an analog in a fuzzy environment called the fuzzy ideal that is characterized linguistically and numerically. The associated fuzzy compromise solutions are derived, and it is shown that there is a set of compromise solutions for each point of the fuzzy ideal. Two algorithmic methods for "de-fuzzification" of the problem are presented that will sometimes yield a point-ideal and point-compromise solution. It is noted that research in dynamic conflict resolution under a fuzzy model is necessary.

We would like to thank all the authors for their fine contributions to this volume and, once again, express our pleasure to be able to contribute our skills to the compilation and editing of a collection of this importance.

Having, at present, reached a conceptual maturity evidenced by the multiplicity of formal notions derived primarily towards application goals, the theory of fuzzy sets has reached a stage where its applicability is the subject of most serious consideration by those interested in the understanding and control of systems specified in imprecise and uncertain terms. With its past primarily characterized by the development of required theoretical notions, its present level of conceptual development is making possible an ever increasing application of fuzzy set concepts to important technological problems. The last few years have seen the crossing of this threshold from directed theoretical development to actual application. At this time we are, as researchers in fuzzy set theory, in the most fortunate position to be the shapers of the new age in the control and understanding of large systems: one based on the use of imprecise concepts and notions as the basis for methods that effectively and efficiently deal with the complexity around us.

REFERENCE

Gaines, B., 1978, Fuzzy and Probability Uncertainty Logics, Inf. Control, 38:154-169.

# ADVANCES IN FUZZY SETS - AN OVERVIEW

Arnold Kaufmann

2, allee du Chene
Corenc - Montfleury
38700 - La Tronche, France

Using a comprehensive study of papers and books published in 1980, the author describes the increasing impact of fuzzy set theory. Many details, both theoretical and applications-oriented, are included. Initial suspicions and criticisms of fuzzy set theory have yielded to growing acknowledgment and acceptance as the theory matures. In 1980 more than 600 scientists produced about 700 papers on fuzzy sets at a high level of technical competence. A concise overview of progress in the theory and application of fuzzy sets is given. In a major development, previous doubts about the applicability of the theory have been answered by the large proportion (50%) of papers devoted to applications published in 1980. Researchers in a wide variety of disciplines now realize that the theory provides a convenient framework for modelling problems that have some inherent imprecision. The boundary between fuzzy set theory and probability theory is precisely delineated; fuzzy sets being based upon subjective valuation (sometimes improperly called fuzzy measure) while probability measure is by necessity objective. Fuzzy logic is a bridge between objects and their perception.

This paper gives current results in applications areas such as management science, information retrieval, process control, clustering, pattern recognition, database, artificial intelligence, decision-making, languages, biology, medicine, architectural science, system theory, operations research, economy, theory of games, automata, man-machine systems, switching, ecology, and many other fields where fuzzy set theory is now being applied.

To give an overview of world-wide research in fuzzy sets I used the following material and communications.

1) Research, published papers, articles, memoranda, books, etc. done or presented in 1980 or a few months earlier.
2) Information from my personal international network of scientists in the field of fuzzy set theory and applications.

I collected information from a 1980 bibliography with approximately 700 titles. We note that from 1965 to 1980 included, about 2,500 titles are available in the field of fuzzy sets. I have on hand about 80% of the total amount. To prepare the present work I analyzed the content of this material. I separated my analysis into two parts: theory and applications. Concept by concept and country by country I tried to obtain a concise summary. From this framework I described an overview.

I also considered a lot of information from many friends all over the world after visiting many countries in 1980 including the People's Republic of China and Australia, to give a clear idea of this activity. These personal interchanges are irreplaceable.

Everything that is considered at the adolescence of fuzzy sets is very promising. Not only has the theory been axiomatized irreproachably but also extended in many directions; also, a lot of applications have been presented in 1980. In fact there are now more papers for applications than papers for theoretical considerations. I remember a few years ago a major criticism from mathematicians and others about fuzzy sets was that there were only a few and poor applications. Now that tendency is reversed. A good equilibrium will be 50-50 between theoretical researches and efficient applications.

Referring to the six page breakdown at the end of my report, the reader will see for each subject the number of papers (of any kind) published, with the nationality of the author and/or country where the research has been done. Each inclined bar corresponds to a work partially or completely concerned with the subject. A very fast analysis is possible.

By structural concepts, I mean all research where the goal is basic axioms and theorems to build the theory. It is interesting to recall there is not a general theory of fuzzy sets but theories which concern each kind of fuzzification. Every theory in classical mathematics can be fuzzified in many ways. The United States of America, the People's Republic of China and France have been specially motivated to consider these theoretical points. But it is normal that a lot of countries have also developed important works. It is a beginning; fuzzification is a natural way to insert uncertainty and imprecision in modelling. The need for imprecise models is increasingly apparent, because formal and stochastic descriptions are less realistic in many cases. Stationary laws in human sciences and others are difficult to check by experiments.

Fuzzy variables and fuzzy functions which generalize boolean variables and functions are a wide domain of strong interest. Important and new properties have been discovered and will be used in the near future for any kind of information processing and treatment. This extension is, in fact, a logical extension of the "Laws of Thought" from G. Boole and later from multi-valued logic variables. Analysis, synthesis and decomposition for boolean functions is a large domain of research, but the problem is much harder in the fuzzy extension. The mathematical literature is already rich in papers addressing this question, but much work remains to be done.

The boolean algebra uses only three main operators, union, intersection and complimentation. In fuzzy set theory we are able to use different and non-denumerable types of algebras. There are roughly two classes of operators: analytic or semantic. The richness of fuzzy algebras is astonishing, too rich for several mathematicians (not in my opinion). In natural law of thought, from primitive thought to abnormal thought, for any kind of thought, we need convenient operators, adjustable, adaptable, controlled. Some new (or rediscovered) operators have been proposed last year or before from Hamacher (W.G.), Yager, Bezdek, Harris, Schwwizer, Sklar (U.S.A.), Dubois, Prade, Kaufmann (France) which generalize or introduce new properties, which parametrize usual connectors. It is a fascinating area of study, and very promising indeed.

It is well known that the adequation of membership functions is a permanent subject or discussion. A fuzzy set is a subjective set, a sensation, not a measure. The function cannot be defined like a formal function, it is imprecise by nature. But some authors gave recently some methodologies in such a way to adapt, at the best, the sensation level with the observed object. These processes include parametrization, a posteriori valuation, or some process already used in the theory of probability. We note many works in the United States of America and the People's Republic of China, for instance. This question will be always open because of the subjective nature of fuzzy sets. But, step by step, we are more able to adapt the model to the individual objective.

After the main work of De Luca and Termini concerning non-probabilistic entropy and my own concept of the index of fuzziness, some other families of entropy have been presented. We need, at each step in a fuzzy process, to be conscious of the entropy change. In fact, a decision is the situation where the entropy is minimal or zero. So in a fuzzy process we are obliged to choose a strategy to decrease (not monotonically) the entropy. I found very good works in Italy, Spain and China on this question.

Extensions from the segment 0-1 to distributive lattice for the membership function is a well known topic addressed in the works of Goguen in 1967. Since this time many researchers have made some interesting investigations in this area. Lattice theory and fuzzy set theory have many points in common.

Partitions, ranking, decompositions in fuzzy sets are generalizations of the concepts of order, and class, and so have fundamental applications. There are also interesting papers on fuzzy constraints, fuzzy domains and integration.

Fuzzy logic in itself is probably the largest concept to consider in theory; main works from the U.S.A., U.K. and China develop some new properties. Associated with the fuzzy logic is the fuzzy reasoning. Some important mathematicians who previously rejected fuzzy logic now seem interested. It is an important development. The mine is a large one and the mineral so rich, new prospectors are welcome.

Fuzzy matrices can be defined in different ways but the combination of fuzzy calculus and matrix calculus is sometimes powerful. Of course, it is a different kind of calculus than the linear calculus with sum-product operator.

Fuzzy measure or valuation is a central issue. Personally I reject the nomenclature fuzzy measure and prefer valuation. A fuzzy measure is not a measure because it is not additive in the sense of measure theory. But the bad habit is almost universal, and I myself sometimes use these words. Valuation plays the same role in fuzzy sets as that of measure plays in ordinary sets and is important for the same reason. Following the work of Sugeno a few years ago, in the 1980 period interesting developments have been proposed. A valuation like a measure is a way to compare, subjectively for valuation.

Relations and equations, after the fundamental work of Sanchez in France, are increasingly important. Systems give a framework to meet the problem of solving equations with fuzzy relations. We now have important material, and a general theory is almost ready to use. During my stay in China I found a lot of people who had on hand many works from Sanchez in this subject. Several papers from D.D.R. and Japan also develop some particular properties, and all of these are very interesting. Some mathematicians criticized the fact that the inverses are not unique, but it is an interesting facet of the theory.

By general properties I categorized some important but not special questions. We can find papers from all over the world. Every day one researcher or another finds something new (but also can open an opened door). A mathematician who is working in the field of fuzzy sets is not a specialist but a generalist in science and he must know all fields of mathematics.

It is not well understood that uncertainty and imprecision must be considered by mathematicians concurrently. Uncertainty can be explained in many ways: for instance by confidence interval, by the Laplace law of equiprobability, etc. In fuzzy set theory we are able to use several kinds of structured uncertainty, in fact an infinity of types. Each one depends subjectively on the perception. Imprecision concerns uncertainty in a particular sense, concentrating often on numerical values. Uncertainty is an epistemologic problem, even a philosophical one. The discussion will be always open (there is a strong connection between freedom and uncertainty!).

Fuzzy set theory is a mathematical concept which needs fuzzy topology to be clearly described. Lowen in Belgium is a master in this direction of studies; at the present time fuzzy topologists number more than one hundred.

When the central figure and also founder of fuzzy sets Lotfi A. Zadeh introduced possibility theory a decade ago, the new theory began to mature. The role possibility theory plays for fuzzy sets is analogous to the same role that mathematical expectancy plays in probability theory. This type of valuation gives the best agreement with information available subjectively. Of course, presently we may find in the research group of Professor Zadeh in Berkeley the best papers on the subject. Discussions between probabilists and possibilists are by no means ended. My personal opinion is that such conversations are unnecessary. Each concept is useful in its own domain, and blending is normal in several situations. We can probabilize the fuzzy sets as well as fuzzify the events. To avoid some confusion I propose to use the word "sensation" when possibility is considered and the word "event" when probability is used. Of course a sensation is something weaker than an event. I hope in the interest of developing the fuzzy set theory that a general study of the terminology used will be reconsidered.

Fuzzy numbers are somewhat recent concepts from the ideas of Nahmias (U.S.A.) and Dubois and Prade (France). What a tool for numerical theories and applications! Fuzzy numbers give the connection between the fuzzy set theory and the confidence theory. It is a confidence theory with $\alpha$ - cuts or levels. For me, a fuzzy number is a couple (interval of confidence of level $\alpha$, level of presumption). This couple associated with max-min convolution gives the extension to classical mono-level confidence theory. I had the personal opportunity to use fuzzy numbers in many applications with efficiency. It is a tool for the pure mathematician, it is a tool for the engineer, the physicist, the economist, the sociologist, it is a tool for any one who is concerned with uncertainty where the interval of confidence can be defined.

I have no room to discuss the connections with other mathematical concepts, so I refer back to my list: fuzzy catastrophic theory, topois, information theory, categories theory, simulation, convexity, cardinality. There are a lot of extensions or connections, an extraordinary amount of materials suitable for many very outstanding theses.

Applications of fuzzy sets have been made everywhere and in many practical problems. Several groups in the university tended to assert the dominance of possibility theory over statistical theory. Fuzzy sets and possibility complete statistics when they are poor, but the domains of concern are different with an ill-defined border. Statistics are dependent on measure, fuzzy sets concern structured uncertainty. Statistics, probability theory and fuzzy set theory are complementary tools. Recent papers show how to combine, at the best, the three.

Management science, from the more general point of view, is a collection of models describing management problems. In the last 30 years, formal or/and statistical data have been used. But the environment becomes more uncertain, flexible, and unforseeable than that which may be addressed by a formal model. Models sometimes become scenarios. In many cases fuzzification can yield better results than simulation. Instead of optimizing strategy, managers prefer a more realistic attitude: to find a domain of acceptance, like in the real life.

Information retrieval, classification, and construction of database, from the pioneer work of S. K. Chang (U.S.A.) are now very important topics in research and applications using fuzzy concepts. Images, sentences, meaning, translations, etc. are inherently fuzzy and cannot be processed as formal data by a computer. Some of those who do not accept fuzzy set concepts use them implicitly when processing this type of data. This domain of computer science in France is dominated in several universities by methodologies derived from fuzzy sets.

Certainly, a domain where fuzziness is widely accepted is process control. We note fundamental results obtained in the U.K. (E. H. Mamdani) and in China. Process control, robotics, automatic manufacturing, etc. must improve their efficiency, and their capabilities. It is not paradoxical to say that better precision needs fuzzy processes. A practical measurement gives a fuzzy number! Not very fuzzy, of course, but fuzzy. Rejection of fuzziness can lead to undesirable results. An acceptable alternative is to take into account fuzziness and use a good strategy to decrease the imprecision.

Identification, clustering, pattern recognition, in our

computer age, are areas of research of major importance. Implicitly or explicitly, the fuzzy set theory is used. If the amount of data is very large, statistical processes are better, but otherwise fuzzy procedures in the man-machine dialogue are available and efficient. In more than 10 international scientific reviews or journals, we can find important papers where fuzzy set theory is employed for identification, clustering or pattern recognition. Sometimes statistical methods are combined with fuzzy methodologies. The concept of a "distance" is fundamental in these questions. The choice of the type of distance is an important point. Fuzziness enlarges the possible choices. A convenient distance and an efficient partition (fuzzy partition) process is an adequate methodology.

In Berkeley, Lotfi A. Zadeh initiated the use of fuzzy sets in artificial intelligence. Machines run with formal (boolean) intelligence and man has a brain that can operate on imprecise data. If, later, we are able to build programs or hardware for true artificial intelligence, we shall use fuzzy mathematics. But already adaptive and learning programs with fuzzy instructions are available. If a marriage between perception and computer is possible, we shall make some progress. It is a promising and exciting idea.

Decision theory from the theoretical point of view and decision making from the practical point of view are domains of research which concern economists and managers. A lot of works have been offered for scientific readers. It is now almost standard to introduce fuzzy set theory in these questions. Connected fields like utility theory, optimization theory, and multicriteria problems, need fuzzy concepts to reflect the real environment. It is important to note a recent interest in socialist countries in the utilization of fuzzy procedures for decision problems, particularly in U.S.S.R. and Poland. Many very interesting papers from researchers of these countries are now available.

In linguistics, fuzzy languages are now better known, but computer scientists generally do not agree with the idea of fuzziness in linguistics, especially in computer languages. It is also important to note that fuzzy languages generally require man-machine dialogue to adjust meaning and connectors. The same word has several meanings and the same object is denoted by several words. With formal language, a word is an instruction for the computer; the correspondence with natural languages, however, is more complicated. Of course such natural languages are in fact, semi-natural. They are also human operator dependent, in that information about the behavior of the operator is indispensable. In spite of these problems, semi-natural languages will be very useful in many applications.

A few words will address spatial economy analysis. In France, at the Institut de Mathematiques Economiques, the team directed in Dijon by Pr. Claude Ponsard has completed many important studies for this type of research using fuzzy concepts extensively.

Biology and medicine are now strongly connected with computer science. But diagnosis is too difficult to do with just measurements, valuation is very important. The valuation is basically a fuzzy concept. Also semiotic relations with diagnosis need fuzziness. At the Faculty of Medicine of Marseilles very interesting thesis has been presented after the master work of E. Sanchez. Phi-fuzzy subsets have been used, also extended modus ponens and modus tollens for decision making in medicine. Because in the future computer will be included inside the attache-case of medical doctors, these efforts to combine measure and valuation with adapted semantics and some special rules for diagnosis will be successful.

Physics, geophysics, geology, will also need the fuzzification in the future. Models of search algorithms can be improved using membership functions. Uncertainty and imprecision are not rare in these sciences. Fuzzy models are beginning to be employed. In China, these models are almost standard.

At Sydney - Department of Architectural Science - on the authority of Pr. J. Gero - fuzzy decision process and evaluation systems for architectural choices are becoming the conventional tools. Some computer programs run with fuzzy algorithms.

Of course, fuzzy set theory is a concept especially useful in system theory. Many papers, thesis and books for this purpose can easily be identified. But during my stay in the People's Republic of China, I found a strong interest in fuzzy system theory. Even a theory which in a way generalizes fuzzy sets, Pan system theory, has been established by Professor Wu Xuemou in Huazhong University of Science and Technology. We probably have a lot of exciting questions posed in Chinese works.

It is known that I am very fond of operations research. Now I think that fuzzy set theory is a good way to introduce uncertainty correctly. O.R. models contain formal data, stochastic data and fuzzy data. It is now possible to combine all these three. Sometimes this needs some caution; one often finds pitfalls. After I collected a lot of works in O.R. on fuzzy concepts (and especially fuzzy numbers) I finished a 400 page text-book presenting the use of fuzzy sets in O.R. and Management. The border between O.R. and economics is not well defined, and my remarks about O.R. are also valid for economics and socio-economics. Many works from all over the world are now available. O.R. analysts now have a good tool for modelling imprecise data and uncertainty in environment.

The max-min rule in theory of games is very similar to the max-min rule in fuzzy relations. A game is a decision process in an environment which, depending on the level of information available, is fuzzy. Fuzzy games are an interesting combination and often more realistic than ordinary games. Some recent papers detail their relationship.

During the decade between 1970 and 1980, a lot of papers have been published on fuzzy automatons. It is now more rare. Probably researchers reported their own interest in a system theory which is more general. But reading some works in automaton theory I discovered that several authors implicitly used fuzzy concepts, without using the word, which is suspicious for some.

In mathematical programming, an important part of O.R., many results from other researchers have been published after the main work of Professor H. Zimmermann in Aachen. I consider personally that the Zimmermann method with linear membership functions is the best process for solving multicriteria linear programming. Extensions to non-linearity are now proposed.

Switching and associated processes are favorite topics of research for A. Kandel (U.S.A.), an outstanding and prolific scientist. F.M.S. (fuzzy switching mechanisms) is now an important subfield with many practical applications.

Among other topics in the large variety of applications, let us point out: military applications, production, ecology, forecasting (Delphi method), reliability (probability of failure are rarely known), psychology (and psychoanalysis), meteorology, didactics (in the University of Louvain - Belgium - Professor A. Jones uses fuzzy programs), agronomy, agriculture, hydrology, transportation, inventory, banking, computer aid to conception, creativity methods (one of my last books published shows how to use fuzzy sets in creativity and innovation methods), coding, literature (remark that the great French poet and philosopher Paul Valery predicted sixty years ago the discovery of a new kind of mathematics with semantics), ... this list is incomplete, unfinishable.

As an example which gives an idea about the exceptional interest all over the world in research in the field of fuzzy set theory and applications, I would like to describe the interest found in the People's Republic of China. While giving a seminar on fuzzy sets in Beijing in the summer of 1974, I found only a small group of scientists attracted by the novel theory, but this group was very active and highly qualified. Returning in the spring of 1980 to give a series of lectures at Huazhong University of Science and Technology in Wuhan (Hubei), I had the opportunity to meet several hundred high level researchers from all the provinces of this great country. I am now a honorary professor of this university and working with a lot of Chinese scientists through correspondence. The People's Republic of China now publishes an important quarterly devoted to this domain, "FUZZY MATHEMATICS". In some other scientific journals like "JOURNAL OF MATHEMATICAL RESEARCH AND EXPOSITION", "SCIENCE EXPLORATION", etc. many important papers on fuzzy set theory have been published. During the last decade, Chinese mathematicians, scientists, and engineers have made important contributions to almost all branches of this theory. They now number several thousand. From a long tradition, Chinese scientists and philosophers know that modelling needs a permanent combination of measurable data and non-measurables, randomness and uncertainty. Now, in each university and research institution, lectures in fuzzy sets are included in regular curriculums. The People's Republic of China will be very soon one of the leaders in this field of mathematics; I strongly believe they already have the leadership.

In the 1980 period more than 600 authors are concerned with fuzzy set theory and applications. I estimate that, all over the world, more than 10,000 scientists know, study, develop, extend, apply, teach, and are devotees of fuzzy sets. Zadeh can be very proud of initiating a new and active area of research and applications, fuzzy sets.

## THEORY OF FUZZY CONCEPTS

### Structural Concepts.

China ////////, Czecho-Slovakia /, Finland /, France ///////////, Hungary ////, Israel /, Italia //, Japan //, D.D.R. /, Nederland / //, Norway //, Poland //, U.S.A. ///////, West Germany /.

### Fuzzy Variables, Fuzzy Functions.

China ///, Israel /, Italy /, Japan //, Norway /, Spain /, U.K. /, U.S.A. ////.

### Operators.

China //, France ///, Norway /, U.K. /, U.S.A. ///////, Spain //, West Germany /.

### Membership Functions.

China /////, Czecho-Slovakia /, Japan //, Nederland //, U.K. /, U.S.A. ///////, U.R.S.S. /.

Non probabilistic Entropy.

China ////, Italia /, Spain ////.

Lattices in Fuzzy Sets.

China ///, Czecho-Slovakia /, Norway /, U.S.A. /.

Partition. Rankink.

France ///, U.S.A. ///.

Constraints - Integration.

China /, D.D.R. /, Japan /.

Logic.

China /////, Finland /, France /, Japan //, D.D.R. /, U.S.A. ///// /////, U.K. /////.

Fuzzy Reasoning.

France //, Japan ///, U.K. ////////, U.S.A. /////, West Germany /.

Matrices.

China //, Japan /, U.S.A. /.

Fuzzy Measures - Valuation.

China //////, France //////, D.D.R. /, Italy /, Poland /, Spain /, U.K. //, U.S.A. ///.

Fuzzy Relations and Equations.

China /////////, France ///, Italy /, Japan /////, D.D.R. /////, U.S.A. ///.

General Properties.

Belgium /, Canada /, China ///////, Finland /, France /////, D.D.R. //, India /, Japan ///, Nederland /, Norway //, Poland //, Romania /, Spain /, U.K. ///, U.S.A. ///////, West Germany /.

Uncertainty - Imprecision.

France ////, Japan /, U.S.A. //, West Germany /.

Fuzzy Topology.

Belgium ///, China /////////////, D.D.R. //, France ///, Hungary /, Italy /, Poland /, Romania /, U.K. //, U.S.A. /////////, West Germany /.

Possibility Theory.

China /, France ////, Norway /, U.S.A. ///////, West Germany /.

Fuzzy Theory and Probability Theory.

China ///, D.D.R. /, Finland //, France /////, Japan ///, Norway /, U.S.A. /////, West Germany //.

Fuzzy Numbers.

China ////, Czecho-Slovakia /, France ////, Japan /, U.S.A. ///, U.S.R.R. /, West Germany /.

Fuzzy Catastrophic Theory.

France ///, Japan /.

Topois.

France //, U.S.A. /, Romania ///.

Information Theory.

China //, France ////, Japan //, U.S.A. /.

Categories Theory.

China /, France ///, U.S.A. ///.

Fuzzy Arithmetic.

China /, France ///, U.S.A. /.

Simulation.

France /, U.S.A. /.

Convexity.

Belgium //.

Cardinality.

D.D.R. //, France //.

APPLICATIONS OF FUZZY CONCEPTS

Statistics.

China ////, France /, Japan /, Spain //, U.S.A. ////, West Germany /.

Management Science.

Denmark /, France /, Japan ///, Sweden /, U.K. //, U.S.A. ////.

Information Retrieval and Classification.

China //, France ////////, Poland /, U.K. ///, U.S.A. ///, U.S.R.R. /, West Germany /.

Process Control - Fuzzy Controllers.

China ////////, Canada //, France /, D.D.R. /, Israel /, Japan //, Poland /, Sweden /, U.K. ////////, U.S.A. //.

Identification.

D.D.R. /, France ///, Poland //, U.K. //, U.S.A. //, U.S.R.R. /, West Germany //.

Clustering.

Belgium /, D.D.R. /, China ///, France /, Italy //, Nederland ////, Sweden /, U.K. /, Spain /, U.S.A. ///, U.S.R.R. /, West Germany /.

Pattern Recognition.

China /////, Czecho-Slovakia /, France //, Hungary /, Italy /, India ///, Poland /, U.S.A. ////.

Databases.

France /, U.K. //, U.S.A. //.

Artificial Intelligence.

France ///, Japan ///, U.S.A. ////, U.K. /.

Optimization Theory.

D.D.R. /, Poland /, Romania /, Spain /, U.K. /, U.S.A. /, West Germany /.

Decision Theory - Decision Making.

Belgium /, China /, Finland /, France ////, Greece /, Hungary /, Japan ///, Poland //////, Nederland //, Spain /, U.K. /, U.S.A. // ///////, U.S.R.R. ///////.

Linguistics - Fuzzy Languages.

Austria /, China ///, France /, D.D.R. /, Italy /, Japan /, Poland //, Spain ///, Norway /, Sweden /, U.K. ////, U.S.A. ///, West Germany /, Yugoslavia /.

Utility Theory.

Sweden /, U.K. /, U.S.A. /.

Spatial Economy Analysis.

France ///////.

Speech Understanding and Synthesis.

India ///, Italy //, Poland /, U.S.A. /.

Biology.

China //, D.D.R. //, Czecho-Slovakia /, France //, India /, Italy /, Nederland /, U.K. /, U.R.S.S. ////.

Medicine.

China /////, D.D.R. /, France ////////////, Japan /, Poland //, Spain /, U.K. ///, U.S.A. //////, U.S.R.R. //.

Physics.

France /.

Geophysics - Geology.

China ////, U.S.A. /.

Picture Treatment - Vision.

China ////, U.S.A. /, U.S.R.R. /.

Architectural Science - City Planning.

Australia /////, U.K. /, U.S.A. //.

System Theory.

China /////////////////, D.D.R. ///, France //, Finland /, India /, Israel /, Italy /, Japan //, Norway /, Poland /, Romania ////, U.S.A. ////, U.S.R.R. //.

Operations Research (general applications).

China //, France //, Israel /, Nederland /, Norway /, U.K. //, West Germany //, U.S.A. //.

Economy - Socio-economy.

D.D.R. /, France ////, Japan ////, Spain /, U.S.A. /, West Germany //.

Theory of Games.

China //, France ///, Romania ///, U.S.A. /, U.S.R.R. /.

Automaton Theory.

D.D.R. /, India /, Italy /, Japan /, U.S.A. /, U.S.R.R. /.

Man-machine Systems.

France /, Japan ////, U.S.A. //, U.K. ////, West Germany /.

Mathematical Programming.

Poland /, U.S.A. //, West Germany //////.

Power Production.

Nederland /, Poland //, U.K. /, U.S.A. ///.

Switching.

China //, Israel /, U.S.A. /////////////.

Social Sciences.

U.S.A. //.

Production Process.

D.D.R. /, France ////, Poland //, U.S.A. /.

Semantics.

Austria /, Poland /, D.D.R. /, Spain /, U.K. /, West Germany /.

Questionaries.

France /, Japan /.

Military Applications.

U.S.A. //.

Ecology - Pollution - Environment.

China ///////, D.D.R. /, France /, West Germany /.

Delphi Method.

Nederland /, U.S.A. /.

Literature.

China /, Japan /.

Reliability.

France /, D.D.R. /, India /, Japan /, U.K. /, U.S.A. //.

Multicriteria Problems.

China //, Japan //, U.K. /, U.S.A. //, U.S.R.R. /.

Psychology.

China ////, U.S.A. //.

Meteorology.

China /////, U.S.A. /.

Computer Science.

China ////, France /, Israel /, D.D.R. /, U.S.A. ///.

<u>Linear Equations</u>.

China /, Romania /, U.S.A. /, West Germany /.

<u>Mechanics</u>.

China /, France /.

<u>Didactics</u>.

Belgium /.

<u>Human Brain</u>.

U.S.R.R. /.

<u>Coding</u>.

U.S.A. /.

<u>Physiology</u>.

China /, U.S.A. /.

<u>Agronomy - Agriculture</u>.

China ////, Japan /.

<u>Hydrology</u>.

Austria /, China /, U.S.A. /.

<u>Queuing Theory</u>.

France /.

<u>Transportation</u>.

China //, D.D.R. /, France ///, U.S.A. /.

<u>Inventory Control</u>.

Poland /.

<u>Banking</u>.

U.K. /, U.S.A. //.

Forecasting.

China /, France //, U.K. /, U.S.A. /.

Drawing.

China /, France /, U.S.A. /.

# A SURVEY OF SOME ASPECTS ON THE RESEARCH WORK OF FUZZY TOPOLOGY IN CHINA

Pu Bao-ming and Liu Ying-ming

Institute of Mathematics
Sichuan University
Chengdu, Sichuan, China

The fundamental concept of a fuzzy set, introduced by Zadeh in 1965 [34] provides a natural foundation for treating mathematically the fuzzy phenomena which exist pervasively in our real world and for building new branches of fuzzy mathematics. In the area of fuzzy topology much research has been carried out since 1968 [1].

Early in the 1970's the authors noticed the study of fuzzy topology and were interested in it. Later on, members of the research group of topology in Sichuan University launched investigation on fuzzy topology under the support and encouragement of Professor Kwan Chao-Chih, the director of Institute of Systems Science and Mathematical Sciences of the Academia Sinica. Today, much research work on fuzzy topology is carried on in many parts of China.

In paper [25], two fundamental problems of fuzzy topology were solved: one problem concerns the proper definition of fuzzy points and their neighborhood structure and the other problem is the establishment of Moore-Smith convergence theory in the fuzzy topology spaces. The authors discovered that for the neighborhood structure of fuzzy points, in addition to the traditional concept of neighborhoods and the usual belonging relation, $\varepsilon$, it was needed to introduce a new kind of neighborhood structure, called Q-neighborhoods, for fuzzy points, and a new kind of relation, called Q-relation (the quasi-coincident relation) between fuzzy points and fuzzy sets. In an ordinary topological space, as a special case of a fuzzy topological space, these concepts: neighborhood system and Q-neighborhood system, $\varepsilon$-relation and Q-relation coincide respectively. With these new concepts as main tools, they generalized all the theorems in the chapters I and II of the celebrated book on general topology [12] to fuzzy topological spaces with exception at most two less important ones. This means that these two problems in fuzzy topology have been solved in almost the same degree as the corresponding problems in General Topology had been solved.

In [22], some mutually equivalent systems of axioms, which seem intuitively to be evident, for establishing the neighborhood structure were given and the theorem that in a L-fuzzy topological space the unique neighborhood structure satisfying any one of these systems of axioms is exactly the Q-neighborhood system given in [25] was proved. This fact thus throws light on the limitation of the traditional neighborhood system and the reasonableness and the intrinsic property of the Q-neighborhood system. Considering those L-fuzzy topological spaces where L is a lattice with some special properties, the paper [15] says that in this class of L-fuzzy topological spaces, there is a certain complementary property between the neighborhood system and the Q-neighborhood system.

In [30, 31], Wang introduced two nice concepts: "fuzzy topological molecular lattice" and "far neighborhood" (which is a complement of Q-neighborhood in some sense) from a more abstract point of view and obtained many interesting results therein. Using Q-neighborhood as the basic tool, the theory of Moore-Smith convergence was established in [25]. But [30] generalized this theory into the topological molecular lattices.

In [14], a new proof of the C. T. Yang's theorem concerning derived sets in the fuzzy topological spaces was given. This proof is much simpler than those given in [25] and [30]. In [29], a revised definition of fuzzy boundary was given in terms of the concept of the Q-neighborhood and was used to study the dimension theory of fuzzy topological spaces [28]. The paper [9] introduced a new definition of fuzzy subspaces. It is more general than that given in [25].

As for the fuzzy product spaces and quotient spaces, [26] completed the generalizations of all the theorems in Chapter III of the famous book [12].

Let us now turn to the aspect of fuzzy compactness. In the literature a lot of different kinds of fuzzy compactness notions have been introduced and studied. (e.g. see [13]) In [7], fuzzy compactness [1] was first characterized in terms of the Moore-Smith convergence of fuzzy nets. In [16] a mistake in [3] was pointed out and a counterexample was constructed. An open problem concerning $\alpha^*$-compactness posed there had been fundamentally solved also. So far as we know, the interesting papers on fuzzy compactness notions may be [17] and [32]. In [32], Wang first defined the notion of $\alpha$-nets and then introduced one new kind of fuzzy compactness, called N-fuzzy compactness (or simply, N-compactness), by using α-net from the point of view of convergence. N-fuzzy compactness has almost all the properties which the ordinary compactness has in general topology. Moreover, this paper gave a clear analysis and a clear description of all the relations and differences among a variety of fuzzy compactness notions given in the foreign literature, inclusive of N-fuzzy compactness. From the standpoint of the net convergence, the definition of N-fuzzy compactness may be considered to be given by analyzing all the α-levels simultaneously. However, the other new fuzzy compactness notion given in [17] was defined in term of the open Q-covers, based upon Q-neighborhoods. This compactness, called Q-compactness, is considered to have less requirements in the sense that it can be characterized in term of the convergence property of "maximum fuzzy point" of the fuzzy set only. Among the nice properties Q-compactness enjoys, the Tychonoff product theorem for Q-compactness is a fundamental one. Within the framework of the better fuzzy compactness notions, i.e., N-compactness and Q-compactness, the fundamental results concerning compactness of Chapter 5 of [12] have been generalized to fuzzy topological spaces.

As to the compactification of fuzzy topological spaces, there were already some works [24, 3] abroad, but the general theory of fuzzy Stone-Cech compactification was recently finished in [21] by using the fuzzy embedding theorem [20] and the N-compactness [30]. Adding a weaker separation axiom, called sub-$T_0$ axiom, to a fuzzy completely regular space [6], the concept of fuzzy Tychonoff space is defined [20]. In [21] it is proved that for every fuzzy Tychonoff space, there exists a fuzzy compactification, called fuzzy Stone-Cech compactification, which has the continuous extension property of continuous mappings similar to that of the usual Stone-Cech compactification in general topology.

In the area of the fuzzy uniformity, the New-Zealand Mathematian Hutton has done a piece of profound and penetrating work [6]. However, the author of [18] has pointed out that the fundamental formula, occuring in Lemma 3 of [6] and often used in this area, does not hold for almost all the completely distributive lattices and proved that it holds good under a simple natural additional condition. From the algebraic point of view, the paper [18] is an investigation concerning the intersection operation on union-preserving mappings in completely distributive lattices. In [19], a further discussion on the inverse operation on the above mentioned mappings was made. These works just mentioned above provide useful algebraic tool for investigation of fuzzy uniformities. The author of [20] contributed three things to fuzzy topology. Firstly, he gave a sound proof of the fuzzy Weil theorem (i.e., a fuzzy topological space is fuzzy uniformizable iff it is fuzzy completely regular.) This theorem was first given in [6], but there are some drawbacks in the original proof. Secondly, he gave a pointwise characterization of fuzzy completely regularity by means of the concept Q-neighborhoods. This characterization thus provides a useful tool for establishing the fuzzy embedding theorem. (Hence it may be seen that in the fuzzy imbedding theory, the concept of fuzzy points cannot be avoided and therefore the point of view of the "point-free" treatment of fuzzy topology has its own limitation). Finally he proved the general fuzzy embedding theorem: a L-fuzzy topological space is a Tychonoff space iff it can be embedded in the product of family of fuzzy interval, i.e., a fuzzy basic cube.

A general theory of fuzzy metric spaces in [2] is considered to be successful in a certain sense. In [5] a kind of metric space for the fuzzy topological spaces was obtained. The fuzzy p. q. metric space has been introduced by Hutton [6] and Erceg [2] respectively. In [37], a connection between these two definitions of fuzzy p. q. metric spaces was completely described. Also, the product space of fuzzy p. q. metric spaces and Q-$C_I$ of p. q. metric space were investigated. Finally, the fuzzy Urysohn metrizable theorem has been successfully established there.

In [36], a theory of fuzzy function spaces has been established. In the space, the fuzzy pointwise topology and fuzzy compact open topology were introduced. The joint continuity and some separation properties of the space, such as the completely regularity, were investigated.

In [11], some basic cardinal functions such as weight, charactor, density and cellularity were introduced in fuzzy topological spaces. Among others, the famous Hewitt-Marczewskipondiczery theorem concerning the density of cartisian products was generalized to fuzzy topological spaces.

As is well-known, fuzzy topology is built up on fuzzy set theory, as its foundation. Now fuzzy topology has developed in such an extent that it can react upon its foundation. That is to say, the results obtained thus far in fuzzy topology have important applications in some other branches of fuzzy set theory. Take for instance, the theory of convex fuzzy sets. In the basic and classical paper [34], Zadeh used almost second half of it to discuss the fuzzy convex sets. There are two main results in this aspect which are worth while to be mentioned. The first one is concerning a property of the shadow of fuzzy convex sets and the other is the separation theorem for fuzzy convex sets. The author of [33] gave a counterexample showing that Zadeh's second result mentioned above cannot be correct even for ordinary (crisp) convex sets and recasted this result by employing the induced fuzzy topology. Later on, the author of [23] gave another counterexample showing that Zadeh's first result is also not true and recasted it by adding some topological conditions to give several positive results.

Finally, we would like to mention briefly two other aspects of research on fuzzy topology in China. Much work on induced fuzzy topology in the sense of Weiss [33] has been done [4, 8, 27, 35]. The dimension theory for fuzzy topology has been initiated [28].

REFERENCES


1. C. L. Chang, Fuzzy topological spaces, J. Math. Anal. Appl., 24 (1968), 191-201.
2. M. A. Erceg, Metric spaces in fuzzy set theory, J. Math. Anal. Appl., 69 (1979, 205-230.
3. T. E. Gantner and R. C. Steinlage and R. H. Warren, Compactness in fuzzy topological spaces, J. Math. Anal. Appl., 62 (1978), 547-562.
4. Hu Cheng-Ming, A class of fuzzy topological spaces I, II, Neimenggu Daxue Xuebao 1 (1981), (in Chinese).
5. Hu Cheng-Ming, A metrization of fuzzy topological spaces, Ziran Zazhi 7 (1981), 554, (in Chinese).
6. B. Hutton, Uniformities on fuzzy topological spaces, J. Math. Anal. Appl., 58 (1977), 559-571.
7. Jiang Ji-Quang, Separation axioms in fuzzy topological spaces and fuzzy compactness, Sichuan Daxue Xuebao 3 (1979), 1-10, (in Chinese).
8. Jiang Ji-Quang, Concerning separation axioms for induced fuzzy topological spaces, Trans. Sichuan Math. Soc., 1 (1981), 25-31, (in Chinese).
9. Jiang Ji-Quang, On fuzzy countably compact spaces and fuzzy sequentially compact spaces, Sichuan Daxue Xuebao, 1 (1980), 37-43, (in Chinese).
10. Jiang Ji-Quang, Fuzzy paracompact spaces, ibid, 3 (1981), 47-52, (in Chinese).
11. Jiang Ji-Quang, Weight, charactor, and density of L-fuzzy topological spaces, Acta Mathematica Scientia, (to appear) (in Chinese).
12. J. L. Kelley, General Topology, Van Nostrand, 1955.
13. R. Lowen, A comparison of different compactness notions in fuzzy topological spaces, ibid., 64 (1978), 446-454.
14. Li Zhongfu, On the generalization of C. T. Yang's theorem concerning derived set in fuzzy topology, Fuzzy Math., 1 (1981), 39-42, (in Chinese).
15. Li Zhongfu, A class of L-fuzzy topological spaces I. Neighborhood structures of L-fuzzy points, J. Math. Research and Exposition, 2 (1982), 1: 37-44, (in Chinese).
16. Liu Ying-Ming, A note on compactness in fuzzy unit interval, Kexux Tongbao 25 (1980), A special issue of Math., Phy., and Chem., 33-35, (in Chinese).
17. Liu Ying-Ming, Compactness and Tychonoff theorem in fuzzy topological spaces, Acta Mathematica Sinica, 24 (1981), 260-268, (in Chinese).

18. Liu Ying-Ming, Intersection operation on union-preserving mappings in completely distributive lattices, J. Math. Anal. Appl., 84 (1981), 249-255.
19. Liu Ying-Ming, Inverse operation on union-preserving mappings in lattices and its applications to fuzzy uniform spaces, Proc. of 12th International Symposium on Multiple-Valued Logic, 280-288.
20. Liu Ying-Ming, A pointwise characterization of fuzzy completely regularity and imbedding theorem in fuzzy topological spaces, Scientia Sinica, 1982, 8:673-680.
21. Liu Ying-Ming, On the fuzzy Stone-Čech compactification, Acta Math. Sinica, (to appear) (in Chinese), for English abstract see Kexue Tongbao, 27 (1982), 7:799.
22. Liu Ying-Ming, Neighborhood structures in fuzzy topological spaces, Kexue Tongbao, 27 (1982), 3:189-190, (abstract in Chinese).
23. Liu Ying-Ming, Some properties of fuzzy convex sets, to appear in JMAA. (A detailed abstract in Chinese has been published in Kexue Tongbao, 27 (1982), 6:328-330).
24. H. W. Martin, A Stone-Čech ultra fuzzy compactification, J. Math. Anal. Appl., 73 (1980), 453-456.
25. Pu Pao-Ming and Liu Ying-Ming, Fuzzy topology I, Neighborhood structure of a fuzzy point and Moore-Smith convergence, J. Math. Annal. Appl., 76 (1980), 571-599. (Also see Sichuan Daxue Xuebao 1 (1977), 31-50, (in Chinese).)
26. Pu Pao-Ming and Liu Ying-Ming, Fuzzy topology II, Product and quotient spaces, J. Math. Anal. Appl., 77 (1980), 20-37. (A detailed abstract in Chinese has also been published in Kexue Tongbao, 24 (1979), No. 3, 97-100.)
27. Pu Si-Li, Mapping properties of induced fuzzy topological spaces, (in Chinese).
28. Pu Si-Li, Zero-dimensional fuzzy topological spaces, (in Chinese).
29. Pu Si-Li, A better definition on fuzzy boundary, Fuzzy Math., 2 (1982), 21-34, (in Chinese).
30. Wang Guo-Jun, Topological molecular lattices (I), Sanshi Sida Xuebao, 1 (1979), 1-15, (in Chinese).
31. Wang Guo-Jun, Separation axioms in topological molecular lattices, J. Math. Research and Exposition, (to appear).
32. Wang Guo-Jun, A new fuzzy compactness defined by fuzzy nets, J. Math. Anal. Appl., (to appear).
33. M. D. Weiss, Fixed points, separation and induced topologies for fuzzy sets, J. Math. Anal. Appl., 50 (1975), 142-150.
34. L. A. Zadeh, Fuzzy sets, Inform. Control, 8 (1965), 338-353.
35. Zhou Hao-Xuan, Relations between topological spaces and fuzzy topological spaces, (in manuscript).
36. Peng Yuwei, Fuzzy function spaces, (to appear), (in Chinese).
37. Liang Jihua, Some problems on fuzzy metric spaces, (to appear), (in Chinese).

# SOME PROPERTIES IN FUZZY CONVEX SETS

Liu Ying-Ming

Department of Mathematics
Sichuan University
Chengdu, Sichuan, China

## 1. INTRODUCTION

In the basic and classical paper [10], where the important concept of fuzzy set was first introduced, L. A. Zadeh developed a basic framework to treat mathematically the fuzzy phenomena or systems which, due to intrinsic indefiniteness, cannot themselves be characterized precisely. He pays special attention to the investigation on the fuzzy convex sets which consists of nearly the second half of the space of his paper. The main results on fuzzy convex sets given in [10] are summarized as follows: (1) The separation theorem; and (2) The theorem on the shadows of fuzzy convex sets. The revised correct version of the separation theorem has been given in [9] by employing induced fuzzy topology. Using the concept of fuzzy hyperplane, Lowen has established some further separation theorem for fuzzy convex sets [6]. Concerning the theorem of shadow of fuzzy convex sets, Zadeh has further investigated in [11]. But in this respect, there exists still some drawbacks which will be shown via a counterexample in the present paper. Perhaps the lack of fuzzy topological assumption in the above mentioned results leads to the appearance of these shortcomings. Such a situation seems to be only natural in the early stage of development of the fuzzy set theory. Adding some assumptions about fuzzy topology, we are able to yield several positive results on the shadows of fuzzy convex sets. Finally we shall give some simple and direct proofs of two theorems that describe the relationships between the fuzzy convex cones and the fuzzy linear subspaces and that originally appeared in [6]. The present proofs do not appeal to the representation theorem established in [6] by R. Lowen.

For simplicity, we shall consider only the fuzzy convex sets defined on the Euclidean space in the paper. But it is not difficult to generalize most of the results obtained in this paper to the case where fuzzy convex set are defined in the linear space over real field or complex field.

## 2. PRELIMINARIES

Throughout this paper I will denote the unit interval [0,1], E the Euclidean space of dimension n, and Y the ordinary (crisp) nonempty set. A map $\lambda$ from Y to I is called a fuzzy set on Y, denoted usually by lower case Greek letter. The ordinary set $\{y \in Y : \lambda(y) > 0\}$ is called the support of $\lambda$ and is to be denoted by supp A. The fuzzy set $\lambda'$ defined by $\lambda'(y) = 1 - \lambda(y)$ is called the complement of $\lambda$. For any family $\beta = \{\lambda_j : j \in J\}$ of fuzzy sets on Y, we further define the intersection inf $\beta$ and the union sup $\beta$ respectively by the following formulae:

$$\inf \beta\ (y) = \inf \{\lambda_j(y) = j \in J\},$$

$$\sup \beta\ (y) = \sup \{\lambda_j(y) = j \in J\}.$$

For real a, $\lambda$-a denotes the map defined as $(\lambda - a)(y) = \lambda(y) - a$. We then define a fuzzy topological space as a pair $(Y, \mathcal{J})$, where $\mathcal{J} \subseteq I^X$ (all maps from X to L) and $\mathcal{J}$ is closed under arbitrary union and finite intersection. In general, the fuzzy topology $\mathcal{J}$ does not include all constant maps, so it is different to the one as given in [5]. A set is called open if it is in $\mathcal{J}$ and closed if its complement is in $\mathcal{J}$. Unless otherwise stated, the fuzzy topology on the Euclidean space E will refer to be induced fuzzy topology [9], i.e. the family of all lower semicontinuous function in E.

For a more detailed account of the concepts just outlined above, the reader is referred to the references [7,10].

<u>Definition 1</u>. The fuzzy set on E is said to be fuzzy convex set iff for all $x, y \in E$ and $a \in I$

$$\lambda(ax + (1-a)y) \geq \lambda(x) \wedge \lambda(y).$$

It is easy to see that $\lambda$ is a fuzzy convex set iff there exists a dense subset D of I and for each $a \in D$, $\lambda^{-1}[a,1]$ ($\lambda^{-1}(a,1]$, respectively) is convex. (This equivalence of fuzzy convexity has essentially been shown in [10], see also [6; Proposition 6.1]).

<u>Definition 2</u>. Suppose $\lambda$ is a fuzzy set on E. The fuzzy convex hull of $\lambda$ is defined by

$$\text{conv } \lambda = \inf \{\nu \geq \lambda : \nu \text{ fuzzy convex}\}$$
$$= \text{smallest convex fuzzy set containing } \lambda.$$

Since the intersection of some fuzzy convex sets is still fuzzy convex, it is obvious that for each $\lambda$, its fuzzy convex hull conv $\lambda$ always exists. Furthermore, as shown in [6], if for any $x \varepsilon E$ and $p \varepsilon N$ (where N denotes the set of positive integer), put

$$C(x,p) = \left\{ (x_1,\ldots,x_p) \subset E: \text{ there exist } a_i \varepsilon I, \sum_1^p a_i = 1,\ x = \sum_1^p a_i x_i \right\},$$

then

$$\text{conv } \lambda\ (x) = \sup_{p \varepsilon N} \sup_{A \varepsilon C(x,p)} \inf\{\lambda(y) : y \varepsilon A\}.$$

Definition 3. A fuzzy set $\lambda$ on E is a fuzzy subspace iff for all $x,y \varepsilon E$ and reals a,b

$$\lambda(ax + by) \geq \lambda(x) \wedge \lambda(y).$$

Lemma 1. $\lambda$ is a fuzzy subspace iff $\lambda$ satisfies the following three conditions:

(1) $\lambda(o) = \sup \{\lambda(x) : x \varepsilon E\}$.
(2) For each $x \varepsilon E$ and real $a \neq o$, $\lambda(ax) = \lambda(x)$.
(3) $\lambda$ is fuzzy convex set.

The proof is trivial, see also the proposition 3.3 of [6].

Definition 4. A fuzzy set $\lambda$ on E is a fuzzy convex cone iff it is convex and for each $x \varepsilon E$ and real $a > o$ $\lambda(ax) = \lambda(x)$.

We can easily verify that $\lambda$ is a fuzzy convex cone iff there exists a dense subset D of I and for each $a \varepsilon D$, $\lambda^{-1}[a,1]$ $(\lambda^{-1}(a,1]$, respectively) is the ordinary convex cone in E. (See also proposition 6.4 of [6]).

Definition 5. Let H be the ordinary hyperplane of the Euclidean space E. The orthogonal projection $p: E \to H$ induces a correspondence $S_H$ from $I^E$ (all maps from E to I) into $I^H$. Then for each fuzzy set $\lambda$ on E, the image $S_H(\lambda)$ is called the shadow of $\lambda$ on H.

Remark. The shadow $S_H(\lambda)$ can be expressed as $S_H(\lambda)(y) = \sup\{\lambda(x) : x \varepsilon E \text{ and } p(x) = y\}$ . (cf. Definition 1.1 of [8]).

Definition 6. Suppose that (Y,J) is a fuzzy topological space.

The fuzzy set $\lambda$ on Y is said to be fuzzy compact iff for all family $\beta \subseteq \mathcal{J}$ satisfying $\sup \beta \geq \lambda$ and for all $\varepsilon > o$, there exists a finite subfamily $\beta_o$ such that $\sup \beta_o \geq \lambda - \varepsilon$.

Lemma 2. If $\lambda$ is fuzzy compact set on E (E equipped with induced fuzzy topology), then for each $a > o$, $\lambda^{-1}[a,1]$ is compact.

Proof. This Lemma has originally been obtained in [6]. Here we shall modify slightly the original proof as given in [6] to apply to other case. Suppose the Lemma is not true for some $a > o$, then by the very structure of the usual Euclidean topology of E there exists a sequence $\{x_n\} \subseteq \lambda^{-1}[a,1]$ $(n=1,2,\ldots)$ such that either the subset $x_n$ of E is discrete or the sequence $x_n \to x \notin \lambda^{-1}[a,1]$; and in both cases for each $x_n$ there exists an open neighborhood $B_n$ of $x_n$ such that the family $B_n$ are pairwise disjoint and each $B_n$ does not contain x. If the limit x does exist, since $\lambda(x) < a$, we shall take b such that it satisfies the following inequality $\lambda(x) < b < a$; otherwise we put $b = o$. Now we further define

$$\nu_o(y) = b \qquad y \in \{x\} \cup \{x_1, x_2, \ldots\}$$
$$= 1 \qquad \text{otherwise}$$

and for $n = 1,2,\ldots$, we also define

$$\nu_n(y) = 1 \qquad y \in \cup \{B_j : j=1,2,\ldots,n\}$$
$$= o \qquad \text{otherwise}$$

Then the family $\beta = \{\nu_m : m=o,1,\ldots\}$ is a family of open fuzzy set and $\sup \beta > \lambda$. Since $b < a$ we choose $\varepsilon > o$ such that $b < a - \varepsilon$. For any finite subfamily $\beta_o = \{\nu_{n_1}, \ldots, \nu_{n_k}\}$ of $\beta$, put $n > \max\{n_1, \ldots, n_k\}$, it is easy then to see that $\sup \beta_o(x_n) = b < a - \varepsilon \leq \lambda(x_n) - \varepsilon$. This result, however, is in contradiction with the fuzzy compactness of $\lambda$. Thus $\lambda^{-1}[a,1]$ is compact.

Remark. Under the assumption of the above Lemma, the set $\lambda^{-1}(a,1]$ is not compact in general.

Definition 7. Let $(Y,\mathcal{J})$ be a fuzzy topological space and $\lambda \in I^Y$. The subfamily $\beta$ of $\mathcal{J}$ is said to be the open Q-cover of $\lambda$ iff for each $y \in \operatorname{supp} \lambda$, $\lambda(y) + \sup \beta(y) > 1$. The fuzzy set $\lambda$ is said to be Q-compact (respect with $\mathcal{J}$) iff for any open Q-cover $\beta$ of $\lambda$, there is a finite subfamily $\beta_o$ of $\beta$ such that $\beta_o$ is Q-cover of $\lambda$. (cf. [4])

## 3. SHADOW OF FUZZY SET

In [10; p. 350] Lotfi Zadeh asserts that supposing $\lambda$ and $\mu$ are fuzzy convex set on E, $S_H(\lambda) = S_H(\mu)$ for each hyperplane H of E implies $\lambda = \mu$. Furthermore he also claims that for a pair of fuzzy sets $\lambda$ and $\mu$, $S_H(\lambda) = S_H(\mu)$ for each hyperplane H of E implies conv $\lambda$ = conv $\mu$. The following counterexample will show that the above assertions are imprecise.

Counterexample. For simplicity, we shall focus on only the case where E is the Euclidean plane. When dim $E \geq 3$, the counterexample may be similarly constructed. As for dim E=1, the above assertion is obviously false. In fact, in the original argument presented in [10], the hypothesis that dim $E \geq 2$ had been implicitly assumed. Let A be a subset of the plane $\{(x,y)$: either $o \leq x < 1$ and $o \leq y < 1$ or $x = 1$ and $o \leq y < 1/2$, the point $p = (1,1/2)$ and $B = A \cup \{P\}$. Suppose that $\lambda$ and $\mu$ denote the characteristic functions of A and B respectively. Then $\lambda \neq \mu$ and both are the (fuzzy) convex sets on E. For any hyperplane H of E, i.e. straight line H, it is quite obvious that any line which goes through the point p and is orthogonal to H will intersect with the subset A. Therefore for each point y of H, the fuzzy set $S_H(\lambda)$ and $S_H(\mu)$ will take the same value (either 1 or o). That is to say, $S_H(\lambda) = S_H(\mu)$.

Adding some hypotheses on fuzzy topology, we are able to recast the assertion in the following more precise form.

Theorem 1. Let $\lambda$ and $\mu$ be fuzzy convex sets on E with dim $E \geq 2$ and $S_H(\lambda) = S_H(\mu)$ for each hyperplane H. If $\lambda$ and $\mu$ are open or closed sets relative to the induced fuzzy topology $\mathcal{J}$ (maybe one is open and another is closed), then $\lambda = \mu$.

Proof. Suppose $\lambda \neq \mu$. There is $x \varepsilon E$ such that $\lambda(x) \neq \mu(x)$. Without loss of generality, we may assume that $a = \lambda(x) > \mu(x) = b$. We then put $c = (a + b)/2$. For the fuzzy open (closed, respectively) set $\mu$, $B = \mu^{-1}(c,1]$ ($B = \mu^{-1}[c,1]$ resp.) is an open (closed, resp.) convex set in E and $x \notin B$. By invoking the convexity theory [2], there is hyperplane F through x such that $F \cap B = \emptyset$. We then choose a hyperplane H which goes through x and is orthogonal to F (Note that dim $E \geq 2$, so such H exists). Then $S_H(\lambda)(x) \geq a$ and since $F \cap B = \emptyset$, max $\{\mu(y)$: $y \varepsilon F\} \leq c$, hence $S_H(\mu)(x) \leq c$. That is to say, $S_H(\lambda) \neq S_H(\mu)$, which apparently contradicts to the hypothesis.

Definition 8. Let $\lambda$ be a fuzzy set on E. The intersection of all the fuzzy convex and open (closed, resp.) sets containing $\lambda$ is called the fuzzy open-convex (closed-convex, resp.) hull of $\lambda$.

The fuzzy closed-convex hull of $\lambda$ is obviously the smallest fuzzy closed convex set containing $\lambda$ but the fuzzy open-convex hull may not be open. Both fuzzy open-convex hull and closed-convex hull of $\lambda$ contain the fuzzy set conv $\lambda$. Their constructions can be described briefly as follows.

Proposition 1. Let $\lambda$ be the fuzzy set on E, $\lambda_a = \lambda^{-1}(a,1]$ for each $a \varepsilon [o,1)$. In the Euclidean space E, we further denote the collection of all the ordinary open and convex subsets containing $\lambda_a$ by

$$\{\lambda(a,j): j \text{ belongs to some index set } J_a\}$$

Furthermore, we designate

$$\Gamma_{a,j}(x) = 1 \qquad x \in \lambda(a,j),$$
$$= a \qquad \text{otherwise},$$

then the fuzzy open-convex hull of $\lambda$ is the fuzzy set

$$\Gamma = \bigcap \{\Gamma_{a,j} : a \in [o,1],\ j \in J_a\}.$$

Proof. Because each $\Gamma_{a,j} \geq \lambda$ and is an open convex set, it is sufficient to show that if a fuzzy open convex set satisfies $\nu \geq \lambda$, then $\nu \geq \Gamma$. In fact, for each $x\varepsilon E$, put $\nu(x) = a$. If $a = 1$, obviously $\nu(x) \geq \Gamma(x)$. If $a < 1$, then $x \notin \nu^{-1}(a,1]$. By $\nu \geq \lambda$, $\nu^{-1}(a,1] \supseteq \lambda_a$ then the open convex set $\nu^{-1}(a,1]$ contains $\lambda_a$. Hence there is $j\varepsilon J_a$ such that $\nu^{-1}(a,1] = \lambda(a,j)$. Since $x \notin \nu^{-1}(a,1]$, so $\Gamma_{a,j}(x) = a$ and $\Gamma(x) \leq a = \nu(x)$.

Proposition 2. Let $\lambda$ be the fuzzy set on E, $\lambda_a = \lambda^{-1}[a,1]$ for each $a\varepsilon I$. In the Euclidean space E, we further denote the smallest closed convex ordinary subset containing $\lambda_a$ by $\tilde{\lambda}_a$. Put

$$\Omega_a(x) = 1 \qquad x \in \tilde{\lambda}_a,$$
$$= a \qquad \text{otherwise},$$

then the fuzzy closed-convex hull of $\lambda$ is the fuzzy set $\Omega = \bigcap\{\Omega_a : a\varepsilon I\}$.

Proof. It is quite clear that each $\Omega_a$ is a fuzzy closed convex set containing $\lambda$. Now it is sufficient to show that if a fuzzy closed convex set $\nu \geq \lambda$, then $\nu \geq \Omega$. In fact, for each $x\varepsilon E$, put $a = \nu(x)$. If $a = 1$, naturally $\nu(x) \geq \Omega(x)$. If $a < 1$, then the set $B = \{b\varepsilon I: b > a\}$ is not empty. For each $b\varepsilon B$, $x \notin \nu^{-1}[b,1] = \lambda_b$. Since $\lambda_b$ is convex closed set in E, so $\lambda_b = \tilde{\lambda}_b$. Hence $\Omega_b(x) = b$ by $x \notin \tilde{\lambda}_b$ and $\Omega(x) \leq b$. Because b may be any number in B, so $\Omega(x) \leq a = \nu(x)$.

Theorem 2. Let $\lambda$ and $\mu$ be fuzzy sets on E with dim $E \geq 2$ and $S_H(\lambda) = S_H(\mu)$ for any hyperplane H of E. If $\tilde{\lambda}$ and $\tilde{\mu}$ denote their fuzzy open-convex (closed-convex) hull respectively, then $\tilde{\lambda} = \tilde{\mu}$.

Proof. Suppose $\tilde{\lambda} \neq \tilde{\mu}$, that is to say, for some $x\varepsilon E$, $\tilde{\lambda}(x) \neq \tilde{\mu}(x)$. Without loss of generality, we may assume $a = \tilde{\lambda}(x) > \tilde{\mu}(x) = b$. Choose $c\varepsilon I$ such that $a > c > b$. From $\tilde{\mu}(x) < c$ and Proposition 1 (using also the notation there), we have $e\varepsilon I$ and $j\varepsilon J_e$ such that $\Gamma_{e,j} \geq \mu$ and $\Gamma_{e,j}(x) < c$. Put $B = \Gamma^{-1}_{e,j}(c,1]$. (From Proposition

2 we have $e \epsilon I$ such that $\Omega_e \geq \mu$ and $\Omega_e(x) < c$. Put $\Omega_e^{-1}[c,1] = B$, respectively). The subset $\bar{B}$ is an ordinary open (closed, resp.) convex set and $x \notin B$. We say that $\sup \{\lambda(y): y \notin B\} > c$. In fact, on the contrary, we get a fuzzy set $\nu$ defined by

$$\nu(y) = 1 \qquad y \epsilon B$$
$$= c \qquad \text{otherwise,}$$

satisfying $\nu \geq \lambda$. Since $\nu$ is fuzzy open (closed, resp.) convex set. So $\nu \geq \bar{\lambda}$. By $x \notin B$ we have $c = \nu(x) \geq \tilde{\lambda}(x) = a$ which contradicts the fact: $a > c$. Now we may choose the point $z \epsilon B$ such that $\lambda(z) > c$. By invoking the convexity theory, in case that B is either open convex set or closed convex set, there exists a hyperplane F through the point z such that $F \cap B = \phi$. Take a hyperplane H through the point z and orthogonal to F. Then $S_H(\lambda)(z) \geq \lambda(z) > c$. On the other hand, since $\mu \leq \Gamma_{e,j}$ ($\mu \leq \Omega_e$, resp.), hence $\mu^{-1}(c,1) \subseteq \Gamma_{e,j}^{-1}(c,1] \subseteq B$ ($\mu^{-1}[c,1] \subseteq \bar{B}$, resp.) Therefore F does not intersect with $\mu^{-1}(c,1]$ ($\mu^{-1}[c,1]$, resp.) and $S_H(\mu)(z) \leq c$. $S_H(\lambda) \neq S_H(\mu)$; a contradiction indeed exists.

Remark. The inverse statement of Proposition 2 is not true. That is to say, the fact that $\tilde{\lambda} = \tilde{\mu}$ does not imply $S_H(\lambda) = S_H(\mu)$. In fact, even in the ordinary convex set theory, the corresponding counterexample is easy to construct.

In the following we shall investigate the relationship between the properties of shadow and the compactness (the fuzzy compactness and Q - compactness).

Theorem 3. Let $\lambda$ and $\mu$ be the fuzzy convex sets on E with $\dim E \geq 2$ and $S_H(\lambda) = S_H(\mu)$ for each hyperplane H. If $\lambda$ and $\mu$ are fuzzy compact relative to the induced topology $\mathcal{J}$, then $\lambda = \mu$.

Proof. From the argument of Theorem 1 it is not difficult to see that if for each $a \epsilon (o,1]$, $\nu^{-1}[a,1]$ is the closed convex subset of E, then the proof presented there can be proceeded throughout regardless of closeness or openness of $\nu$. Now, by invoking Lemma 2 for each $a \epsilon (o,1]$, $\nu^{-1}[a,1]$ is compact, hence it is a closed set. So the present proof can be given similarly to that of Theorem 1.

As to the concept of fuzzy topological subspace (simply, subspace), we shall refer to [7].

Proposition 3. Let $(Y, \mathcal{J})$ be a fuzzy topological space, $\lambda$ the fuzzy set on Y. If there exists an increasing positive real sequence $\varepsilon_n \to o$ such that each fuzzy set $\lambda' + \varepsilon_n$ is Q-compact in the subspace supp $\lambda$, then $\lambda$ is fuzzy compact, where $\lambda' + \varepsilon_n$ is defined by $(\lambda' + \varepsilon_n)(y) = \min \{1, 1 + \varepsilon_n - \lambda(y)\}$, $y \epsilon$ supp $\lambda$.

Proof. Suppose that the subfamily $\beta \subseteq \mathcal{J}$ satisfies sup $\beta \geq \lambda$ and $\varepsilon > o$. We may choose $\varepsilon_n$ such that $\varepsilon_n < \varepsilon$. For each fuzzy set $\mu \varepsilon \beta$ the restriction of $\mu$ to the subspace supp $\lambda$ can be denoted by $\tilde{\mu}$. So we have the family

$$\tilde{\beta} = \{\tilde{\mu} : \mu \varepsilon \beta\}.$$

We say that $\tilde{\beta}$ is an open Q-cover of $\lambda'+\varepsilon_n$ in the subspace supp $\lambda$. In fact, we may take any $x \varepsilon$ supp $\lambda$. If $(\lambda'+\varepsilon_n)(x) = 1$, then by sup $\tilde{\beta}$ $(x) \geq \lambda(x) > o$ we have sup $\tilde{\beta}$ $(x) + (\lambda'+\varepsilon_n)(x) > 1$; If $(\lambda'+\varepsilon_n)(x) < 1$, then $(\lambda'+\varepsilon_n)(x) = 1 - \lambda(x) + \varepsilon_n$. Since sup $\tilde{\beta} \geq \lambda$, hence sup $\tilde{\beta}(x) + (\lambda'+\varepsilon_n)(x) =$ sup $\tilde{\beta}(x) - \lambda(x) + 1 + \varepsilon_n \geq 1 + \varepsilon_n > 1$. Thus $\tilde{\beta}$ indeed is an open Q-cover of $\lambda' + \varepsilon_n$. By the Q-compactness of $\lambda' + \varepsilon_n$ there exists a finite subfamily $\tilde{\beta}_o$ of $\tilde{\beta}$ such that for each $x \varepsilon$ supp $\lambda$, sup $\tilde{\beta}_o(x)+(\lambda'+\varepsilon_n)(x) > 1$. Now, when $1 - \lambda(x) + \varepsilon_n \leq 1$ we have $(\lambda'+\varepsilon_n)(x) = 1 - \lambda(x) + \varepsilon_n$, thus sup $\tilde{\beta}_o(x) > \lambda(x)-\varepsilon_n$; When $1 - \lambda(x) + \varepsilon_n > 1$, we have $\lambda(x) < \varepsilon_n$, hence sup $\tilde{\beta}_o(x) \geq o > \lambda(x) - \varepsilon_n$. To sum up, we always have sup $\tilde{\beta}_o(x) \geq \lambda(x) - \varepsilon_n$ for each $x \varepsilon$ supp $\lambda$. Put $\beta_o = \{\mu\varepsilon\beta : \tilde{\mu}\varepsilon\tilde{\beta}_o\}$. Then $\beta_o$ is a finite subfamily of $\beta$ and sup $\beta_o \geq \lambda - \varepsilon_n > \lambda - \varepsilon$. By Definition 6, $\lambda$ is a fuzzy compact set in $(Y,\mathcal{J})$.

From Proposition 3 and Theorem 3, we can establish the following corollary.

Corollary. Let $\lambda$ and $\mu$ be fuzzy convex sets on E with dim E $\geq 2$. If there exists an increasing positive real sequence $\varepsilon_n \to o$ such that each $\lambda' + \varepsilon_n$ and each $\mu' + \varepsilon_n$ are Q-compact in subspace supp $\lambda$ and subspace supp $\mu$ respectively, then $S_H(\lambda) = S_H(\mu)$ for each hyperplane H implies $\lambda = \mu$.

Definition 9. Let $(Y, \mathcal{J})$ be a fuzzy topological space and $\lambda$ the fuzzy set on Y. The subfamily $\beta$ of $\mathcal{J}$ is said to be the complete Q-cover of $\lambda$ iff for each $y \varepsilon$ supp $\lambda$, sup $\beta(y) + \lambda(y) > 1$ and for each $y \notin Y \setminus$ supp $\lambda$, there is a $\nu\varepsilon\beta$ such that $\nu(y) = 1$. The fuzzy set $\lambda$ on Y is said to be complete Q-compact iff for each complete Q-cover $\beta$ of $\lambda$, there is a finite subfamily $\beta_o$ of $\beta$ such that $\beta_o$ is complete Q-cover of $\lambda$.

Remark. When $\lambda = \phi$, the complete Q-compactness of $\lambda$ is equivalent to 1*-compactness [1] of $(Y, \mathcal{J})$.

Proposition 4. If the fuzzy set $\lambda'$ is complete Q-compact on E (E equipped with induced fuzzy topology), then for each $a > o$, $\lambda^{-1}[a,1]$ is compact.

Proof. The argument of Lemma 2 can be applied to the present case with some trivial modification.

Theorem 4. Let $\lambda$ and $\mu$ be fuzzy convex sets on E with dim E $\geq 2$.

If $\lambda'$ and $\mu'$ are complete Q-compact and for each hyperplane H, $S_H(\lambda) = S_H(\mu)$, then $\lambda = \mu$.

Proof. Analogonous to the proof of Theorem 3.

## 4. FUZZY CONVEX CONE AND FUZZY SUBSPACE

In this section we shall give a simple and direct proof of two theorems describing the relation between the fuzzy convex cone and fuzzy subspace which has been previously established in [6] via so called representation theorem.

Theorem 5. Suppose that $\lambda$ is a fuzzy convex cone and $\lambda(o) = \sup\{\lambda(x): x\varepsilon E\}$. Then there exists a smallest fuzzy subspace $\tilde{\lambda}$ containing $\lambda$ and $\tilde{\lambda}$ can be defined by

$$\tilde{\lambda}(x) = \sup\{\lambda(z) \wedge \lambda(y) : z-y=x\}.$$

Proof. We first point out that the fuzzy set $\tilde{\lambda}$ defined by the above-mentioned formula is a subspace. In fact, by definition of $\tilde{\lambda}$ we have $\tilde{\lambda}(x) = \tilde{\lambda}(-x)$. Now we need to prove the following formula holds for any reals $a_1$ and $a_2$ and any points $x_1$ and $x_2$:

$$\tilde{\lambda}(a_1x_1 + a_2x_2) \geq \tilde{\lambda}(x_1) \wedge \tilde{\lambda}(x_2). \qquad (*)$$

If some $a_i = o$, the formula (*) follows directly from the definition of the fuzzy convex cone. If $a_1 \cdot a_2 \neq o$, put $y_i = (\operatorname{sgn} a_i)\, x_i$, then $a_ix_i = |a_i|y_i$ (i=1,2) and

$$\tilde{\lambda}(a_1x_1 + a_2x_2) = \tilde{\lambda}(|a_1|y_1 + |a_2|y_2)$$

$$= \sup\{\lambda(w) \wedge \lambda(z) : z - w = a_1x_1 + a_2x_2\}$$

$$= \sup\{\lambda(|a_1|w_1+|a_2|w_2) \wedge \lambda(|a_1|z_1+|a_2|z_2): z_1-w_1=y_1,\ z_2-w_2=y_2\}.$$

In view of the definition of the fuzzy convex cone

$$\lambda(|a_1|w_1 + |a_2|w_2) \geq \lambda(w_1) \wedge \lambda(w_2) \text{ and}$$

$$\lambda(|a_1|z_1 + |a_2|z_2) \geq \lambda(z_1) \wedge \lambda(z_2), \text{ we have}$$

$$\lambda(a_1x_1 + a_2x_2) \geq \sup\{\lambda(w_1) \wedge \lambda(w_2) \wedge \lambda(z_1) \wedge \lambda(z_2): z_1-w_1=y_1,\ z_2-w_2=y_2\}$$

$$= \sup\{\lambda(w_1) \wedge \lambda(z_1): w_1-z_1=y_1\} \wedge \sup\{\lambda(w_2) \wedge \lambda(z_2): z_2-w_2=y_2\}$$

$$= \tilde{\lambda}(y_1) \wedge \tilde{\lambda}(y_2) = \tilde{\lambda}(x_1) \wedge \tilde{\lambda}(x_2).$$

i.e. the formula (*) still holds. Next, since $\lambda(o)$ is the greatest $\lambda(x)$, hence

$$\tilde{\lambda}(x) = \sup\{\lambda(z) \wedge \lambda(y) : z-y=x\}$$

$$\geq \lambda(x) \wedge \lambda(o)$$

$$= \lambda(x),$$

i.e. $\tilde{\lambda} \geq \lambda$. Finally, let $\nu$ be any fuzzy subspace containing $\lambda$. For each $x\varepsilon E$, we take the points y and z such that $y-z=x$. Then

$$\nu(x) = \nu(z-y) \geq \nu(z) \wedge \nu(y) \geq \lambda(z) \wedge \lambda(y).$$

Based upon the definition of $\tilde{\lambda}$, we have $\nu(x) \geq \tilde{\lambda}(x)$. Thus $\tilde{\lambda}$ is exactly the smallest fuzzy subspace which contains $\lambda$.

Theorem 6. Suppose that $\lambda$ is a fuzzy convex cone and $\lambda(o) = \sup\{\lambda(x): x\varepsilon E\}$, then there exists the largest fuzzy subspace $\tilde{\lambda}$ which is contained in $\lambda$ such that

$$\tilde{\lambda}(x) = \lambda(x) \wedge \lambda(-x), \qquad x \varepsilon E.$$

Proof. The fuzzy set $\tilde{\lambda}$ defined by the above-mentioned formula is a fuzzy subspace. In fact, for any reals $a_1$ and $a_2$ and any points $x_1$ and $x_2$, put $y_1 = (\text{sgn } a_i)\ x_i$, $i=1,2$, then $a_i x_i = a_i y_i$. If some $a_i=o$, then $\tilde{\lambda}(a_1x_1+a_2x_2)$ is greater than or equal to $\tilde{\lambda}(x_1) \wedge \tilde{\lambda}(x_2)$. If $a_1 \cdot a_2 \neq o$, we have

$$\tilde{\lambda}(a_1x_1+a_2x_2) = \tilde{\lambda}(|a_1|y_1 + |a_2|y_2)$$

$$= \lambda(|a_1|y_1 + |a_2|y_2) \wedge \lambda(|a_1|(-y) + |a_2|(-y_2))$$

$$\geq \lambda(y_1) \wedge \lambda(y_2) \wedge \lambda(-y_1) \wedge \lambda(-y_2)$$

$$= \tilde{\lambda}(y_1) \wedge \tilde{\lambda}(y_2)$$

$$= \tilde{\lambda}(x_1) \wedge \tilde{\lambda}(x_2).$$

Hence, we claim that $\tilde{\lambda}$ is a fuzzy subspace. Next, it is quite obvious that $\tilde{\lambda} \leq \lambda$. Finally, let $\nu$ be a fuzzy subspace contained in $\lambda$, then $\nu(x) = \nu(x) \wedge \nu(-x) \leq \lambda(x) \wedge \lambda(-x) = \tilde{\lambda}(x)$. Hence, we conclude

$$\nu \leq \tilde{\lambda}.$$

REFERENCES

1. T. E. Gantner, R. C. Steinlage and R. H. Warren, Compactness in

fuzzy topological spaces, J. Math. Anal. Appl., 62(1978), 547-562.
2. J. L. Kelley and I. L. Namioka, "Linear Topological Spaces," Van Nostrand, Princeton (1963).
3. Ying-ming Liu, A note on compactness in fuzzy unit interval, Kexue Tongbao 25(1980). A special issue of Math., Phy., and Chem., 33-35 (in Chinese).
4. Ying-ming Liu, Compactness and Tychonoff theorem in fuzzy topological spaces, Acta Math. Sinica, 24(1981), 260-268 (in Chinese).
5. R. Lowen, Fuzzy topological spaces and fuzzy compactness, J. Math. Anal. Appl., 56(1976), 621-633.
6. R. Lowen, Convex fuzzy sets, Fuzzy Sets and Systems, 3(1980), 291-310.
7. Pao-ming Pu and Ying-ming Liu, Fuzzy topology I, neighborhood structure of a fuzzy point and Moore-Smith convergence, J. Math. Anal. Appl., 76(1980), 571-599.
8. Pao-ming Pu and Ying-ming Liu, Fuzzy topology II, product and quotient spaces, J. Math. Anal. Appl., 77(1980), 20-37.
9. M. D. Weiss, Fixed points, separation and induced topologies for fuzzy sets, J. Math. Anal. Appl., 50(1975), 142-150.
10. L. A. Zadeh, Fuzzy sets, Inform. Contr., 8(1965), 338-353.
11. L. A. Zadeh, Shadows of fuzzy sets, Problems of Information Transmission, 2(1) (1966), 37-44 (MR 35#4817).

"NON-STANDARD" CONCEPTS IN FUZZY TOPOLOGY

Umberto Cerruti

Istituto di Geometria
Università di Torino
10123 Torino, Italy

## 1. INTRODUCTION

The principal concept introduced in this paper is that of admissible extension $(Y,\phi)$ of a set X. Here $\phi$ is a function $\phi : \mathcal{P}(X) \to \mathcal{P}(Y)$ which preserves finite boolean operations and determines each filter $\mathcal{F}$ on X be means of its nucleus $\bigcap_{B\in\mathcal{F}} \phi(B)$.

As examples we can think of Y as the space of ultrafilters on X or as the scope of X in a non-standard enlargement.

The advantage is that we do not need to explicitly use ultrafilters or non-standard models but only some simple formal properties of them. In a following paper this concept will be used to set up what I call Topology Admissing Universes, in which results from non-standard topology can be obtained and extended in an easy way.

In this paper my intention is more restricted: I want to characterize compactness and ultracompactness in fuzzy topological spaces by means of a nearness relation between points of X and of an admissible extension Y of X. This is done in the main Theorems of §3. In §2, I give preliminary definitions, examples, and the construction of the extension $\tilde{\phi}$ of $\phi$, $\tilde{\phi} : I^X \to I^Y$.

I hope that admissible extensions can also be useful to relate concepts from classical topology to fuzzy topology, through non-standard topology.

## 2. ADMISSIBLE EXTENSIONS

### 2.1 Definition

Given a set X we call a pair $(Y,\phi)$ <u>admissible</u> <u>extension</u> of X if Y is a set, $\phi : \mathcal{P}(X) \to \mathcal{P}(Y)$ is a function and the following axioms are satisfied:

<u>2.1.1.</u> $\phi$ is a boolean algebra morphism.

<u>2.1.2.</u> If $\mathcal{F}$ is a filter on X then

$$\forall B\varepsilon\mathcal{P}(X),\ \phi(B) \supseteq \bigcap_{A\varepsilon\mathcal{F}} \phi(A) \Rightarrow B\varepsilon\mathcal{F}\ .$$

### 2.2 Remarks

The definition above has three immediate consequences:

<u>2.2.1.</u> If $\mathcal{F}$ is a filter on X then $\mathcal{F} = \{B\varepsilon\mathcal{P}(x)\colon \phi(B) \supseteq \bigcap_{A\varepsilon\mathcal{F}} \phi(A)\}$.

<u>2.2.2.</u> $\phi$ is a boolean algebra monomorphism.

<u>2.2.3.</u> If a family $\mathcal{A} \subseteq \mathcal{P}(X)$ has the FIP (finite intersection property) the intersection $\bigcap_{A\varepsilon\mathcal{A}} \phi(A)$ is non-empty.

### 2.3 Examples

<u>2.3.1.</u> For notations and definitions of this example we refer to (3). Let X be a set in some universe of discourse $\mathcal{U}$ and let $^*\mathcal{U}$ be a non-standard enlargement of $\mathcal{U}$. We define $Y = \{y\colon y{*}\varepsilon X\}$. So Y is the collection of *members of X, called "the scope of X". Given now $A \subseteq X$ we define $\phi(A) = \{z\colon z{*}\varepsilon A\} \subseteq Y$. Then $\phi$ is a boolean morphism. We define now for a filter $\mathcal{F}$ in X $\text{Nuc}\ \mathcal{F} = \bigcap_{A\varepsilon\mathcal{F}} \phi(A)$ and for $D \subseteq T$ $\text{Fil}\ D = \{A\varepsilon\mathcal{P}(X)\colon D \subseteq \phi(A)\}$; then $\text{Fil}\ (\text{Nuc}\ \mathcal{F}) = \mathcal{F}$ and 2.1.2. is verified. So $(Y,\phi)$ is an admissible extension of X.

<u>2.3.2.</u> Given a set X we define $Y = \beta X$ (the Stone-Cech compactification of the discrete topology on X), and for $A \subseteq X$, $\phi(A) = \bar{A} =$ {ultrafilters on X to which A belongs} (so $\phi(A)$ is the closure of A in $\beta X$). We verifiy only 2.1.2. In effect $\bigcap_{A\varepsilon\mathcal{F}} \phi(A) =$ {ultrafilters containing $\mathcal{F}$} and the statement "B belongs to $\mathcal{F}$ iff B belongs to every ultrafilter containing $\mathcal{F}$" is clearly true.

If $(Y,\phi)$ is an admissible extension of X, every $A \subseteq X$ has a correspondent $\phi(A) \subseteq Y$. We now should like to find a correspondent for every <u>fuzzy</u> subset $\mu$ of X. As a guide we take the example 2.3.2.

An ultrafilter $G$ is in $\phi(A)$ iff A belongs to $G$, and A belongs to an ultrafilter $G$ iff A has non-empty intersection with every element of $G$. If we denote by $1_K$ the characteristic function of a set K, this can be expressed as:

$$\forall G\varepsilon\beta X \,\forall A\varepsilon P(X) \left( 1_{\phi(A)}(G) = \bigwedge_{H\varepsilon G} \bigvee_{x\varepsilon X} 1_H(x)\, 1_A(x) = \bigwedge_{H\varepsilon G} \bigvee_{x\varepsilon H} 1_A(x). \right)$$

So, given a fuzzy subset $\mu$ of X it is natural to define the <u>fuzzy</u> subset $\tilde{\phi}(\mu)$ of $\beta X$ in this way:

$$\forall \mu\varepsilon I^X \forall G\varepsilon\beta X \left( <\tilde{\phi}(\mu),G> = \bigwedge_{H\varepsilon G} \bigvee_{x\varepsilon H} \mu(x), \text{ where } H \subseteq X. \right)$$

Recalling that $H\varepsilon G$ iff $G\varepsilon\phi(H)$ this leads us - in the general case - to the following:

<u>2.4 Definition</u>

Let $(Y,\phi)$ be an admissible extension of X. We denote by $\tilde{\phi}$ the function $\tilde{\phi} : I^X \to I^Y$ so defined:

$$\forall \mu\varepsilon I^X \forall q\varepsilon Y \left( <\tilde{\phi}(\mu),q> = \bigwedge_{q\varepsilon\phi(H)} \bigvee_{x\varepsilon H} \mu(x), \text{ where } H \subseteq X. \right)$$

Every $A\varepsilon P(X)$ can be identified by its characteristic function $1_A\varepsilon I^X$. So in order that definition 2.4 makes sense we need that the image of $1_A$ be the characteristic function of $\phi(A)$ _ $P(Y)$. This is indeed our case:

<u>2.5 Property</u>

$$\forall A\varepsilon P(X) \left( \tilde{\phi}(1_A) = 1_{\phi(A)}. \right)$$

<u>Proof</u>: From the definition

$$<\tilde{\phi}(1_A),q> = \bigwedge_{q\varepsilon\phi(H)} \bigvee_{x\varepsilon H} 1_A(x).$$

If $q\varepsilon\phi(A)$, given any H s.t. $q\varepsilon\phi(H)$, $q\varepsilon\phi(A)\ \ \phi(H) = \phi(A\ \ H)$. So A  H cannot be empty ($\phi(K) = \emptyset \Leftrightarrow K = \emptyset$ because $\phi$ is a boolean monomorphism) and there exists $t\varepsilon A\cap H$. So if $q\varepsilon\phi(H)$,

$$\bigvee_{x\varepsilon H} 1_A(x) \geq 1_A(t) = 1 \text{ and } <\tilde{\phi}(1_A), q> = 1.$$

If $q\notin\phi(A)$ then $q\varepsilon\phi(X\setminus A)$; so

$$\bigwedge_{q\varepsilon\phi(H)} \bigvee_{x\varepsilon H} 1_A(x) \leq \bigvee_{x\varepsilon X\setminus A} 1_A(x) = 0.$$

As it is well known (4) a fuzzy set $\mu$ can be identified in a canonical way with the family $\{\mu_\alpha\}_{\alpha\varepsilon I}$ of its level sets $\mu_\alpha = \{x\varepsilon X: \mu(x) \geq \alpha\}$. So, given $\mu\varepsilon I^X$, it is also natural to construct an image of $\mu$ in $I^Y$ by means of the images of the level sets:

2.6 Definition

Given $\mu\varepsilon I^X$ we define $\tilde{\mu}\varepsilon I^Y$ in this way:

$$\forall q\varepsilon Y, \quad \tilde{\mu}(q) = \bigvee\{\alpha : q\varepsilon\phi(\mu_\alpha)\}.$$

It is important to note that:

2.7 Property

$$\forall \mu\varepsilon I^X \left( \tilde{\phi}(\mu) = \tilde{\mu}. \right)$$

Proof - I) $\forall q\varepsilon Y, \quad \tilde{\mu}(q) \leq \langle\tilde{\phi}(\mu), q\rangle$.

Let us take $\gamma\varepsilon I$ and $H\varepsilon P(X)$ s.t. $q\varepsilon\phi(\mu_\gamma)$ and $q\varepsilon\phi(H)$. Then there exists $t\varepsilon H\cap\mu_\gamma$. So for each $\gamma$ s.t. $q\varepsilon\phi(\mu_\gamma)$ we have

$$\bigwedge_{q\varepsilon\phi(H)} \bigvee_{x\varepsilon H} \mu(x) \geq \gamma \text{ and } \langle\tilde{\phi}(\mu), q\rangle \geq \bigvee\{\alpha : q\varepsilon\phi(\mu_\alpha)\} = \tilde{\mu}(q).$$

II) $\forall q\varepsilon Y \langle\tilde{\phi}(\mu), q\rangle \leq \tilde{\mu}(q)$.

It is easy to see that $\tilde{\mu}(q) = \bigvee\{\alpha : q\varepsilon\phi(\mu_\alpha)\} = \bigwedge\{\beta : q\not\varepsilon\phi(\mu_\beta)\}$.
Let us take any $\gamma$ s.t. $q\not\varepsilon\phi(\mu_\gamma)$; then $q\varepsilon\phi(X\setminus\mu_\gamma)$ and $\bigwedge_{q\varepsilon\phi(H)} \bigvee_{x\varepsilon H} \mu(x) \leq \bigvee_{x\varepsilon X\setminus\mu_\gamma} \mu(x) \leq \gamma$. Then
$\langle\tilde{\phi}(\mu), q\rangle \leq \bigwedge\{\beta : q\not\varepsilon\phi(\mu_\beta)\} = \tilde{\mu}(q)$.

2.8 Remark

We have $\bigvee\{\alpha : q\varepsilon(\tilde{\mu})_\alpha\} = \bigvee\{\beta : q\varepsilon\phi(\mu_\beta)\}$ because they are both equal to $\tilde{\mu}(q)$; but $(\tilde{\mu})_\alpha \overset{\supseteq}{\neq} \phi(\mu_\alpha)$ and the equality does not hold in general; furthermore the LHS of the equality is always a maximum while the RHS need not to be it.

$I^X$ (and $I^Y$) is a complete browerian lattice (see (1)) and $2^X$ can be thought as a sublattice which is a boolean algebra; $\phi$ is a boolean monomorphism from $2^X$ to $2^Y$ and $\tilde{\phi}$ is a function from $I^X$ to $I^Y$ which extends $\phi$. Now we shall see that $\tilde{\phi}$ is a lattice morphism.

2.9 Lemma

$\tilde{\phi} : I^X \to I^Y$ preserves the order.

Proof: If $\nu \leq \mu$, for every $\alpha$, $\nu_\alpha \subseteq \mu_\alpha$ and then $\phi(\nu_\alpha) \subseteq \phi(\mu_\alpha)$. So

$$\langle\tilde{\phi}(\nu), q\rangle = \bigvee\{\beta : q\varepsilon\phi(\nu_\beta)\} \leq \bigvee\{\gamma : q\varepsilon\phi(\mu_\gamma)\} = \langle\tilde{\phi}(\mu), q\rangle$$

<u>2.10 Theorem</u>

$\tilde{\phi} : I^X \to I^Y$ is a lattice morphism.

<u>Proof</u> - I) $\tilde{\phi}(\mu\wedge\gamma) = \tilde{\phi}(\mu)\wedge\tilde{\phi}(\nu)$.

By Lemma 2.9 it is sufficient to prove that $\tilde{\phi}(\mu)\wedge\tilde{\phi}(\nu) \leq \tilde{\phi}(\mu\wedge\nu)$.

$$\langle\tilde{\phi}(\mu\wedge\nu), q\rangle = \bigwedge\{\gamma : q\varepsilon\phi((\mu\wedge\nu)_\gamma)\} = \{\gamma : q\notin\phi(\mu_\gamma)\cap\phi(\nu_\gamma)\}.$$

Let us take $\gamma$ s.t. $q\notin\phi(\mu_\gamma)\cap\phi(\nu_\gamma)$. Then for example $q\notin\phi(\mu_\gamma)$ and $\gamma \geq \bigwedge\{\alpha : q\notin\phi(\mu_\alpha)\}$. Then $\gamma \geq \langle\tilde{\phi}(\mu), q\rangle \geq \langle\tilde{\phi}(\mu), q\rangle\wedge\langle\tilde{\phi}(\nu), q\rangle$.

II) $\tilde{\phi}(\mu\vee\nu) = \tilde{\phi}(\mu)\vee\tilde{\phi}(\nu)$.

It is sufficient to prove that $\tilde{\phi}(\mu\vee\nu) \leq \tilde{\phi}(\mu)\vee\tilde{\phi}(\nu)$.

$$\langle\tilde{\phi}(\mu\vee\nu), q\rangle = \bigvee\{\alpha : q\varepsilon\phi((\mu\vee\nu)_\alpha)\} = \bigvee\{\alpha : q\varepsilon\phi(\mu_\alpha)\cup\phi(\nu_\alpha)\}.$$

Let us take $\alpha$ s.t. $q\varepsilon\phi(\mu_\alpha)\cup\phi(\nu_\alpha)$. Then for example $q\varepsilon\phi(\mu_\alpha)$. So

$$\alpha \leq \bigvee\{\gamma : q\varepsilon\phi(\mu_\gamma)\} = \langle\tilde{\phi}(\mu), q\rangle \text{ and } \alpha \leq \langle\tilde{\phi}(\mu), q\rangle\vee\langle\tilde{\phi}(\nu), q\rangle.$$

<u>2.11 Remarks</u>

2.11.1. It is easy to see that $\tilde{\phi}$ does not preserve pseudo-complements.

2.11.2. For every $\alpha\varepsilon I$ $\tilde{\phi}(\underline{\alpha}) = \underline{\alpha}$.

## 3. NEARNESS AND COMPACTNESS IN FUZZY TOPOLOGICAL SPACES

In non-standard topology the basic concepts are those of <u>nearness</u> and of <u>monad of a point</u> (see (3)).

The monad $M(p)$ of a point $p\varepsilon X$ is the set $\{q{*}\varepsilon X$: q is near to p$\}$. Here "q near to p" means that q is a *member of every neighborhood of p, i.e. an element of the nucleus of the neighborhood filter of p.

This leads to the following:

<u>3.1 Definition</u>

Let $(X,\mathcal{T})$ be a topological space and $(Y,\phi)$ an admissible extension of X. We define a <u>nearness relation</u> $N \subseteq X \times Y$ in this way: $(x,y)\varepsilon N \Leftrightarrow y\varepsilon \bigcap_{A\varepsilon F[x]} \phi(A)$, where $F[x]$ is the neighborhood filter of x in $\mathcal{T}$. If $(x,y)\varepsilon N$, we say that <u>y is near to x</u>.

3.2 Remark

In the case of example 2.3.2 $(x,\mathcal{U})\varepsilon N$ iff x is a limit point of $\mathcal{U}$ i.e. an ultrafilter $\mathcal{U}$ is near to x iff it contains $F[x]$.

The following statement is a generalization of the famous Robinson's theorem about compact spaces:

3.3 Theorem

Let $(X,\mathcal{T})$ be a topological space, $(Y,\phi)$ any admissible extension of X. $(X,\mathcal{T})$ is compact iff every point of Y is near to some point of X.

Proof - I) Let us suppose that $(X,\mathcal{T})$ is compact and that there exists a point $y\varepsilon Y$ which is not near to any point of X. Then $\forall x\varepsilon X\, \exists A_x\varepsilon F[x] : y\notin\phi(A_x)$. We can choose every $A_x$ open. Let $\{A_{x_1}, \ldots, A_{x_k}\}$ be a finite subcover of the open cover $\{A_x\}_{x\varepsilon X}$. Then $\bigcup_{i=1}^{k} \phi(A_{x_i}) = Y$ and $y\varepsilon\phi(A_{x_j})$ for some j, which is a contradiction.

II) Let us suppose now that $(X,\mathcal{T})$ is not compact. Then there exists an open cover $\{A_i\}_{i\varepsilon J}$ of X which has no finite subcover. Then the family $\{B_i\}_{i\varepsilon J}$ where $B_i = X\setminus A_i$ has the FIP and there exists $y\varepsilon \bigcap_i \phi(B_i)$. Taken any $x\varepsilon X$, $x\varepsilon A_k$ for some $k\varepsilon J$ and $A_k\varepsilon F[x]$. But $y\varepsilon\phi(B_k) = Y\setminus\phi(A_k)$. Then $y\notin\phi(A_k)$ and $(x,y)\notin N$.

3.4 Remark

In the case of Example 2.3.2 Theorem 3.3 says: "$(X,\mathcal{T})$ is compact iff every ultrafilter on X is convergent" (see Remark 3.2).

Our intention is now to introduce a nearness relation in f.t.s.'s in such a way that (fuzzy) compactness and ultracompactness can be characterized by means of theorems of the kind of Theorem 3.3.

We have two problems in the fuzzy case: 1) the extension $\tilde{\phi}(\mu)$ of a fuzzy subset of X is a fuzzy subset of Y; 2) we have no neighborhood filter of x in a f.t.s. and only a membership degree of x to an open set $\mu$. So we must expect that: 1) the nearness relation will be a fuzzy relation; 2) we shall have a concept of $\alpha$-nearness for each membership degree $\alpha$.

Let $I_0$ be - as usual - the set $]0,1]$; given the f.t.s. $(X,\mathcal{T})$ and $p\varepsilon X$ let $\mathcal{T}_\alpha(p)$ be the family $\mathcal{T}_\alpha(p) = \{\mu\varepsilon\mathcal{T}: \mu(p) \geq \alpha\}$. So in $\mathcal{T}_\alpha(p)$ we find the open sets which contain p with a membership degree at least equal to $\alpha$. The nearness relation in $(X,\mathcal{T})$ is a ternary fuzzy relation:

3.5 Definition

We call nearness relation on $(X,\mathcal{T})$ the fuzzy relation $N: I_0 \times X \times Y \to I$ so defined:

$$\forall\alpha\varepsilon I_0 \;\; \forall p\varepsilon X \;\; \forall q\varepsilon Y, \; N(\alpha,p,q) = \bigwedge_{\mu\varepsilon\mathcal{T}_\alpha(p)} \tilde{\mu}(q).$$

3.6 Remark

From the fact that $\underline{\alpha}\varepsilon\mathcal{T}_\alpha(p)$ and from Remark 2.11.2 we have $\forall\alpha\;\forall p\;\forall q,\; N(\alpha,p,q) \leq \alpha$.

3.7 Definitions

Given $\alpha\varepsilon I_0$, $q\varepsilon Y$, $p\varepsilon X$ we say that q is $\alpha$-near to p if $N(\alpha,p,q)$ has its maximum possible value in $(p,q)$, i.e. if $N(\alpha,p,q) = \alpha$. We say that $q\varepsilon Y$ is $\alpha$-near to the set X if $\bigvee_{p\varepsilon X} N(\alpha,p,q) = \alpha$. We say that $q\varepsilon Y$ is near to $p\varepsilon X$ if $\forall\alpha\varepsilon I_0\; N(\alpha,p,q) = \alpha$. We say that $q\varepsilon Y$ is near to the set X if $\forall\alpha\varepsilon I_0 \bigvee_{p\varepsilon X} N(\alpha,p,q) = \alpha$.

We now are ready to characterize ultracompactness and compactness in f.t.s.'s (for definitions see (2)).

3.8 Theorem

Let $(X,\mathcal{T})$ be a f.t.s. and $(Y,\phi)$ any admissible extension of X. $(X,\mathcal{T})$ is ultracompact iff every q in Y is near to some p in X.

Proof - I) Let us suppose $(X,\mathcal{T})$ ultracompact. Then the initial topology is compact and given q in Y there exists - by Theorem 3.3 - an element p of X s.t. $q\varepsilon \bigcap_{B\varepsilon F[p]} \phi(B)$. We prove that q is near to p (as in Definition 3.7). Supposing on the contrary that q is not near to p, there exists $\alpha > 0$ s.t. $N(\alpha,p,q) < \alpha$, i.e.

$$\bigwedge_{\nu\varepsilon\mathcal{T}_\alpha(p)} \; \bigwedge_{q\varepsilon\phi(H)} \; \bigvee_{t\varepsilon H} \nu(t) < \alpha; \text{ then}$$

$$\exists\, \nu\varepsilon\mathcal{T} \;\exists H \subseteq X,\; \nu(p) \geq \alpha \;\; q\varepsilon\phi(H) \;\; \bigvee_{t\varepsilon H} \nu(t) = \delta < \alpha.$$

So $\nu(p) > \delta$, $p\varepsilon\nu^\delta = \{x\varepsilon X: \nu(x) > \delta\}\varepsilon F[p]$ and $q\varepsilon\phi(\nu^\delta)$. Now q is in

$\phi(H)$; so there exists r in $H \cap \nu^{\delta}$. Since $r \in H$, $\nu(r) \leq \delta$ and since $r \in \nu^{\delta}$, $\nu(r) > \delta$, and this is a contradiction.

II) Let us suppose $(X, \mathcal{T})$ not ultracompact. Then the initial topology is not compact and - by Theorem 3.3 - there exists an element $q \in Y$ s.t. $\forall\, p \in X\ q \notin \bigcap_{B \in F[p]} \phi(B)$. So, for each p in X we can find a subbasic open set $\mu^{\alpha}(\mu \in \mathcal{T},\ 0 \leq \alpha < 1)$ s.t. $p \in \mu^{\alpha}$ and $q \notin \phi(\mu^{\alpha})$. Then $q \in \phi(X \setminus \mu^{\alpha})(X \setminus \mu^{\alpha} = \{x \in X\colon \mu(x) \leq \alpha\})$. Now $\mu(p) = \varepsilon > 0$ (remember that $\mu(p) > \alpha$) and $N(\alpha,p,q) \leq \bigvee_{t \in X \setminus \mu^{\alpha}} \mu(t) \leq \alpha < \varepsilon$. Then q is not near to p.

## 3.9 Theorem

Let $(X, \mathcal{T})$ be a f.t.s. and $(Y, \phi)$ any admissible extension of X. $(X, \mathcal{T})$ is compact iff every q in Y is near to X.

Proof - I) Let us suppose that $(X, \mathcal{T})$ is not compact. Then there exist $\alpha, \beta \in I$ - with $\beta < \alpha$ - and $\mathcal{B} \subseteq \mathcal{T}$ s.t. $\bigvee \mathcal{B} \geq \underline{\alpha}$ and for every finite $\mathcal{B}_0 \subseteq \mathcal{B}$, $\bigvee \mathcal{B}_0 \not\geq \underline{\beta}$. From this it follows that, calling $\mathcal{H}$ the family $\mathcal{H} = \{\mu_{(\beta)} : \mu \in \mathcal{B}\}$ with $\mu_{(\beta)} = X \setminus \mu_{\beta} = \{x \in X\colon \mu(x) < \beta\}$, $\mathcal{H}$ has the FIP. Then there exists $q \in Y$ s.t. $q \in \bigcap_{H \in \mathcal{H}} \phi(H)$. Let us take now $\varepsilon$ such that $\beta < \varepsilon < \alpha$ (so in particular $\varepsilon > 0$). By the hypothesis $\bigvee \mathcal{B} \geq \underline{\alpha}$, given any $p \in X$ we can find $\nu \in \mathcal{B}$ s.t. $\nu(p) \geq \varepsilon$. So we have $\nu \in \mathcal{T}_{\varepsilon}(p)$ and $q \in \phi(\nu_{(\beta)})$. Then $N(\varepsilon,p,q) \leq \bigvee_{t \in \nu_{(\beta)}} \nu(t) \leq \beta$. This is true for every $p \in X$; then $\bigvee_{p \in X} N(\varepsilon,p,q) \leq \beta < \varepsilon$. So q is not near to X.

II) Let $(X, \mathcal{T})$ be compact. Let us suppose that there exists $q \in Y$ not near to X. So there exists $\alpha > 0$ s.t. $\bigvee_{p \in X} N(\alpha,p,q) < \alpha$. Then we can find $\varepsilon$

$$0 < \varepsilon < \alpha \quad \text{s.t.} \quad \bigvee_{p \in X} N(\alpha,p,q) < \varepsilon \quad \text{i.e.}$$

$$\bigvee_{p \in X}\ \bigwedge_{\mu \in \mathcal{T}_{\alpha}(p)}\ \bigwedge_{q \in \phi(H)}\ \bigvee_{t \in H} \mu(t) < \varepsilon .$$

Then

$$(*) \qquad \forall\, p \in X\ \exists\, \mu^{p} \in \mathcal{T}_{\alpha}(p)\ \exists\, H^{p} \subseteq X\ \forall\, t \in H^{p}\ \ q \in \phi(H^{p}) \text{ and } \mu^{p}(t) < \varepsilon .$$

We take the family $\mathcal{B} = \{\mu^{p}\}_{p \in X}$; of course $\bigvee \mathcal{B} \geq \underline{\alpha}$. Since $(X, \mathcal{T})$ is compact and $\varepsilon < \alpha$ there exists $\mathcal{B}_0$ finite $\mathcal{B}_0 \subseteq \mathcal{B}$ s.t. $\mathcal{B}_0 \geq \underline{\varepsilon}$. Let us suppose $\mathcal{B}_0 = \{\mu^{x_1}, \ldots, \mu^{x_n}\}$ with $x_1, \ldots, x_n \in X$. The family $\{\mu_{\varepsilon}^{x_i}\}_{1 \leq i \leq n}$ is a cover of Y and there exists k, $1 \leq k \leq n$ s.t.

$q \varepsilon \phi(\mu_\varepsilon^{x_k})$. So $q \varepsilon \phi(\mu_\varepsilon^{x_k}) \cap \phi(H^{x_k})$ (see (*)) and there exists $r \varepsilon \mu_\varepsilon^{x_k} \cap H^{x_k}$. But this is a contradiction because $r \varepsilon \mu_\varepsilon^{x_k}$ implies $\mu^{x_k}(r) \geq \varepsilon$ while $r \varepsilon H^{x_k}$ implies (see (*)) $\mu^{x_k}(r) < \varepsilon$.

REFERENCES

1. A. Deluca and S. Termini, Algebraic properties of fuzzy sets, J. Math. Anal. Appl., 40, 373/386 (1972).
2. R. Lowen, A comparison of different compactness notions in fuzzy topological spaces, J. Math. Anal. Appl., 64, 446/454, (1978).
3. M. Machover and J. Hirschfeld, Lectures on Non-Standard Analysis, L.N.M. 94, Springer-Verlag (1969).
4. C. V. Negoita and D. A. Ralescu, Application of Fuzzy Sets to Systems Analysis, John Wiley, New York (1975).

THE SPACES OF FUZZY PROBABILITY AND POSSIBILITY (I)

Zhende Huang[1] and Tong Zhengxiang[2]

Research Section of Fuzzy Mathematics[1]
Huazhong Institute of Technology
Wuhan, China
Research Section of Multi-Logic and Fuzzy System[2]
Shanghai Railway Institute
Shanghai, China

## 1. INTRODUCTION

Zadeh[1], 1978, pointed out that Fuzzy Sets' theory can be a basis for a theory of possibility, which is similar to the role of measure theory for probability theory. He put emphasis on the fact that a variable usually relates with both probability distribution and possibility distribution. Though he developed the possibility/probability consistency principle, he found it difficult to investigate all the relations between probability distribution and possibility distribution[2]. Ellen Hisdal[3], 1979, considered the relations between conditional possibilities independence and noninteraction in the case of 2-dimensional possibility distribution. Her results are not very satisfactory because the definition of possibility independence developed in [3] is analogous to that of probabilities independence. We feel that the results in [3] didn't make the relations between independence and noninteraction of two fuzzy events clear. Following the basic definitions in [2], this paper discusses fuzzy events, develops the concept of fuzzy probability and possibility spaces and investigates some properties and weak relations between p-independence and $\pi$-noninteraction of two fuzzy events, setting the stage for further research.

## 2. FUZZY EVENTS AND FUZZY PROBABILITY AND POSSIBILITY SPACES

Let $\Omega$ be a set, $(\Omega,\Sigma,P)$ be a probability space, $BL([0,1])$ be a system of Borel subsets in $[0,1]$, $F(\Omega)$ be a lattice of fuzzy subsets on $\Omega$.

DEFINITION 1. The mapping $\mu_B:\Omega\to[0,1]$ is called a $\Sigma$-m, iff

$$\forall\alpha\varepsilon[0,1], \quad \{\omega|\mu_B(\omega)\varepsilon[0,\alpha]\}\varepsilon\Sigma \tag{1}$$

PROPOSITION 1. The mapping $\mu_B:\Omega\to[0,1]$ is a $\Sigma$-m, iff

$$\forall A\varepsilon BL([0,1]), \quad \{\omega|\mu_B(\omega)\varepsilon A\}\varepsilon\Sigma \tag{2}$$

PROOF. Let $A = [0,\alpha]$ in (2), then we obtain (1) immediately.

$$\text{Let } M = \{A|A\ [0,1], \quad \{\omega|\mu_B(\omega)\varepsilon A\}\varepsilon\Sigma\} \tag{3}$$

First, it is easy to show that $M$ is a $\sigma$-algebra in $[0,1]$. In fact,

1. $[0,1]\varepsilon M$, because $\{\omega|\mu_B(\omega)\varepsilon[0,1]\} = \Omega\varepsilon\Sigma$

2. if $A\varepsilon M$, then $A^c = [0,1]-A\subset[0,1]$ and $\{\omega|\mu_B(\omega)\varepsilon A^c\} = \{\omega|\mu_B(\omega)\varepsilon A\}^c\varepsilon\Sigma$, namely, $A^c\varepsilon M$.

3. if $A_i\varepsilon M$, $i=1,2,\ldots$, then $\bigcup_{i=1}^{\infty} A_i\subset[0,1]$ and $\{\omega|\mu_B(\omega)\varepsilon\bigcup_{i=1}^{\infty} A_i\} = \bigcup_{i=1}^{\infty}\{\omega|\mu_B(\omega)\varepsilon A_i\}\varepsilon\Sigma$ namely, $\bigcup_{i=1}^{\infty} A_i\varepsilon M$. Secondly, we can show that $\forall\alpha\varepsilon[0,1]$, $[0,\alpha]\varepsilon M$. In fact, it is evident that $[0,\alpha]\subset[0,1]$ and $\{\omega|\mu_B(\omega)\varepsilon[0,\alpha]\}\varepsilon\Sigma$, since $\mu_B$ is a $\Sigma$-m.

We have now proved that $M$ is a $\sigma$-algebra and $BL([0,1])\subset M$. So $A\varepsilon BL([0,1])$, $\{\omega|\mu_B(\omega)\varepsilon A\}\varepsilon\Sigma$. The proposition is true. Q.E.D.

DEFINITION 2. Let $\beta = \{B|B\varepsilon F(\Omega),\mu_B \text{ is a } \Sigma\text{-m}\}$. If $B\varepsilon\beta$, then $B$ is called a fuzzy event on $\Omega$.

PROPOSITION 2. All fuzzy events in $(\Omega,\Sigma,P)$, $\beta$, constitute a fuzzy $\sigma$-algebra in $\Omega$.

PROOF. Corresponding to the three conditions of fuzzy $\sigma$-algebra, we prove the proposition as follows.

1. $\Omega\varepsilon\beta$, since

$$\{\omega|\mu_\Omega(\omega)\varepsilon[0,\alpha]\} = \begin{cases}\Omega\varepsilon\Sigma & \alpha=1\\ \phi\varepsilon\Sigma & \alpha\varepsilon[0,1)\end{cases}$$

2. If $B\varepsilon\beta$, it is evident from PROP. 1 that for any $\alpha\varepsilon[0,1]$,

$$\{\omega|\mu_{B^c}(\omega)\varepsilon[0,\alpha]\} = \{\omega|1-\mu_B(\omega)\varepsilon[0,\alpha]\}$$
$$= \{\omega|\mu_B(\omega)\varepsilon[1-\alpha,1]\}\varepsilon\Sigma$$

namely, $B^c\varepsilon\beta$.

3. If $B_i\varepsilon\beta$, $i=1,2,\ldots$, then for any $\alpha\varepsilon[0,1]$,

$$\{\omega|\mu_{\bigcup_{i=1}^{\infty} B_i}(\omega)\varepsilon[0,\alpha]\} = \{\omega \sup_i\{\mu_{B_i}(\omega)\}\varepsilon[0,\alpha]\}$$
$$= \bigcap_{i=1}^{\infty}\{\omega|\mu_{B_i}(\omega)\varepsilon[0,\alpha]\}\varepsilon\Sigma$$

namely $\bigcup_{i=1}^{\infty} B_i\varepsilon\beta$. Q.E.D.

COROLLARY 1. 1° $\phi\varepsilon\beta$

2° If $B_i\varepsilon\beta$, $i=1,2,\ldots$, then $\bigcap_{i=1}^{\infty} B_i\varepsilon\beta$.

PROPOSITION 3. Let $(\Omega,\Sigma,P)$ be a probability space, $B\varepsilon F(\Omega)$. The following statements are equivalent to each other.

1. B is a fuzzy event on $\Omega$.
2. $\mu_B$ is a $\Sigma$-m.
3. $\forall\alpha\varepsilon[0,1]$, $B_\alpha\varepsilon\Sigma$, where $B_\alpha$ is the strong $\alpha$-cut of B, i.e., $B_\alpha = \{\omega|\mu_B(\omega)\geq\alpha\}$.
4. $\forall\alpha\varepsilon[0,1]$, $B_{\alpha\omega}\varepsilon\Sigma$, where $B_{\alpha\omega}$ is the weak $\alpha$-cut of B, i.e., $B_{\alpha\omega} = \{\omega|\mu_B(\omega)>\alpha\}$.
5. $\forall c\varepsilon(-\infty,\infty)$, $\{\omega|\mu_B(\omega)>c\}\varepsilon\Sigma$
6. $\forall c\ (-\infty,\infty)$, $\{\omega|\mu_B(\omega)<c\}\varepsilon\Sigma$

PROOF. Note that $\Sigma$ is a $\sigma$-algebra. Then it is clear from definitions and some simple sets operations that the proposition is true.

PROPOSITION 4. If $A,B\varepsilon\beta$, then $A\oplus B\varepsilon\beta$ and $A\otimes B\varepsilon\beta$, where

$A\oplus B$: $\mu_{A\oplus B}(\omega) = \min(1,\ \mu_A(\omega)+\mu_B(\omega))$;

$A\otimes B$: $\mu_{A\otimes B}(\omega) = \max(0,\ \mu_A(\omega)+\mu_B(\omega)-1)$.

PROOF. First, we assert that $\mu_A(\omega)+\mu_B(\omega)$ is a $\Sigma$-m. In order to prove this assertion, it is sufficient to show that for any $c\varepsilon(-\infty,\infty)$, we have $\{\omega|\mu_A(\omega)+\mu_B(\omega)>c\}\varepsilon\Sigma$. Now, let $\{\gamma_n\}$ be that sequence of all rational numbers. It is clear that

$$\{\omega|\mu_A(\omega)+\mu_B(\omega)>c\} = \bigcup_{i=1}^{\infty}\{\omega|\mu_A(\omega)>\gamma_i\}\cap\{\omega|\mu_B(\omega)>c-\gamma_i\}.$$

Since $\forall i=1,2,\ldots$, $\{\omega|\mu_A(\omega)>\gamma_i\}\in\Sigma$, $\{\omega|\mu_B(\omega)>c-\gamma_i\}\in\Sigma$ and $\Sigma$ is a $\sigma$-algebra, we have $\{\omega|\mu_A(\omega)+\mu_B(\omega)>C\}\in\Sigma$.

Now that

$$\{\omega|\mu_{A\oplus B}(\omega)>c\} = \{\omega|\mu_\Omega(\omega)>c\}\cap\{\omega|\mu_A(\omega)+\mu_B(\omega)>c\},$$

and

$$\{\omega|\mu_{A\otimes B}(\omega)>c\} = \{\omega|\mu_\phi(\omega)>c\}\cup\{\omega|\mu_A(\omega)+\mu_B(\omega)>1+c\},$$

we have $\{\omega|\mu_{A\oplus B}(\omega)>c\}\in\Sigma$ and $\{\omega|\mu_{A\otimes B}(\omega)>c\}\in\Sigma$ namely $A\oplus B\in\beta$ and $A\otimes B\in\beta$. Q.E.D.

PROPOSITION 5. If $A,B\in\beta$, then $A\cdot B\in\beta$ and $A\hat{+}B\in\beta$, where

$$A\cdot B: \quad \mu_{A\cdot B}(\omega) = \mu_A(\omega)\cdot\mu_B(\omega)$$

$$A\hat{+}B: \quad \mu_{A\hat{+}B}(\omega) = \mu_A(\omega)+\mu_B(\omega)-\mu_A(\omega)\mu_B(\omega).$$

PROOF. First, we assert that $\mu_A(\omega)-\mu_B(\omega)$ is a $\Sigma$-m. In fact, let $\{\gamma_i\}$ be the sequence of all rational numbers, it is clear that for any $c\in(-\infty,\infty)$

$$\{\omega|\mu_A(\omega)-\mu_B(\omega)>c\} = \bigcup_{i=1}^{\infty}\{\omega|\mu_A(\omega)>\gamma_i\}\cap\{\omega|\mu_B(\omega)<\gamma_i-c\}.$$

Since $\Sigma$ is a $\sigma$-algebra, $\mu_A(\omega)-\mu_B(\omega)$ is a $\Sigma$-m.

Furthermore we can prove that if $\mu_A(\omega)$ is a $\Sigma$-m, then so is $\mu_{A\cdot A}(\omega)$. In fact, for any $\alpha\in[0,1]$, we have $\sqrt{\alpha}\in[0,1]$ and $(A\cdot A)_\alpha=A_{\sqrt{\alpha}}$. It is clear from PROP. 3 that $(A\cdot A)_\alpha\in\Sigma$, namely, $\mu_{A\cdot A}(\omega)$ is a $\Sigma$-m.

Now that $\mu_{A\cdot B}(\omega) = \left[\frac{1}{4}\,(\mu_A(\omega)+\mu_B(\omega))^2-(\mu_A(\omega)-\mu_B(\omega))\right]^2$ and $(\mu_A(\omega)+\mu_B(\omega))^2-(\mu_A(\omega)-\mu_B(\omega))^2$ is a $\Sigma$-m, we have that $\mu_{A\cdot B}(\omega)$ is a $\Sigma$-m, i.e., $A\cdot B\in\beta$. Of course, $\mu_{A\hat{+}B}(\omega)$ is a $\Sigma$-m too. Q.E.D.

DEFINITION 3. If B is a fuzzy event in the probability space $(\Omega,\Sigma,P)$, then the mathematical expectation of $\mu_B$, i.e., $E[\mu_B]$, is called the fuzzy probability measure of B, denoted by P(B).

PROPOSITION 6. Fuzzy probability measure possesses the following properties:

1. $\forall B\in\beta$, $0 \leq \tilde{P}(B) \leq 1$;
2. $\tilde{P}(\Omega) = 1$;
3. $A,B\in\beta$, $A\subset B \Rightarrow \tilde{P}(A) \leq \tilde{P}(B)$;
4. if there are such countable fuzzy events $B_n$ that for any $i\neq j$, $B_i\cap B_j=\phi$, then

$$\tilde{P}\left(\bigcup_{n=1}^{\infty} B_n\right) = \sum_{n=1}^{\infty} \tilde{P}(B_n).$$

PROOF. Properties 1, 2 and 3 are very simple. So we only prove properties 4. Under the given conditions,

$$\mu_{\bigcup_{n=1}^{m} B_n}(\omega) = \sum_{n=1}^{m} \mu_{B_n}(\omega) \qquad m=1,2,\ldots$$

$$\mu_{\bigcup_{n=1}^{\infty} B_n}(\omega) = \sum_{n=1}^{\infty} \mu_{B_n}(\omega) .$$

Then

$$\begin{aligned}\tilde{P}\left(\bigcup_{n=1}^{\infty} B_n\right) &= E\left[\mu_{\bigcup_{n=1}^{\infty} B_n}\right] \\ &= E\left[\sum_{n=1}^{\infty} \mu_{B_n}\right] \\ &= \sum_{n=1}^{\infty} E[\mu_{B_n}] \\ &= \sum_{n=1}^{\infty} \tilde{P}(B_n)\end{aligned}$$

Q.E.D.

DEFINITION 4. The mapping $\pi:\Omega\to[0,1]$ is called a possibility distribution on $\Omega$, if $\forall\omega\in\Omega$, $\pi(\omega)\geq P(\omega)$.

DEFINITION 5. If $B\in\beta$, then $\bigvee_{\omega\in\Omega}(\pi(\omega)\wedge\mu_B(\omega))$ is called the possibility measure of fuzzy event B, denoted by $\pi(B)$.

DEFINITION 6. By a fuzzy prpbability and possibility space we mean a 5-tuple $(\Omega,\Sigma,\beta,P,\pi)$ in which $(\Omega,\Sigma,P)$ is a probability space, $\beta$ is the family of all fuzzy events in $(\Omega,\Sigma,P)$ and $\pi$ is a possibility distribution.

PROPOSITION 7. If A, $B\in\beta$, $A\subset B$, then $\pi(A)\leq\pi(B)$.

## 3. P-INDEPENDENCE AND $\pi$-NONINTERACTION BETWEEN TWO FUZZY EVENTS

DEFINITION 7. Let A, $B\in\beta$. If

$$\tilde{P}(A\cdot B) = \tilde{P}(A)\cdot\tilde{P}(B),$$

then A and B are said to be independent to each other under the probability distribution P, or, simply, P-independent.

DEFINITION 8. Let A, B$\epsilon\beta$. If

$$\pi(A \cap B) = \pi(A) \wedge \pi(B),$$

then A and B are said to be noninteraction to each other under the possibility distribution $\pi$ or, simply, $\pi$-noninteraction.

The following two propositions are well known.

PROPOSITION 8. If $\mu_A(\omega) \equiv$ const, then $\forall B\epsilon\beta$, A and B are P-independent.

PROPOSITION 9. If A, B$\epsilon\beta$, then the following statements are equivalent to each other.

1. A and B are P-independent.
2. A and $B^c$ are P-independent.
3. $A^c$ and $B^c$ are P-independent.

DEFINITION 9. Let $\Omega$ be a finite set $\{\omega_1,\omega_2,\ldots,\omega_n\}$. The probability distribution P is said to be undegenerative, if for any $i\epsilon\{1,2,\ldots,n\}$, $P(\omega_i)>0$.

In order to make clear the properties of P-independence and $\pi$-noninteraction and their relations, we consider at present the simplest cases, namely the cases in which $\Omega$ contains two or three elements.

Let $\Omega$ be the set $\{\omega_1,\omega_2\}$, $\Sigma$ consists of all subsets of $\Omega$. The probability P, possibility distribution $\pi$, fuzzy sets A and B are as follows:

| $\Omega$ | $\omega_1$ | $\omega_2$ |
|---|---|---|
| P | x | 1-x |
| $\pi$ | $\pi_1$ | $\pi_2$ |
| A | $a_1$ | $a_2$ |
| B | $b_1$ | $b_2$ |

where, $\mu_A(\omega_i) = a_i$, $\mu_B(\omega_i) = b_i$, i=1,2, $0\le x\le 1$. Without loss of generality, we assume that $a_1 \ge a_2$.

PROPOSITION 10. A and B are $\pi$-noninteraction iff $\pi_1 \wedge \pi_2 \wedge a_1 \wedge b_2 \le \pi(A \cap B)$.

PROOF. From Definition 5, we have

$$\pi(A) \wedge \pi(B) = [(\pi_1 \wedge a_1) \vee (\pi_2 \wedge a_2)] \wedge [(\pi_1 \wedge b_1) \vee (\pi_2 \wedge b_2)]$$

$$= (\pi_1 \wedge \pi_2 \wedge a_1 \wedge b_2) \vee (\pi_1 \wedge \pi_2 \wedge a_2 \wedge b_1) \vee \pi(A \cap B).$$

Since

$$\pi_1 \wedge \pi_2 \wedge a_2 \wedge b_1 \leq \pi_1 \wedge a_1 \wedge b_1 \leq \pi(A \cap B),$$

the proposition is true. Q.E.D.

COROLLARY 1. If $a_1 = a_2$, then A and B are $\pi$-noninteraction.

PROOF. In this case, we have $\pi_1 \wedge \pi_2 \wedge a_1 \wedge b_2 \leq \pi_2 \wedge a_2 \wedge b_2 \leq \pi(A \cap B)$.

COROLLARY 2. If $b_1 \geq b_2$, then A and B are $\pi$-noninteraction.

PROOF. In this case, we have

$$\pi_1 \wedge \pi_2 \wedge a_1 \wedge b_2 \leq \pi_1 \wedge a_1 \wedge b_1 \leq \pi(A \cap B).$$

COROLLARY 3. If $\pi_1 \leq a_2$, then A and B are $\pi$-noninteraction.

PROOF. In this case, we have

$$\pi_1 \wedge \pi_2 \wedge a_1 \wedge b_2 = \pi_1 \wedge \pi_2 \wedge b_2 \leq \pi(A \cap B).$$

COROLLARY 4. If $a_2 \vee b_1 < \pi_1 \wedge \pi_2 \wedge a_1 \wedge b_2$, then A and B are not $\pi$-noninteraction.

PROOF. In this case, it is clear that

$$\pi_1 \wedge \pi_2 \wedge a_1 \wedge b_2 > \pi(A \cap B).$$

PROPOSITION 11. If $a_1 = a_2$ or $b_1 = b_2$, then A and B are P-independent, where P is any probability distribution on $\Omega$.

This proposition is the special case of PROP. 8.

PROPOSITION 12. If $a_1 \neq a_2$, $b_1 \neq b_2$, then A and B are not P-independent, where P is any undegenerative probability distribution on $\Omega$.

PROOF. We have

$$\begin{aligned} P(A)P(B)-P(A\cdot B) &= [xa_1+(1-x)a_2][xb_1+(1-x)b_2]-[xa_1b_1+(1-x)a_2b_2] \\ &= x(1-x)(a_1-a_2)(b_1-b_2). \end{aligned}$$

Under the conditions of $a_1 \neq a_2$ and $b_1 \neq b_2$, the function $P(A)P(B)-P(A\cdot B)$ has only two zero points, i.e., $x=0$ and $x=1$. So under an undegenerative probability distribution P, i.e., $0<x<1$, we have $P(A\cdot B) \neq P(A)P(B)$. Q.E.D.

PROPOSITION 13. If $a_1 = a_2$ or $b_1 = b_2$, then for any possibility distribution $\pi$ and probability distribution P, A and B are $\pi$-independent. If $a_1 > a_2$, $b_1 > b_2$ then A and B are noninteraction under any possibility distribution $\pi$ but are not independent under any undegererative probability distribution P.

This is the combination of PROP. 11, 12 and CORO. 1, 2.

Now let $\Omega$ be the set $\{\omega_1, \omega_2, \omega_3\}$, $\Sigma$ consists of all subsets in $\Omega$. The probability P, possibility distribution $\pi$, fuzzy sets A and B are as follows:

| $\Omega$ | $\omega_1$ | $\omega_2$ | $\omega_3$ |
|---|---|---|---|
| P | x | y | 1-x-y |
| $\pi$ | $\pi_1$ | $\pi_2$ | $\pi_3$ |
| A | $a_1$ | $a_2$ | $a_3$ |
| B | $b_1$ | $b_2$ | $b_3$ |

where $x \geq 0$, $y \geq 0$, $x+y \leq 1$. Without loss of generality, we assume that $a_1 \geq a_2 \geq a_3$.

PROPOSITION 14. A and B are $\pi$-noninteraction iff the following inequalities hold.

$$\pi_1 \wedge \pi_2 \wedge a_1 \wedge b_2 \leq \pi(A \cap B)$$

$$\pi_1 \wedge \pi_3 \wedge a_1 \wedge b_3 \leq \pi(A \cap B)$$

$$\pi_2 \wedge \pi_3 \wedge a_2 \wedge b_3 \leq \pi(A \cap B)$$

PROOF. From DE. 5, we have

$$\begin{aligned}\pi(A) \wedge \pi(B) &= [(\pi_1 \wedge a_1) \vee (\pi_2 \wedge a_2) \vee (\pi_3 \wedge a_3)] \wedge [(\pi_1 \wedge b_1) \\ &\quad \vee (\pi_2 \wedge b_2) \vee (\pi_3 \wedge b_3)] \\ &= (\pi_1 \wedge \pi_2 \wedge a_1 \wedge b_2) \vee (\pi_1 \wedge \pi_3 \wedge a_1 \wedge b_3) \vee \\ &\quad (\pi_2 \wedge \pi_3 \wedge a_2 \wedge b_3) \vee (\pi_2 \wedge \pi_1 \wedge a_2 \wedge b_1) \\ &\quad \vee (\pi_3 \wedge \pi_1 \wedge a_3 \wedge b_1) \vee (\pi_3 \wedge \pi_2 \wedge a_3 \wedge b_2) \vee \pi(A \cap B) \\ &= (\pi_1 \wedge \pi_2 \wedge a_1 \wedge b_2) \vee (\pi_1 \wedge \pi_3 \wedge a_1 \wedge b_3) \\ &\quad \vee (\pi_2 \wedge \pi_3 \wedge a_2 \wedge b_3) \vee \pi(A \cap B)\end{aligned}$$

So the proposition is true. Q.E.D.

COROLLARY 5. If $b_1 \geq b_2 \geq b_3$, then A and B are $\pi$-noninteraction.

PROOF. In this case, we have

$$\pi_1 \wedge \pi_2 \wedge a_1 \wedge b_2 \leq \pi_1 \wedge a_1 \wedge b_1 \leq \pi(A \cap B)$$

$$\pi_1 \wedge \pi_3 \wedge a_1 \wedge b_3 \leq \pi_1 \wedge a_1 \wedge b_1 \leq \pi(A \cap B)$$

$$\pi_2 \wedge \pi_3 \wedge a_2 \wedge b_3 \leq \pi_2 \wedge a_2 \wedge b_2 \leq \pi(A \cap B)$$ Q.E.D.

COROLLARY 6. If $b_1 \geq \max\{b_2,b_3\}$, $\pi_1 \geq \min\{\pi_2,\pi_3\}$, then A and B are $\pi$-noninteraction.

PROOF. In this case, we have

$$\pi_1 \wedge \pi_2 \wedge a_1 \wedge b_2 \leq \pi_1 \wedge a_1 \wedge b_1 \leq \pi(A \cap B)$$

$$\pi_1 \wedge \pi_3 \wedge a_1 \wedge b_3 \leq \pi_1 \wedge a_1 \wedge b_1 \leq \pi(A \cap B)$$

$$\pi_2 \wedge \pi_3 \wedge a_2 \wedge b_3 \leq \pi_1 \wedge a_1 \wedge b_1 \leq \pi(A \cap B)$$ Q.E.D.

COROLLARY 7. If $b_i \geq a_i, \forall i=1,2,3$, then A and B are $\pi$-noninteraction.

PROOF. In this case, $B \supseteq A$, $\pi(B) \geq \pi(A)$. Then

$$\pi(A \cap B) = \pi(A) = \pi(A) \wedge \pi(B)$$ Q.E.D.

DEFINITION 10. By the judge function of P-independence for A and B we mean the function $\Delta(x,y)$

$$\Delta(x,y) = x(x-1)(a_1-a_3)(b_1-b_3)+y(y-1)(a_2-a_3)(b_2-b_3)$$
$$+xy[(a_1-a_3)(b_2-b_3)+(a_2-a_3)(b_1-b_3).$$

PROPOSITION 15. The fuzzy events A and B are P-independent if and only if $\Delta(x,y) = 0$.

PROOF. It is easy to show that

$$\tilde{P}(A)\tilde{P}(B)-\tilde{P}(A \cdot B) \equiv \Delta(x,y).$$ Q.E.D.

PROPOSITION 16. If $a_1 = a_3$, then A and B are independent to each other under any probability distribution.

This proposition is the special case of PROP. 8.

PROPOSITION 17. Assume that $a_1 = a_2 > a_3$. Then if and only if $b_1 \wedge b_2 < b_3 < b_1 \vee b_2$, A and B can be independent to each other under some undegenerative probability distribution P. In this case, the probability distribution P(x,y,1-x-y) satisfies the following linear equation

$$x(b_1-b_3)+y(b_2-b_3) = 0.$$

PROOF. From the given condition, we have

$$\Delta(x,y) = (a_1-a_3)[x(x-1)(b_1-b_3)+y(y-1)(b_2-b_3)+xy(b_1+b_2-2b_3)]$$
$$= (a_1-a_3)(x+y-1)[x(b_1-b_3)+y(b_2-b_3)].$$

Since P is undegenerative, the sufficient and necessary condition for $\Delta(x,y) = 0$ is that

$$x(b_1-b_3)+y(b_2-b_3) = 0.$$

This equation has positive roots if and only if

$$(b_1-b_3)(b_2-b_3) < 0$$

or, in other terms, $b_1 \wedge b_2 < b_3 < b_1 \vee b_2$. Q.E.D.

PROPOSITION 18. Assume that $a_1 > a_2 = a_3$. Then if and only if $b_2 \wedge b_3 < b_1 < b_2 \vee b_3$, A and B can be independent to each other under some undegenerative probability distribution P. And in this case, the probability distribution P(x,y,1-x-y) satisfies the following linear equation

$$(1-x-y)(b_3-b_1)+y(b_2-b_1) = 0.$$

PROOF. By PROP. 9, A and B are P-independent iff $A^c$ and $B^c$ are P-independent. Since $1-a_3 = 1-a_2 > 1-a_1$, it is clear from PROP. 17 that $A^c$ and $B^c$ are P-independent if and only if

$$(1-b_2) \wedge (1-b_3) < 1-b_1 < (1-b_2) \vee (1-b_3),$$

$$(1-x-y)[(1-b_3)-(1-b_1)]+y[(1-b_2)-(1-b_1)] = 0,$$

or, in other words,

$$b_2 \wedge b_3 < b_1 < b_2 \vee b_3$$

$$(1-x-y)(b_3-b_1)+y(b_2-b_1) = 0. \qquad \text{Q.E.D.}$$

PROPOSITION 19. Assume that $a_1 > a_2 > a_3$ and $\prod_{1 \le i \le j \le 3} (b_i-b_j) \neq 0$. Then if and only if either $b_2 > b_1 \vee b_3$ or $b_2 < b_1 \wedge b_3$, A and B can be independent to each other under some undegenerative probability distribution $P(x,y,1-x-y)$.

PROOF. Using the point $(x,y)$ to represent the probability distribution $P(x,y,1-x-y)$, we can see that the undegenerative probability distribution P is nothing but a point $(x,y)$ located inside the triangle T (Fig. 1). All the vertices of T are zero points of $\Delta(x,y)$. On each side of T, the changing values of $\Delta(x,y)$ remain certainly positive or negative. In fact,

$$\ell_1: \quad \Delta(x,y) = y(y-1)(a_2-a_3)(b_2-b_3)$$

$$\mathrm{Sgn}(\Delta(x,y)) = -\mathrm{Sgn}(b_2-b_3)$$

$$\ell_2: \quad \Delta(x,y) = x(x-1)(a_1-a_3)(b_1-b_3)$$

$$\mathrm{Sgn}(\Delta(x,y)) = -\mathrm{Sgn}(b_1-b_3)$$

$$\ell_3: \quad \Delta(x,y) = x(x-1)(a_1-a_2)(b_1-b_2)$$

$$\mathrm{Sgn}(\Delta(x,y)) = -\mathrm{Sgn}(b_1-b_2).$$

Under the given condition, the equation $\Delta(x,y) = 0$ is obviously hyperbolic type. It has real roots inside T if and only if the function $\mathrm{Sgn}(\Delta(x,y))$ takes different values on $\ell_1$, $\ell_2$ and $\ell_3$, i.e.

$$\left| \sum_{1\leq i<j\leq 3} \mathrm{Sgn}(b_i-b_j) \right| = 1,$$

or, in other words, $b_2 \neq b_1 \vee b_3$. Q.E.D.

The locus of the point $(x,y)$ is a segment of hyperbola.

PROPOSITION 20. If $\forall\, i\neq j$, $(a_i-a_j)(b_i-b_j) > 0$, then A and B are noninteraction under any possibility distribution but aren't independent under any undegenerative probability distribution P.

This proposition is the combination of CORO. 5 and PROP. 19.

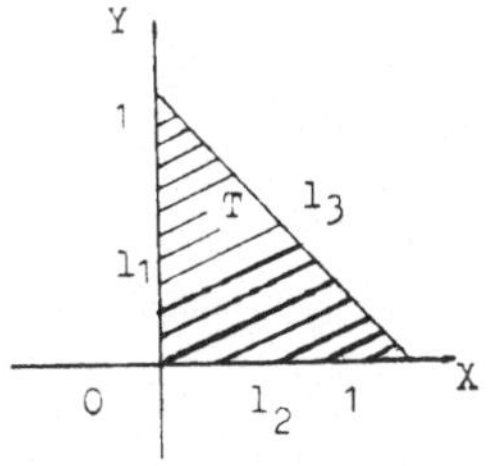


Fig. 1

It shows us that the relation between P-independence and $\pi$-non-interaction is indeed very weak.

REFERENCES

1. L. A. Zadeh, Fuzzy sets as a basis for a theory of possibility, Fuzzy Sets and Systems, 1 (1978), 3-28.
2. L. A. Zadeh, Probability measures of fuzzy sets, JMAA, 23 (1968).
3. E. Hisdal, Possibilically Dependent Variables and a General Theory of Fuzzy Sets, Advances in Fuzzy Set Theory and Applications, 1979, 215-234.
4. S. Khalili, Independent Fuzzy Events, JMAA, 67 (1979), 412-420.

# AN ALGEBRAIC SYSTEM GENERALIZING THE FUZZY SUBSETS OF A SET

Cao Zhi-Qiang

Institute of Automation
Academia Sinica
Beijing, China

## I. INTRODUCTION

In this paper we introduce a new algebraic system. The theoretical and practical background of it is very interesting.

In the last ten years many operations of fuzzy sets were introduced for various applications, some algebraic properties of the fuzzy sets with these operations and some related theories of fuzzy matrices with these operations were discussed by a series of papers. In fact, these operations have some common properties. Hence, it will be helpful to abstract their common properties in algebraic language for the further investigation of the properties of these operations. This is just the main reason we introduce the new algebraic system.

Second, in the research of some social and psychological pheenomena we found a kind of new type of mathematical quantities which, however, were commonly met in daily life. Therefore, it is necessary to develop a suitable mathematical structure for the study of the algebraic properties of these new mathematical quantities. In fact, by means of the algebraic system introduced in this paper the author has established a new system theory in which many dynamic processes in social and psychological phenomena can be described very well [1].

As an introduction of this algebraic system here we only give some fundamental properties of it and several important theorems related to the structure of module and matrices over it.

## II. SLOPE

Definition 2.1 A slope is a structure consisting of a non-empty set P together with two binary operations "+" and "·", such that

(1) (P,+) is a semilattice

(2) (P,·) is a commutative semigroup

(3) $a\cdot(b+c)=a\cdot b+a\cdot c$

(4) $a+a\cdot b=a$

for all $a,b,c\in P$.

Here, a semilattice is a non-empty set with a binary operation "+" such that (1) $a+b=b+a$ (2) $a+(b+c)=(a+b)+c$ (3) $a+a=a$ hold for all elements a,b,c of the set. Therefore, we can define a partially ordered structure on a semilattice by its addition. The relation $a+b=a$ implies $a \geq b$ (or $a \leq b$). Thus a semilattice is a partially ordered set in which each pair of elements a, b have a least upper bound (or greatest lower bound) $a+b$.

The word SLOPE used in this definition is the acronym for "Semilattice with another operation". Another reason we adopt the term SLOPE is that the assumption (4) $a+a\cdot b=a$ may be figuratively thought as "sliding down" property, since the relation $a+a\cdot b=a$ implies $a\cdot b \leq a$ and $a\cdot b \leq b$ (or $a\cdot b \geq a$ and $a\cdot b \geq b$) so that the multiplication seems to be defined on a slope. Hence, a slope is actually a semilattice-semigroup with "sliding down" property.

Example 2.1 $Mm=([0,1],\vee,\wedge)$, with $a\vee b=\max\{a,b\}$ and $a\wedge b=\min\{a,b\}$, is a slope.

Example 2.2 $Mp=([0,1],\vee,\cdot)$, with the ordinary multiplication of real numbers "·", is a slope.

Example 2.3 $Mc=([0,1],\wedge,\circ)$, where $a\circ b=a+b-ab$, is a slope.

Example 2.4 $My=([0,1],\vee,*)$ is a slope, here * is a general class of fuzzy connectives suggested by R. R. Yager in [2] i.e., $a*b=1-\min\{1,[(1-a)^p+(1-b)^p]^{1/p}\}$ $(p \geq 1)$.

Example 2.5 Every distributive lattice is a slope.

From the above examples we see that the operations in the definition 2.1 cover many useful operations of fuzzy sets.

Proposition 2.1 A slope is a lattice iff $a\cdot a=a$ for any a in the slope.

The symbols 0 and 1 will be denoted the additive and multiplicative identities respectively, in case the slope has these identities. These identities are obviously unique.

Proposition 2.2 A slope with the identities 0 and 1 has the following properties:

(1) $a+1=1$ (2) $a\cdot 0=0$

(3) If $a\cdot b=1$, then $a=b=1$.

(4) If $a+b=0$, then $a=b=0$.

Proof. (1) By Definition 2.1, we have $b+a\cdot b=b$. Let $b=1$, then $a+1=1$.

(2) Since $a\cdot 0+b=a\cdot 0+a\cdot b+b=a\cdot(0+b)+b=b$ for any b.

(3) If $a\cdot b=1$, then $a=a\cdot b+a=1+a=1$. Similarly $b=1$.

(4) If $a+b=0$, then $a=(a+b)+a=a+b=0$. And $b=0$.

Definition 2.2 A slope is an integral slope iff for any $a, b\in P$ the following conditions are held:

(1) If $a+b=1$, then $a=1$ or $b=1$.

(2) If $a\cdot b=0$, then $a=0$ or $b=0$.

Examples 2.6 The slopes Mm, Mp, Mc and My all are integral slopes.

## III. SLOPE MODULE

Definition 3.1 A slope module over a slope P is a commutative semigroup M together with a map $(a,x) \to ax$ of $P\times M$ into M such that for $x,y\in M$; $a,b\in P$

$$a(x+y)=ax+ay$$

$$a(bx)=(a\cdot b)x$$

$$(a+b)x=ax+bx$$

for P with the identities 0,1;

$$1x=x$$

$$0x=\theta$$

where $\theta$ is the additive identity of M.

Example 3.1 Let X be a universe of discourse and $[0,1]^X$ be denoted as the fuzzy subsets of X. For $f(x)$, $g(x)\varepsilon[0,1]^X$ and $a\varepsilon[0,1]$ we define $(f+g)(x)=\max\{f(x),g(x)\}$ and $(af)(x)=\min\{a,f(x)\}$, then the fuzzy subsets $[0,1]^X$ is a slope module over Mm.

Example 3.2 Let $MpV_2=(\{(a,b):a,b\varepsilon Mp\},+)$ and $(a_1,b_1)+(a_2,b_2)=$ $=(a_1\ \ a_2,b_1\ \ b_2)$; $c(a,b)=(c\cdot a,c\cdot b)$. Then $MpV_2$ is a slope module over Mp.

Proposition 3.1 Let M be a slope module over P. Then

(1) $ax+x=x$

(2) $a\theta=\theta$

(3) If $x+y=\theta$, then $x=y=\theta$

hold for all $a\varepsilon P$; $x,y\varepsilon M$.

Proof. (1) $ax+x=(a+1)x=x$.

(2) $a\theta=x=(a\theta+ax)+x=ax+x=x$ for all $x\varepsilon M$. So $a\theta=\theta$.

(3) If $x+y=\theta$, then $x=x+y+x=x+y=\theta$. Similarly $y=\theta$.

Corollary. A module, as a commutative semigroup, is a semilattice and is a partially ordered set.

The following concepts related to the module theory over a slope have similar definitions to those in the theory of module over a ring: finitely generated module, free module, independent, basis and module-homomorphism. Now we discuss some properties of module over a slope.

In general, a free module over a slope may have several bases, but it is different from the case of module over a ring that the cardinals of these bases may be inequal. However, some modules have unique bases.

Theorem 3.1 If P is an integral slope and M is a finitely generated free module over P, then M has a unique basis.

Proof. Let both $\{e_i\}$ and $\{f_j\}$ be bases of M. Then there are $\{a_{ij}\}$ and $\{b_{jk}\}$ such that

$$e_i = \sum_j a_{ij} f_j \qquad i=1,2,\ldots,m. \tag{3.1}$$

$$f_j = \sum_k b_{jk} e_k \qquad j=1,2,\ldots,n. \tag{3.2}$$

Suppose $m \geq n$. From (3.1) and (3.2) we have

$$e_i = \sum_k (\sum_j a_{ij} b_{jk}) e_k$$

Since $\{e_i\}$ is a basis, for every i we have

$$\sum_j a_{ij} b_{ji} = 1$$

$$\sum_j a_{ij} b_{jk} = 0 \qquad (k \neq i).$$

Here P is an integral slope, by Definition 2.2 we have

$$a_{ij}=0 \quad \text{or} \quad b_{jk}=0 \qquad (k \neq i),$$

and there is at least a j such that

$$a_{ij} b_{ji} = 1.$$

By Proposiiton 2.2 we have

$$a_{ij} = b_{ji} = 1.$$

Thus

$$b_{jk} = 0 \qquad (k \neq i).$$

Hence, for every i there is at least a j such that

$$f_j = e_i \; ,$$

i.e. $\{e_i\} \subseteq \{f_j\}$, and $m \leq n$. Since we have already supposed $m \geq n$, we have $m=n$ and $\{e_i\}=\{f_j\}$.

It is easy to show that a submodule of a finitely generated free module may not be finitely generated, and a module generated by an independent set may have no bases. The concept of basis is only applicable to free module. In general, a finitely generated module may have no basis. The concept of standard basis which is due to K. H. Kim and F. W. Routh [3] is very useful for the discussion of a finitely generated module.

Definition 3.2 A finite set S is a standard basis of the module generated by S iff whenever

$$s_i = \sum_j a_{ij} s_j ,$$

where $\{s_k\} \subseteq S$, then $s_i = a_{ii} s_i$.

Theorem 3.2 Let M be a finitely generated module. If for any spanning set $\{s_1, s_2, \ldots, s_n\}$ of M the set

$$S(i) = \{a_{ii} : s_i = \sum_j a_{ij} s_j\}$$

has at least a minimal element, then there exists at least a standard basis of M.

Proof. Let $\{s_1, s_2, \ldots, s_n\}$ be an independent spanning set of M. First, we show that there exists an $e_1$ such that the set $\{e_1, s_2, \ldots, s_n\}$ is a spanning set of M and $b_{11}e_1 = e_1$ whenever $e_1 = b_{11}e_1 + b_{12}s_2 + \ldots + b_{1n}s_n$. Suppose that the set S(1) has a minimal element c and denote $cs_1$ by $e_1$. We have

$$s_1 = e_1 + \sum_j a_{1j} s_j .$$

For every $x \in M$,

$$x = \sum_i d_i s_i = d_1 e_1 + (d_1 a_{12} + d_2) s_2 + \ldots + (d_1 a_{1n} + d_n) s_n ,$$

hence $\{e_1, s_2, \ldots, s_n\}$ is a spanning set of M. If $e_1 = b_{11}e_1 + b_{12}s_2 + \ldots + b_{1n}s_n$, then $s_1 = b_{11}e_1 + (b_{12} + a_{12})s_2 + \ldots + (b_{1n} + a_{1n})s_n = b_{11}cs_1 + \ldots + (b_{1n} + a_{1n})s_n$. Since $b_{11}c \leq c$ and c is a minimal element, we have $b_{11}c = c$. Hence, $b_{11}e_1 = b_{11}cs_1 = cs_1 = e_1$.

Repeating the above process we can get another set $\{e_1, e_2, s_3, \ldots, s_n\}$ such that it is also a spanning set of M and $b_{22}e_2 = e_2$ whenever $e_2 = b_{21}e_1 + b_{22}e_2 + b_{23}s_3 + \ldots + b_{2n}s_n$. Since $e_2 = c_2 s_2$, whenever $e_1 = b_{11}e_1 + b_{12}e_2 + \ldots + b_{1n}s_n$ we have $e_1 = b_{11}e_1 + b_{12}c_2 s_2 + \ldots + b_{1n}s_n$. Hence, $e_1 = b_{11}e_1$.

Repeating the above process until getting the set $\{e_1, e_2, \ldots, e_n\}$, we obtain a standard basis of M, and complete the proof.

In [4] we proved that the condition of Theorem 3.2 is satisfied in the case of a finitely generated submodule of a finitely generated free module over the slope Mm. In fact, the result of [4] is applicable to any totally ordered set yet.

Theorem 3.3 Let P be a totally ordered set. Any finitely generated submodule of a finitely generated free module over P has a unique standard basis.

Proof. Let M be a submodule satisfying the assumption of the theorem and $\{e_i\}$ be a standard basis of M.

Suppose that we have

$$e_i = \sum_j y_j$$

where $e_i \varepsilon \{e_k\}$ and $\{y_j\} \subset M$. From

$$y_j = \sum_k a_{jk} e_k$$

we have

$$e_i = \sum_k (\sum_j a_{jk}) e_k \ .$$

As an element of a standard basis

$$e_i = (\sum_j a_{ji}) e_i \ .$$

Since P is a totally ordered set, there exists an n such that

$$a_{ni} = \sum_j a_{ji} \ ,$$

that is, $e_i = a_{ni} e_i$. By Corollary of Proposition 3.1,

$$a_{ni} e_i = e_i \leq y_n \quad \text{and} \quad e_i = \sum_j y_j \geq y_n \ .$$

Hence, $e_i = y_n$. Let $\{f_j\}$ be another standard basis of M. By the above proof we have $e_i = c_{ij} f_j$ and $f_j = d_{jk} e_k$ for every i and j. Thus $e_i = c_{ij} d_{jk} e_k$. Since $\{e_i\}$ is a standard basis, we have $e_i = e_k$. Thus there exists a one to one correspondence between $\{e_i\}$ and $\{f_j\}$, that is, $|\{e_i\}| = |\{f_k\}|$. Since P is a lattice, we have $c_{ij} c_{ij} = c_{ij}$ and $e_i = c_{ij} d_{ji} f_j = f_j$.

This proves the theorem.

## IV. SLOPE FUZZY SUBSETS

Now the interval [0,1] with some certain operations can be referred to a slope. Hence the concept of slope fuzzy subsets arises naturally.

Definition 4.1 A slope fuzzy subset is a map from X to P, where X is the universe of discourse and P is a slope. The set of all of the slope fuzzy subsets of X is denoted by $P^X$.

Theorem 4.1 Let $P^X$ be the set of all of the slope fuzzy subsets of X, $(f+g)(x)=f(x)+g(x)$ and $(fg)(x)=f(x)g(x)$ for all $f,g\epsilon P^X$. Then $P^X$ is a slope.

In some practical applications we have to discuss the sets $(P^X)^X,\ldots,(..(P^X)^X..)^X$. Theorem 4.1 shows that the algebraic properties of these sets are identical.

Theorem 4.2 The set $(P^X,+)$, with $(f+g)(x)=f(x)+g(x)$ and $(af)(x)=a\cdot f(x)$ for $a\epsilon P$, f, $g\epsilon P^X$, is a module over P.

Theorem 4.3 If X is a finite set and P is a slope, then there exists a finitely generated free module M over P such that $P^X$ and M are isomorphic.

The proofs of the above theorems are immediate.

## V. SLOPE MODULE-HOMOMORPHISMS AND SLOPE MATRICES

Theorem 5.1 Let H(M,N,P) be the set consisting of all of homomorphisms from a module M to a module N. The set H(M,N,P), with $(G+K)x=Gx+Kx$ and $(aG)x=a\cdot Gx$, is a module over P.

Every element of a finitely generated free module can be represented as a vector under a certain basis, while every module-homomorphism as a matrix. Converseley, every slope matrix is just a slope module-homomorphism. We define the composition of two homomorphisms as $(GK)x=G(K)x$. Then the following properties are obvious:

Proposition 5.1 For any $n\times m$ matrices $A=(a_{ij})$, $B=(b_{ij})$ and $a\epsilon P$ we have

(1) $A+B=(a_{ij}+b_{ij})$

(2) $AB=(\sum_j a_{ij}b_{jk})$

(3) $aA=(a\cdot a_{ij})$.

Definition 5.1 An integral slope matrix T is invertible iff there exists a matrix $T^{-1}$ such that $TT^{-1}=T^{-1}T=I$.

Theorem 5.2 An integral slope matrix T is invertible iff the matrix T can be obtained through interchanging several rows of identity matrix I. And $T'=T^{-1}$.

Proof. A matrix obtained through interchanging several rows of identity matrix I is obviously invertible. Conversely, if there exists a matrix $T^{-1}$ such that $TT^{-1}=T^{-1}T=I$, then the image of $TT^{-1}$, as a module-homomorphism from M to M, must be M. Hence, from $Ix=TT^{-1}x=T(T^{-1}x)=x$ we know that the set of all of columns of T is a spanning set of M. By Theorem 3.1 T has the same columns as I.

Definition 5.2 An integral slope matrix A is similar to a matrix B iff there is an invertible matrix T such that $A=TBT'$.

Theorem 5.3 For integral slope matrix A there is an n such that $A^n=0$ iff the matrix A has a similar matrix which has the following structure:

$$\begin{bmatrix} oxx\ldots x \\ oox\ldots x \\ ooo\ldots x \\ \ldots \quad . \\ ooo\ldots x \end{bmatrix}$$

where x stands for an arbitrary element in the slope.

Proof. Suppose $A^n=0$ and $A^{n-1}\neq 0$. Let A be a module-homomorphism of M and $\{e_i\}$ be the unique basis of M.

First, since $A^n=0$ and $A^{n-1}\neq 0$, there exists at least an $x\varepsilon M$ such that $A^{n-1}x\neq 0$ and $A^n_x=0$. Suppose $Ae_i\neq 0$ for any $e_i\varepsilon\{e_i\}$. Then

$$Ay = \sum_i b_i Ae_i \neq 0$$

when

$$y = \sum_i b_i e_i \neq 0,$$

this is contrary to $A(A^{n-1}x)=0$. Hence, there exists at least an $e_i$ such that $Ae_i=0$, this means that there exists at least a column $a_i$ in the matrix A, $a_i=0$. We can find an invertible matrix T such that the first column of $TAT'$ is 0.

It is easy to show that if a matrix A has the following form:

$$A = \begin{bmatrix} C_1 & B_1 \\ 0 & A_1 \end{bmatrix}$$

where $A_1$ and $C_1$ are square matrices, then

$$A^n = \begin{bmatrix} C_1^n & B_n \\ 0 & A_1^n \end{bmatrix}.$$

Now we have shown that the matrix TAT' has the following form:

$$TAT' = \begin{bmatrix} 0 & B_1 \\ 0 & A_1 \end{bmatrix}.$$

Since $A_1^n=0$, repeating the above proof we prove the theorem.

Theorem 5.4 Every slope module-homomorphism is a monotone mapping.

Proof. Since $H(x+y)=Hx+Hy$, if $x+y=x$ then $H(x+y)=Hx$.

For the particular slope: distributive lattice, the following result given by Y. Give'on in [5] is very important.

Theorem 5.5 For any matrix A over a distributive lattice the sequence $A, A^2, A^3, \ldots$ is ultimately periodic.

## VI. ACKNOWLEDGMENTS

The author is indebted to Professor K. H. Kim for his helpful suggestions.

## VII. REFERENCES

1. Zhi-Qiang Cao, Modern control theory of psychological phenomena, Proceedings of China-IEEE Bilateral Meeting on Control Systems, (1981) Shanghai, China.
2. R. R. Yager, On a general class of connectives, Fuzzy Sets and Systems, 4 (1980) 235-242.
3. K. H. Kim and F. W. Routh, Generalized fuzzy matrices, Fuzzy Sets and Systems, 4 (1980) 293-315.
4. Zhi-Qiang Cao, Solution of fuzzy relation equations, ACTA AUTOMATICA SINICA, to appear.
5. Y. Give'on, Lattice matrices, Information and Control, 7 (1964) 477-484.

FROM THE FUZZY STATISTICS TO THE FALLING RANDOM SUBSETS

Wang Pei-zhuang

Department of Mathematics
Beijing Normal University
Beijing, China

In this paper we shall discuss some problems concerning the fuzzy statistics and the random subsets. The Rain space, a framework treating the fuzzy subsets as random subsets will be introduced. In this space, the measurability of random subsets is equivalent to the strong measurability. Applying the graph of random subsets we will give a correspondence theorem which combines the falling random subsets and the measurable fuzzy subsets. Finally, we will discuss the operations of fuzzy subsets, some interesting results will be found.

INTRODUCTION

An important and difficult problem in the fuzzy sets theory and its applications is how to treat the relationship between the objectivity and the subjectivity. The results of fuzzy statistical experiments obtained by Zhang Nan-lun show that there are some objective rules hiding in the subjectivity experiments here.

A fuzzy statistics experiment concerning a fuzzy concept $\underset{\sim}{A}$ consists of three factories as shown in the following:

1. An universe of discourse U;
2. A fixed element $u_o \in U$;
3. The trials on $\underset{\sim}{A}$, which yields a movable set $A^*$. The set is variable in different trials but must be fixed once a trial is completed.

We say that $u_o$ belongs to $\underset{\sim}{A}$ in one trial if $A^*$ covers $u_o$ at that moment. Let m be the number of that $u_o$ belongs to $\underset{\sim}{A}$ in n trials, we call m/n the belonging frequance of $u_o$. Zhang's work shows that there is a rule of stability on the belonging frequances in some situations. When n increases the belonging frequance of $u_o$ will steady near some real value $\lambda \in [0,1]$, we describe the degree of membership of $u_o$ for $\underset{\sim}{A}$ as $\lambda$;

$$\mu_{\underset{\sim}{A}}(u_o) = \lambda \ .$$

Fuzzy statistics is not the same as the probability statistics. A probabilility statistical experiment consists of three factories as shown in the following:

1. A fundamental space $\Omega$;
2. A fixed event $A_o$;
3. A point $w \in \Omega$ being constrained by the condition S where w is the variable in different trials but it must be fixed once a trial is performed.

The main difference between the fuzzy statistics and the probability statistics is: In fuzzy statistical experiments the element $u_o$ is fixed and the subset $A^*$ is variable, but in probability statistical experiments the subset (event) $A_o$ is fixed and the element w is the variable.

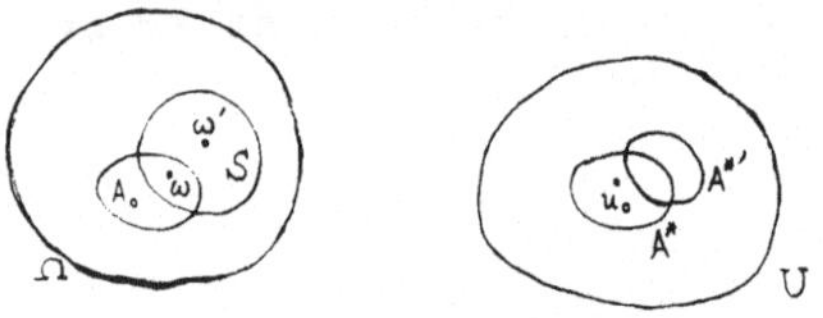


(Graph 1)

The fact that the stability of belonging frequancy does exist leads us to consider the $A^*$ as a random subset which depends upon some hidden variable acting in a fundamental space $\Omega$. The element $w \in \Omega$ maybe determined by many subjective factors. The relationship between fuzzy subsets and the random subsets have been studied by I. R. Goodman and others. Wang Pei-zhuang and E. Sanchez have formalized the concept of falling random subset, we will explore further this novel concept in this paper.

<u>Preliminary Knowledge</u>

Let U be a set, given

$$\mathcal{B} \subset \mathcal{P}(U) = \{B \mid B \subset U\}. \qquad (2.1)$$

An $\mathcal{B}$-(almost numerable) division of U is a subclass

$$d = \{D_i \mid i \in I\}, \quad \text{(I almost numerable)} \qquad (2.2)$$

satisfying that

$$D_i \in \mathcal{B} \quad (i \in I); \tag{2.3}$$

$$\bigcup_{i \in I} D_i = U; \tag{2.4}$$

$$D_i \cap D_j = \phi \quad (i \neq j). \tag{2.5}$$

The collection of all $\mathcal{B}$-division on U is denoted by

$$\mathcal{D} = \mathcal{D}(U,\mathcal{B}).$$

We further define that $d_1 \prec d_2$ iff for any $D_i^1 \in d_1$, there is $D_j^2 \in d_2$ such that

$$D_i^1 \subset D_j^2 . \tag{2.6}$$

Hence, $(\mathcal{D},\prec)$ is a poset.

DEFINITION 2.1. We call $\mathcal{B}$ is thin if there exists a sequance of divisions $\{d^{(n)}\}$ satisfying that

1) $d^{(1)} \succ d^{(2)} \succ \cdots \succ d^{(n)} \succ \cdots$ (2.7)

2) for any $d \in \mathcal{D}(U,\mathcal{B})$, there is an $n_o$ such that

$$d \succ d^{(n_o)} . \tag{2.8}$$

DEFINITION 2.2. Let $(\Omega,A,\mu)$ be a probability field, a normal net $\mathcal{D}^* = \{d^{(n)}\} \subset \mathcal{D}(\Omega,A)$ is such a sequance of division on $(\Omega,A)$:

$$d^{(1)} = \{D_1,D_2\},$$

$$\mu(D_1) = \mu(D_2) = 1/2;$$

$$d^{(2)} = \{D_{11},D_{12},D_{21},D_{22}\},$$

$$\mu(D_{11}) = \mu(D_{12}) = \mu(D_{21}) = \mu(D_{22}) = 1/2^2,$$

$$D_{11} \cup D_{12} = D_1 ,$$

$$D_{21} \cup D_{22} = D_2 ; \tag{2.9}$$

$$\cdots \quad \cdots \quad \cdots$$

$$d^{(n)} = \{D_{i_1 \cdots i_n} \mid i_k = 1,2;\ k = 1,\ldots,n\}$$

$$D_{i_1 \cdots i_n} \in A \quad (i_k = 1,2;\ k = 1,\ldots,n)$$

$$\mu(D_{i_1 \cdots i_n}) = 1/2^n ,$$

$$D_{i_1 \cdots i_{n-1}1} \cup D_{i_1 \cdots i_{n-1}2} = D_{i_1 \cdots i_{n-1}} ; \quad (2.10)$$

Let $(U,\mathcal{B})$ be a measurable space, i.e., $\mathcal{B}$ is an $\sigma$-algebra on U. For any $E \in \mathcal{P}(U)$, we set

$$\check{E} = \check{E}_{\mathcal{B}} = \{B \mid B \in \mathcal{B},\ B \cap E \neq \phi\}; \quad (2.11)$$

$$\underset{\vee}{E} = \underset{\vee}{E}_{\mathcal{B}} = \{B \mid B \in \mathcal{B},\ B \subset E\}. \quad (2.12)$$

Taking the set operations on the set $\mathcal{B}$ yields;

$$(\check{E})^{\overset{c}{\cdot}} = (E^c)_{\vee}; \quad (2.13)$$

$$(\underset{\vee}{E})^{\overset{c}{\cdot}} = (E^c)^{\vee}, \quad (2.14)$$

where $\overset{c}{\cdot}$ is the operation of the complement on $\mathcal{B}$. For any $\mathcal{C} \subset \mathcal{P}(U)$, denote that

$$\check{\mathcal{C}} = \check{\mathcal{C}}_{\mathcal{B}} = \{\check{c} \mid c \in \mathcal{C}\} \quad (2.15)$$

$$\underset{\vee}{\mathcal{C}} = \underset{\vee}{\mathcal{C}}_{\mathcal{B}} = \{\underset{\vee}{c} \mid c \in \mathcal{C}\}. \quad (2.16)$$

DEFINITION 2.3. A falling measurable space is a triple $(U,\mathcal{B},\hat{\mathcal{B}})$ where $(U,\mathcal{B})$ is a measurable space, $\hat{\mathcal{B}}$ is an $\sigma$-algebra on $\mathcal{B}$ satisfying the following inclusion operation.

$$\hat{\mathcal{B}} \supset [\check{\mathcal{C}}_o]_{\mathcal{B}} , \quad (2.17)$$

where

$$\mathcal{C}_o = \{\{u\} \mid u \in U\}, \quad (2.18)$$

and $[\check{\mathcal{C}}_o]_{\mathcal{B}}$ is the smallest $\sigma$-algebra on $\mathcal{B}$ containning $\check{\mathcal{C}}_o$.

DEFINITION 2.4. $\hat{\mathcal{B}}$ is called c-closed if for any $E \in \mathcal{P}(U)$, we have that

$$\check{E} \in \hat{\mathcal{B}} \Longleftrightarrow (E^c)^{\vee} \in \hat{\mathcal{B}}, \quad (2.19)$$

or equivalently, we are able to say:

$$\check{E} \in \hat{\mathcal{B}} \Longleftrightarrow \underset{\vee}{E} \in \hat{\mathcal{B}} . \quad (2.20)$$

A multivalued mapping from $\Omega$ to U is a mapping

$$\Gamma: \Omega \to \mathcal{P}(U). \quad (2.21)$$

Given $\Gamma$, the upper (lower) inverse of $\Gamma$ is such a mapping:

$$\Gamma^*: \mathcal{P}(U) \to \mathcal{P}(\Omega)$$

$$E \mapsto E^*_\Gamma \triangleq \Gamma^{-1}(\check{E}) \triangleq \{w | \Gamma(w) \in \check{E}\}. \quad (2.22)$$

$$(\Gamma_*: \mathcal{P}(U) \to \mathcal{P}(\Omega)$$

$$E \mapsto E_{*\Gamma} \triangleq \Gamma^{-1}(\underset{\vee}{E}) \triangleq \{w | \Gamma(w) \in \underset{\vee}{E}\}) \quad (2.23)$$

Clearly, we have that

$$(E^*)^c = (E^c)_*; \quad (E_*)^c = (E^c)^*; \quad (2.24)$$

$$(\bigcup_{t\in T} E_t)^\vee = \bigcup_{t\in T} \check{E}_t; \quad (\bigcup_{t\in T} E_t)^* = \bigcup_{t\in T} E^*_t; \quad (2.25)$$

$$(\bigcap_{t\in T} E_t)_\vee = \bigcap_{t\in T} \underset{\vee}{E}_t; \quad (\bigcap_{t\in T} E_t)_* = \bigcap_{t\in T} E_{t*}; \quad (2.26)$$

$$E_1 \subset E_2 \Rightarrow \check{E}_1 \subset \check{E}_2,\ \underset{\vee}{E}_1 \subset \underset{\vee}{E}_2,\ E^*_1 \subset E^*_2,\ E_{1*} \subset E_{2*}. \quad (2.27)$$

Let $(\Omega,\mathcal{A})$, $(U,\mathcal{B})$ both be measurable spaces, we call the mapping $\Gamma$ as $\mathcal{A}$-$\mathcal{B}$ strong measurable if

$$\Gamma^*(\mathcal{B}) \triangleq \{B^* | B \in \mathcal{B}\} \subset \mathcal{A}, \quad (2.28)$$

or equivalently, if

$$\Gamma_*(\mathcal{B}) \triangleq \{B_* | B \in \mathcal{B}\} \subset \mathcal{A} \quad (2.29)$$

Let $(\Omega,\mathcal{A},\mu)$ be a probability field, suppose that mapping $\Gamma$ is $\mathcal{A}$-$\mathcal{B}$ strong measurable, we have the following mapping

$$\mu^*: \mathcal{B} \to [0,1]$$
$$\mu^*(B) = \mu(B^*). \quad (2.30)$$

$$(\mu_*: \mathcal{B} \to [0,1]$$
$$\mu_*(B) \triangleq \mu(B_*)) \quad (2.31)$$

and we designate this mapping as the upper (lower) probability measure on $\mathcal{B}$.

Let $\Gamma_t (t \in T)$ be a family of multivalued mappings, we may define the operations as following:

$$(\Gamma_t: \Omega \to \mathcal{P}(U)) \quad (2.32)$$

$$(\bigcup_{t\in T} \Gamma_t)(w) \triangleq \bigcup_{t\in T} \Gamma_t(w); \quad (2.33)$$

$$(\bigcap_{t\in T}\Gamma_t)(w) = \bigcap_{t\in T}\Gamma_t(w); \tag{2.34}$$

$$(\Gamma^c)(w) = (\Gamma(w))^c. \tag{2.35}$$

Let $\Gamma$ be a multivalued mapping as defined in (2.20).

DEFINITION 2.5. The graph of $\Gamma$ is defined as

$$G_\Gamma \triangleq \{(w,u) \mid \Gamma(w) \ni u\}. \tag{2.36}$$

Clearly, we have the following identities;

$$G_{(\bigcup_{t\in T}\Gamma_t)} = \bigcup_{t\in T} G_{\Gamma_t}; \tag{2.37}$$

$$G_{(\bigcap_{t\in T}\Gamma_t)} = \bigcap_{t\in T} G_{\Gamma_t}; \tag{2.38}$$

$$G_{\Gamma^c} = (G_\Gamma)^c. \tag{2.39}$$

Let $G \subset \Omega \times U$, the section and projection of it are designated as follows:

$$G|_{w\cdot} \triangleq \{u \mid (w,u) \in G\} \quad (G|_{\cdot u} \triangleq \{w \mid (w,u) \in G\}) \tag{2.40}$$

$$\mathrm{Pro}_\Omega(G) = \{w \mid \exists u: (w,u) \in G\} \quad (\mathrm{Pro}_U(G) = \{u \mid \exists w: (w,u) \in G\}) \tag{2.41}$$

DEFINITION 2.6. Let $(\Omega,\mathcal{A})$, $(U,\mathcal{B})$ are both the measurable space, we call $\mathcal{A}$ is enough for $\mathcal{B}$ if

$$\mathrm{Pro}_\Omega(\mathcal{A}\times\mathcal{B}) \triangleq \{\mathrm{Pro}_\Omega D \mid D \in \mathcal{A}\times\mathcal{B}\} = \mathcal{A} \tag{2.42}$$

where $\mathcal{A}\times\mathcal{B}$ is the product measurable space:

$$\mathcal{A}\times\mathcal{B} \triangleq [\{A\times B \mid A \in \mathcal{A},\ B \in \mathcal{B}\}]. \tag{2.43}$$

$\underset{\sim}{A} \in \mathcal{F}(U)$ is called $\mathcal{B}$-measurable fuzzy subset on $(U,\mathcal{B})$ if its membership function $\mu_{\underset{\sim}{A}}$ is $\mathcal{B}$-measurable. All of them may be denoted as $\mathcal{F}_o = \mathcal{F}_o(U,\mathcal{B})$.

DEFINITION 3.1. $\mathcal{R} = (\Omega,\mathcal{A},\mu;U,\mathcal{B},\hat{\mathcal{B}})$, the combination of a probability field $(\Omega,\mathcal{A},\mu)$ and a falling measurable space $(U,\mathcal{B},\hat{\mathcal{B}})$ is called a rain space if

(R.1) there is a fixed subclass

$$\mathcal{C} \supset \mathcal{C}_o = \{\{u\} \mid u \in U\} \tag{3.1}$$

such that

$$\mathcal{B} = [\mathcal{C}]; \quad \hat{\mathcal{B}} = [\check{\mathcal{C}}]_{\mathcal{B}}. \tag{3.2}$$

(R.2) $\mathcal{B}$ is thin;

(R.3) $\hat{\mathcal{B}}$ is c-closed;

(R.4) $A$ has a fixed normal net $\mathcal{D}^*$;

(R.5) $A$ is enough for $\mathcal{B}$.

DEFINITION 3.2. Given a rain space $\mathcal{R} = (\Omega, A, \mu; U, \mathcal{B}, \hat{\mathcal{B}})$, we call a mapping

$$s: \Omega \to \mathcal{B} \tag{3.3}$$

the falling random subset if it is $A - \hat{\mathcal{B}}$ measurable, i.e.,

$$\Gamma^{-1}(\hat{\mathcal{B}}) = \{\Gamma^{-1}(\mathcal{E}) \mid \mathcal{E} \in \hat{\mathcal{B}}\} = \{\{w \mid \Gamma(w) \in \mathcal{E}\} \mid \ \in \hat{\mathcal{B}}\} \subset A. \tag{3.4}$$

the collection of all falling random subsets on $\mathcal{R}$ is denoted by $\mathcal{S} \triangleq \mathcal{S}(\mathcal{R}) \triangleq \mathcal{S}(\Omega, A, \mu; U, \mathcal{B}, \hat{\mathcal{B}}) \triangleq \mathcal{S}(A, \hat{\mathcal{B}})$.

THEOREM 3.1. $s \in \mathcal{S}(\mathcal{R})$ iff $s$ is $A - \mathcal{B}$ strong measurable.

PROOF. Set

$$\mathcal{B}' = \{E \mid \check{E} \in \hat{\mathcal{B}}\} . \tag{3.5}$$

Because $\hat{\mathcal{B}}$ is $\check{c}$-closed, we may establish the following identity by invoking (2.20).

$$\mathcal{B}' = \{E \mid \underset{\vee}{E} \in \hat{\mathcal{B}}\} . \tag{3.6}$$

From (2.25) and (2.26) we have that

$$\begin{aligned}
& E_n \in \mathcal{B}'(n \geq 1) \Rightarrow \check{E}_n \in \hat{\mathcal{B}} \\
& \Rightarrow \left(\bigcup_{n=1}^{\infty} E_n\right)^{\vee} = \bigcup_{n=1}^{\infty} \check{E}_n \in \hat{\mathcal{B}} \\
& \Rightarrow \bigcup_{n=1}^{\infty} E_n \in \mathcal{B}', \\
& E_n \in \mathcal{B}'(n \geq 1) \Rightarrow \underset{\vee}{E}_n \in \hat{\mathcal{B}} \\
& \Rightarrow \left(\bigcap_{n=1}^{\infty} E_n\right)_{\vee} = \bigcap_{n=1}^{\infty} \underset{\vee}{E}_n \in \hat{\mathcal{B}} \\
& \Rightarrow \bigcap_{n=1}^{\infty} E_n \in \mathcal{B}'.
\end{aligned}$$

So that $\mathcal{B}'$ is an $\sigma$-algebra.

Because of $\mathcal{C} \subset \mathcal{B}'$, $\mathcal{B} = [\mathcal{C}]$, hence we have

$$\mathcal{B}' \supset \mathcal{B} \tag{3.7}$$

If s is $\mathcal{A}-\hat{\mathcal{B}}$ measurable, then from (3.7) we have that

$$B \in \mathcal{B} \Rightarrow B \in \mathcal{B}' \Rightarrow \check{B} \in \hat{\mathcal{B}} \Rightarrow B^* \in s^{-1}(\hat{\mathcal{B}}) \subset \mathcal{A}. \tag{3.8}$$

So that $s^*$ is $\mathcal{A}-\mathcal{B}$ measurable, i.e., s is $\mathcal{A}-\mathcal{B}$ strong measurable. If s is $\mathcal{A}-\mathcal{B}$ strong measurable, then

$$\mathcal{E} \in \check{\mathcal{C}} \Rightarrow \mathcal{E} = \check{E} \Rightarrow s^{-1}(\mathcal{E}) = s^{-1}(\check{E}) = E^* \in \mathcal{A},$$

so that

$$s^{-1}(\check{\mathcal{C}}) \subset \mathcal{A}. \tag{3.9}$$

From (3.2) we have that

$$s^{-1}(\hat{\mathcal{B}}) \subset \mathcal{A}$$

consequently s is $\mathcal{A}-\hat{\mathcal{B}}$ measurable. End.

DEFINITION 3.2. Let $s \in \mathcal{S}(\Omega,\mathcal{A},\mu;U,\mathcal{B},\hat{\mathcal{B}})$, we call $\underset{\sim}{A} \in \mathcal{F}(U)$ the down-fallen of s if its membership function satisfying:

$$\mu_{\underset{\sim}{A}}(u) = \mu(w|s(w) \in \{\check{u}\}) = \mu(w|s(w) \ni u). \tag{3.10}$$

## The Graph of a Random Subset

Let $\mathcal{R} = (\Omega,\mathcal{A},\mu;U,\mathcal{B},\hat{\mathcal{B}})$ be a rain space, consider the mapping

$$\phi\colon \mathcal{S} \to \mathcal{P}(\Omega \times U)$$

$$s \mapsto G_s \tag{4.1}$$

here, $G_s$ is the graph of s defined in (2.36).

THEOREM 4.1. The mapping $\phi$ is a bijection between $\mathcal{S}$ and $\mathcal{A} \times \mathcal{B}$.

PROOF. 1. Suppose that $s \in \mathcal{S}$, we need to prove that

$$G_s \in \mathcal{A} \times \mathcal{B}. \tag{4.2}$$

Consider arbitrary division $d \in \mathcal{D}(U,\mathcal{B})$:

$$d = \{D_i\} \; (i \in I) \quad (D_i \in \mathcal{B}) \tag{4.3}$$

Set

$$G_s^d = \bigcup_{i \in I} (D_i^* \times D_i), \tag{4.4}$$

where

$$D_i^* = s^{-1}(\check{D}_i).$$

Because of $s \in \mathcal{S}(\mathcal{R})$, from theorem 3.1, $s$ is $\mathcal{A}-\mathcal{B}$ measurable, so that $D_i^* \in \mathcal{A}$ and then

$$D_i^* \times D_i \in \mathcal{A} \times \mathcal{B} , \tag{4.5}$$

therefore

$$G_s^d \in \mathcal{A} \times \mathcal{B} . \tag{4.6}$$

From (2.6) and (2.27), it is clear that if $d_1 \prec d_2$, then

$$G_s^{d_1} \subset G_s^{d_2} . \tag{4.7}$$

We also have that

$$G_s = \bigcap_{d \in \mathcal{D}} G_s^d . \tag{4.8}$$

In fact, on the one hand, we have the inclusion identity as shown in the following.

$$G_s \subset G_s^d \qquad (d \in \mathcal{D}), \tag{4.9}$$

so that

$$G_s \subset \bigcap_{d \in \mathcal{D}} G_s^d , \tag{4,10}$$

But on the other hand, suppose that $(w_o,u_o) \notin G_s$, we may take

$$d_o = \{\{u_o\}, \{u_o\}^c\} \in \mathcal{D} , \tag{4.11}$$

but $(w_o,u_o) \notin G_s$ implies that $s(w_o) \notin \{\check{u}_o\}$, i.e., $w_o \notin \{u_o\}^*$, and then

$$(w_o,u_o) \notin (\{u_o\}^* \times \{u_o\}) \cup ((\{u_o\}^c)^* \times \{u_o\}^c) = G_s^{d_o}, \tag{4.12}$$

therefore we have

$$(w_o,u_o) \notin \bigcap_{d \in \mathcal{D}} G_s^d , \tag{4.13}$$

so that

$$G_s \supset \bigcap_{d \in \mathcal{D}} G_s^d \tag{4.14}$$

from (4.10) and (4.14) we have proved (4.8).

Because of $\mathcal{B}$ is thin, there are $d^{(n)} \in \mathcal{D}(U,\mathcal{B})$ satisfying

(2.7) and (2.8). It is quite straight forward to prove that

$$\bigcap_{d\varepsilon\mathcal{D}} G_s^d = \bigcap_{n=1}^{\infty} G_s^{d^{(n)}} \tag{4.15}$$

From (4.6), (4.8) and (4.15), we obtain (4.2). This is

$$\phi(s) \in \mathcal{A} \times \mathcal{B}. \tag{4.16}$$

2. Given $G \varepsilon \mathcal{A} \times \mathcal{B}$, we define the mapping

$$s: \Omega \rightarrow \mathcal{B}$$

$$w \mapsto G|_{w.} = \{u | (w,u) \varepsilon G\}. \tag{4.17}$$

for any $B \varepsilon \mathcal{B}$, we have that

$$B_G \triangleq (\Omega \times B) \cap G \varepsilon \mathcal{A} \times \mathcal{B}. \tag{4.18}$$

Because of $\mathcal{A}$ is enough for $\mathcal{B}$, so that

$$Pro_{\Omega}(B_G) \varepsilon \mathcal{A}. \tag{4.19}$$

But

$$Pro_{\Omega}(B_G) = s^{-1}(\check{B}) = s^*(B) = B_s^*. \tag{4.20}$$

So that $s$ is $\mathcal{A}-\mathcal{B}$ strong measurable. From theorem 3.1, we have that $s \varepsilon \mathcal{S}(\mathcal{R})$. This implies that the mapping $\phi$ is a surjection.

3. It is clear that $\phi$ is injective, i.e., if $s_1, s_2 \varepsilon \mathcal{S}$ and if

$$\phi(s_1) = \phi(s_2) \tag{4.21}$$

then we have that

$$s_1 = s_2. \qquad \text{End.} \tag{4.22}$$

THEOREM 4.2. $\mathcal{S}$ is closed under set operations. $(\mathcal{S},\cup,\cap,c)$ form an σ-complete Boolean algebra. The mapping $\phi$ is an isomorphism between $(S,\cup,\cap,c)$ and $(\mathcal{A} \times \mathcal{B},\cup,\cap,c)$:

$$\mathcal{S} \cong \mathcal{A} \times \mathcal{B} \tag{4.23}$$

From (2.37) - (2.39), and by means of theorem 4.1, this theorem is obvious.

THEOREM 4.3. Let $\underset{\sim}{A}_s \varepsilon \mathcal{F}(U)$ be a downfallen of $s \varepsilon \mathcal{S}$, $G_s$ is the graph of $s$, then

$$\mu_{\underset{\sim}{A}}(u) = \mu(G_{x|\cdot u}). \tag{4.24}$$

This theorem can be proved immediately from that fact:

$$G_{s|\cdot u} = s^{-1}(\{\check{u}\}) = \{u\}^*_s . \tag{4.25}$$

This permits us to call a downfallen of s the downfallen of $G_s$.

<u>The Correspondence Between $S$ and $F_o$.</u>

Let $R = (\Omega, A, \mu; U, B, \hat{B})$ be a rain space, $S = S(R)$, consider the mapping

$$\nu: S \to F(U)$$

$$s \mapsto \underset{\sim}{A}_s \quad \text{(downfallen of s)} \tag{5.1}$$

By invoking the measure theory, we know that

$$\mu_{\underset{\sim}{A}_s}(u) = \mu(G_{s|\cdot u}) \text{ is } B\text{-measurable,} \tag{5.2}$$

i.e., $\nu(s) = \underset{\sim}{A} \in F_o$ , so that the mapping $\nu$ is a mapping from $S$ to $F_o$:

$$\nu: S \to F_o \tag{5.3}$$

For any $s_1, s_2 \in S$ , we denoted that

$$s_1 \sim s_2 \iff (\forall u \in U)(\mu(G_{s_1|\cdot u}) = \mu(G_{s_2|\cdot u})). \tag{5.4}$$

Obviously, "$\sim$" is an equivalent relation on $S$ , set

$$\tilde{S} = S/\sim \tag{5.5}$$

Because that

$$s_1 \sim s_2 \Rightarrow \nu(s_1) = \nu(s_2), \tag{5.6}$$

so that we can write $\nu$ into mapping $\tilde{\nu}$:

$$\tilde{\nu}: \tilde{S} \to F_o \tag{5.7}$$

For any $ss \in \tilde{S}$, define that

$$\tilde{\nu}(ss) = \nu(s) \quad (s \in ss) \tag{5.8}$$

THEOREM 5.1. $\tilde{\nu}$ is a bijection between $\tilde{S}$ and $F_o$.

PROOF 1. $\tilde{\nu}$ is injective.

2. We want to prove that $\tilde{\nu}$ is surjective, i.e., for any $\underset{\sim}{A} \in F_o(U,\mathcal{B})$, there is an $ss \in \tilde{S}$ such that $\tilde{\nu}(ss) = \underset{\sim}{A}$.

Because of $\underset{\sim}{A} \in F_o$, so that there are $\mathcal{B}$-simple functions $\{f_n\}$ defined as following:

$$f_n(u) \triangleq \sum_{k=1}^{2^n} k/2^n \, \chi_{[\mu_{\underset{\sim}{A}}(u) \geq k/2^n]}. \tag{5.9}$$

Clearly

$$f_n(u) \nearrow \mu_{\underset{\sim}{A}}(u) \quad (n \to \infty). \tag{5.10}$$

Note that $\mathcal{A}$ has a fixed normal net $\mathcal{D}^*$ defined in (2.10), for any fixed $n \geq 1$, define mapping on the set of n-array:

$$\theta_1: (i_1,\ldots,i_n) \to (i_1-1,\ldots,i_n-1); \tag{5.11}$$

$$\theta_2: (j_1,\ldots,j_n) \to k \triangleq (j_1 \cdot 2^{n-1} + j_2 \cdot 2^{n-2} + \ldots + j_n \cdot 2)^o) + 1. \tag{5.12}$$

Obviously, $\theta_2 \circ \theta_1$ is a bijection between $\{(i_1,\ldots,i_n) | i_k = 1,2; k=1,\ldots,n\}$ and $\{1,2,\ldots,2^n\}$. Set

$$\zeta = \theta_1^{-1} \circ \theta_2^{-1}. \tag{5.13}$$

Denote that

$$B_k^{(n)} = \bigcup_{i=1}^{k} D_{\zeta(i)} \quad (k=1,\ldots,2^n), \tag{5.14}$$

clearly, $B_k^{(n)} \in \mathcal{A}$, and

$$\mu(B_k^{(n)}) = k/2^n. \tag{5.15}$$

Define that

$$G^n = \bigcup_{k=1}^{2^n} (B_k^{(n)} \times \chi_{[\mu_{\underset{\sim}{A}} \geq k/2^n]}). \tag{5.16}$$

Obviously, $G^n \in \mathcal{A} \times \mathcal{B}$, and

$$f_n(u) = \mu(G^n | \cdot u). \tag{5.17}$$

It is easy to prove that if $n_1 < n_2$, then

$$G^{n_1} \subset G^{n_2}. \tag{5.18}$$

By setting

$$G^* = \lim_{n\to\infty} G^n = \bigcup_{n=1}^{\infty} G^n. \tag{5.19}$$

It is clear that $G^* \in \mathcal{A} \times \mathcal{B}$, and

$$\mu(G^* | \cdot u) = \mu_{\underset{\sim}{A}}(u). \tag{5.20}$$

From theorem 4.1 set that

$$s^* = \phi^{-1}(G^*), \tag{5.21}$$

suppose that $s^* \varepsilon$ ss, then

$$\tilde{\nu}(ss) = \underset{\sim}{A}. \qquad \text{End.} \tag{5.22}$$

DEFINITION 5.1. The falling random subset $s^*$ (resp. its graph $G^*$) defined in (5.21) (resp. (5.19)) is called the represent of ss (resp. $G = \{G_s | s \varepsilon ss\}$), or the represent falling random subset (resp. graph) of $\underset{\sim}{A}$. The collection of represent random subsets in $S$ is denoted by $S_o$, hence we have

$$G_o = \{G_s | s \varepsilon S_o\}. \tag{5.23}$$

There are bijections between $S_o$, $G_o$ and $F_o$.

<u>The Operations Of Fuzzy Subsets</u>

Because there are many different falling random subsets correspond to the same fuzzy subset, so we can not define the operations of fuzzy subset uniquely. However, it does not prevent us in investigating some special cases.

Case 1. Represent in $S_o$

THEOREM 6.1. Denoting the downfallen of s by $\underset{\sim}{A}_s$, then for any $s_1, s_2 \varepsilon S_o$, we have that $s_1 \cup s_2$, $s_1 \cap s_2 \varepsilon S_o$, where

$$\mu_{\underset{\sim}{A}_{(s_1 \cup s_2)}}(u) = \max(\mu_{\underset{\sim}{A}_{s_1}}(u), \mu_{\underset{\sim}{A}_{s_2}}(u)); \tag{6.1}$$

$$\mu_{\underset{\sim}{A}_{(s_1 \cap s_2)}}(u) = \min(\mu_{\underset{\sim}{A}_{s_1}}(u), \mu_{\underset{\sim}{A}_{s_2}}(u)). \tag{6.2}$$

PROOF. Suppose that $s_1, s_2 \varepsilon S_o$, $\nu(s_i) = \underset{\sim}{A}_i (i=1,2)$, we have that

$$G_{s_i} = \lim_{n\to\infty} G_{s_i}^{(n)} \quad (i=1,2) \tag{6.3}$$

where

$$G_{s_i}^{(n)} = \bigcup_{k=1}^{2^n} (B_k^{(n)} \times \{u | \mu_{\underset{\sim}{A}_i}(u) \geq k/2^n\}) \quad (i=1,2) \tag{6.4}$$

Set $\underset{\sim}{A} \varepsilon F_o$:

$$\mu_{\underset{\sim}{A}}(u) = \max(\mu_{\underset{\sim}{A}_1}(u), \mu_{\underset{\sim}{A}_2}(u)). \tag{6.5}$$

Suppose that s is represent of $\underset{\sim}{A}$, then

$$G_s = \lim_{n\to\infty} G_s^{(n)} \tag{6.6}$$

where

$$G_s^{(n)} = \bigcup_{k=1}^{2^n} (B_k^{(n)} \times \{u \mid \mu_{\underset{\sim}{A}}(u) \geq k/2^n\}) \tag{6.7}$$

We can see that

$$\begin{aligned} G_s^{(n)} &= \bigcup_{k=1}^{2^n} (B_k^{(n)} \times (\{u \mid \max(\mu_{\underset{\sim}{A}_1}(u), \mu_{\underset{\sim}{A}_2}(u)) \geq k/2^n\})) \\ &= \bigcup_{k=1}^{2^n} (B_k^{(n)} \times (\{u \mid \mu_{\underset{\sim}{A}_1}(u) \geq k/2^n\} \cup \{u \mid \mu_{\underset{\sim}{A}_2}(u) \geq k/2^n)) \\ &= (\bigcup_{k=1}^{2^n} (B_k^{(n)} \times \{u \mid \mu_{\underset{\sim}{A}_1}(u) \geq k/2^n\})) \cup (\bigcup_{k=1}^{2^n} (B_k^{(n)} \times \{u \mid \mu_{\underset{\sim}{A}_2}(u) \geq k/2^n)) \\ &= G_{s_1}^{(n)} \cup G_{s_2}^{(n)} , \end{aligned} \tag{6.8}$$

so that

$$G_s = G_{s_1} \cup G_{s_2} \tag{6.9}$$

However, $G_{s_1} \cup G_{s_2} = G_{(s_1 \cup s_2)}$, so that

$$s_1 \cup s_2 = s \tag{6.10}$$

It is quite straight forward to obtain (6.1). Analogousely, (6.2) is true.

Case 2. Represented apart

DEFINITION 6.1. $s_1, s_2 \in S$ are said to be separate if for any $u \in U$ we have that

$$\mu(G_{s_1}|_{\cdot u}) + \mu(G_{s_2}|_{\cdot u}) \leq 1 \Longleftrightarrow G_{s_1}|_{\cdot u} \cap G_{s_2}|_{\cdot u} = \phi \; ; \tag{6.11}$$

$$\mu(G_{s_1}|_{\cdot u}) + \mu(G_{s_2}|_{\cdot u}) > 1 \Longleftrightarrow G_{s_1}|_{\cdot u} \cup G_{s_2}|_{\cdot u} = \Omega \; . \tag{6.12}$$

THEOREM 6.2. If $s_1$, $s_2$ are seperate then

$$\mu_{\underset{\sim}{A}(s_1 \cup s_2)}(u) = \min(\mu_{\underset{\sim}{A}_{s_1}}(u) + \mu_{\underset{\sim}{A}_{s_2}}(u), 1); \tag{6.13}$$

$$\mu_{\underset{\sim}{A}(s_1 \cap s_2)}(u) = \max(0, \mu_{\underset{\sim}{A}_{s_1}}(u) + \mu_{\underset{\sim}{A}_{s_2}}(u) - 1). \tag{6.14}$$

PROOF.

$$\mu_{\underset{\sim}{A}(s_1 \cup s_2)}(u) = \mu(G_{(s_1 \cup s_2)} | \cdot u) = \mu((G_{s_1} \cup G_{s_2}) | \cdot u).$$

Based Upon (6.11), if $\mu(G_{s_1} | .u) + \mu(G_{s_2} | .u) \leq 1$, then we have

$$\mu_{\underset{\sim}{A}(s_1 \cup s_2)}(u) = \mu(G_{s_1} | .u) + \mu(G_{s_2} | \cdot u) = \mu_{\underset{\sim}{A}_{s_1}}(u) + \mu_{\underset{\sim}{A}_{s_2}}(u).$$

else $\mu_{\underset{\sim}{A}(s_1 \cup s_2)}(u) = 1$, so that (6.13) is true. Analogously, (6.14) is true.

Case 3. Represented independently

DEFINITION 6.2. $s_1$, $s_2$ is called independent if $s_i^*(\mathcal{B})$ and $s_2^*(\mathcal{B})$ are independent, i.e., for any $B_1$, $B_2 \in \mathcal{B}$

$$\mu(s_1^*(B_1) \cap s_2^*(B_2)) = \mu(s_1^*(B_1) \cdot \mu(s_2^*(B_2)). \tag{6.15}$$

THEOREM 6.3. If $s_1$, $s_2$ are independent, then

$$\mu_{\underset{\sim}{A}(s_1 \cap s_2)}(u) = \mu_{\underset{\sim}{A}_{s_1}}(u) \cdot \mu_{\underset{\sim}{A}_{s_2}}(u); \tag{6.16}$$

$$\mu_{\underset{\sim}{A}(s_1 \cap s_2)}(u) = \mu_{\underset{\sim}{A}_{s_1}}(u) + \mu_{\underset{\sim}{A}_{s_2}}(u) \div \mu_{\underset{\sim}{A}_{s_1}}(u) \cdot \mu_{\underset{\sim}{A}_{s_2}}(u). \tag{6.17}$$

PROOF. We have established that;

$$\mu_{\underset{\sim}{A}(s_1 \cap s_2)}(u) = \mu(G_{(s_1 \cap s_2)} \cdot u) = \mu(G_{s_1|\cdot u} \cap G_{s_2|\cdot u})$$

$$= (s_1^*(\{u\}) \cap s_2^*(\{u\})).$$

Due to the fact that $s_1$, $s_2$ are independent, so that

$$\mu(s_1^*(\{u\}) \cap s_2^*(\{u\}) = \mu(s_1^*(\{u\})) \cdot \mu(s_2^*(\{u\})).$$

Consequently, (6.16) is true. We can easily prove (6.17) by invoking (6.16).

REFERENCES

1. Choquet, G., Theorem of capacities, Ann. Inst. Four., U. Grenoble, V. 131-296, 1954.
2. Goodman, I. R., Identification of fuzzy sets with a class of canonically induced random sets and some applications, Proc. 19th I.E.E.E. confer. on Desision and Control (Albuquerque, NM) and (Longer version) Naval research laboratory report 8415, 1980.
3. Goodman, I. R., Fuzzy sets as equivalence class of random

sets, Recent Developments in Fuzzy set and Possibility Theory, Edited by Yager, R. T., 1981.

4. Hirota, K., Extended fuzzy expression of Probabilistic sets. In advance. In fuzzy Set Theory and Appl., M. M. Gupta, R. K. Ragade, R. R. Yager (eds). North-Holland Publishing, 1979.
5. Kendall, D. G., Fundations of a theory of random sets, In Stochastic Geometry, New York, 322-376, 1974.
6. Klement, E. P., Characterizations of finite fuzzy measures using Markoff-kernals, Journal of Math. Anal. Applic., 75 (2) 330-339, 1980.
7. Kwakernaak, H. Fuzzy random variables-I. Info. Sci. 15, 1-29, 1978.
8. Nguyen, H. T., On random sets and belieg functions, J. Math. Anal. & Applic., 65, 531-542, 1978.
9. Shafer, G., Allocations of probability, Annals of Prob., 7 (5), 827-839, 1979.
10. Stallings, W., Fuzzy sets theory versus Bayesion statistics, I.E.E.E. Trans. Sys. Man, Cybernetics, SMC-7, 216-219, 1977.
11. Wang, P. P., Chang, S. K., Fuzzy Sets: Theory and Application to Policy Analysis and Information Systems, Plenum Press, New York, 1981.
12. Wang Pei-zhuang, Sanchez, E., Treating a Fuzzy subset as a fallable random subset, to be published.
13. Wang Pei-zhuang, Sugeno, M., The factories space and the fuzzy background structure, to be published.
14. Zadeh, L. A., Fuzzy sets as a basis for a possibility, Fuzzy Sets and Systems, 1, 3-28, 1978.
15. Zhang Nan-lun, The Membership and probability characteristics of random appearance, Journal of Wuhan Institute of Building Materials, No. 1, 1981.

ON FUZZY RELATIONS AND PARTITIONS

B. Bouchon[1] and G. Cohen[2]

CNRS[1]
University of Paris VI
Tour 45, 4 Place JUSSIEU
75230 PARIS CEDEX 05
France

ENST[2]
46 rue Barrault
75634 PARIS CEDEX 13
France

## I. INTRODUCTION

Let E be a finite set, of cardinality N. A fuzzy binary relation R on E, is a mapping from E x E to [0,1] R.

In the sequel, R is a *-relation, i.e.

R is:

- definite: for x, y in E, $R(x,y) = 1$ if and only if $x = y$.
- symmetric: $R(x,y) = R(y,x)$ $\forall x, \forall y$.
- *-transitive: $R(x,y) \geq \underset{z \in E}{\vee} (R(x,z) * R(y,z))$ with * being an operation from [0,1] x [0,1] to [0,1]. (v denotes the supremum).

Proposition 1. If R verifies $\forall x, y, z$

$$R(x,z) * R(y,z) \geq (R(x,z) + R(y,z) - 1) \vee 0,$$

then $d(x,y) = 1-R(x,y)$ is a T-distance (or distance), i.e., d is definite, symmetric and satisfies the triangular inequality.

This will be the case in this paper: d and R will be called associated.

## II. R-HOMOGENOUS DISTANCES

For a distance d and $x \in E$, we denote by $B_x(r)$ the r-sphere of radius r and center x, i.e., $B_x(r) = \{y \in E, d(x,y) \leq r\}$, and by C(r) a r-clique, characterized by $C(r) \subset E$, $\forall$ x, $y \in C(r)$, $d(x,y) \leq r$, and C(r) maximal for inclusion. Obviously for all x in C(r), $C(r) \subset B_x(r)$.

We consider the following examples:

Let $*$ be the product ($a*b = ab$). The associated distance satisfies $\forall$ x, y, $z \in E$

$$d(x,y) \leq d(x,z) + d(z,y) - d(x,z).\, d(z,y). \tag{i}$$

Let $*$ be the infimum ($a*b = a \wedge b$). The associated distance satisfies $\forall$ x, y, $z \in E$

$$d(x,y) \leq d(x,z) \vee d(z,y). \tag{ii}$$

For more details on $*$-relations, see [2], [3].

Definition 1. A distance d satisfying (i) is called a probabilistic distance (P-distance or P). If it satisfies (ii), d is called ultrametric (UM-distance or UM).

Proposition 2. UM $\Rightarrow$ P $\Rightarrow$ T.

The proof follows from $a \wedge b \geq ab \geq (a+b-1) \vee 0$ for all a,b in [0,1].

Definition 2. A distance d is said r-homogenous if

$$B_x(r) \cap B_y(r) = \emptyset \text{ or } B_x(r) = B_y(r) \quad \forall x,y.$$

Proposition 3. d is r-homogenous iff for every r-clique C(r) and any x in C(r), $C(r) = B_x(r)$.

Proof. Suppose d is r-homogenous and let y, z be in $B_x(r)$. Then $B_y(r) \cap B_z(r)$ contains x; hence $B_y(r) = B_z(r)$, $d(y,z) \leq r$ and y, z are in any r-clique containing x. So $C(r) \supset B_x(r)$ and $C(r) = B_x(r)$. Conversely, suppose $B_x(r) \cap B_y(r) \neq \emptyset$. Let $z \in B_x(r) \cap B_y(r)$ and C(r) be a r-clique containing z. Then x, $y \in B_z(r)$. Now by hypothesis $B_z(r) \subset C(r)$ so $d(x,y) \leq r$ and x, y, $z \in C(r)$. Hence $B_x(r) = B_y(r) = B_z(r)$.

Proposition 4. d is UM-distance iff d is r-homogenous $\forall r \in [0,1]$.

Proof. If d is UM, let x and y be in E, and z in $B_x(r) \cap B_y(r)$. Then $d(x,y) \leq d(x,z) \vee d(y,z) \leq r$ and for any t in $B_x(r)$, $d(x,t) \leq r$; hence $d(y,t) \leq r$, t is in $B_y(r)$ and $B_x(r) = B_y(r)$. Conversely, if d is not UM, there exist x, y, z, such that $d(x,y) > d(x,z) \vee d(y,z)$. Suppose $d(z,y) \leq d(z,x)$, then x and y are in $B_z(d(z,x))$, and $B_x(d(z,x))$ contains z but not y. Hence two spheres are different but not disjoint, and d is not d(z,x) - homogenous.

Proposition 5. d is a non trivial P-distance $\Rightarrow \exists r < 1$ such that d is r-homogenous.

Proof. Let R be associated with d, and

$$r = 1 - \bigwedge_{\{(v,w),\ R(v,w) \neq 0\}} R(v,w)$$

For all x, $B_x(r) = \{z | R(x,z) \geq 1-r\} = \{z | R(x,z) > 0\}$. Let x and y be in E and z belong to $B_x(r) \cap B_y(r)$. For all t in $B_x(r)$, $R(t,z) \geq R(t,x) \cdot R(x,z) > 0$ and $R(t,y) \geq R(z,y) \cdot R(t,z) > 0$; so t is in $B_y(r)$ and $B_x(r) = B_y(r)$.

Remarks. 1) The converse is wrong. Consider E = {x, y, z} $d(x,y) = 0.2$, $d(x,z) = 0.8$, $d(y,z) = 0.7$. Then d is not P $(0.8 > 0.2 + 0.7 - 0.14)$ but d is 0.1-homogenous. 2) A r-homogenous distance is not in general r'-homogenous for $r' \neq r$ : in the previous example d is 0.2- and 0.8-homogenous but not 0.7-homogenous.

## III. PARTITIONS OF E

### III.1. Classical Case

$P = \{P_i\}$ is a hard partition of E if $P_i \subset E$, $\bigcup_i P_i = E$ and $i \neq j \Rightarrow P_i \cap P_j = \emptyset$.

Definition 3. P will be called homogenous for a distance d if for all x, y in $P_i$, z in $P_j$, $i \neq j \Rightarrow d(x,y) < d(x,z)$. P will be called centered for d if for all i, there is a $c_i$ in $P_i$ (a center) such that $x \in P_i$ and $i \neq j \Rightarrow d(x,c_i) < d(x,c_j)$.

Remarks. The center of a $P_i$ is not necessarily unique; if P is homogenous, then it is centered, every element of $P_i$ being a center. It is easy to prove the following.

Proposition 6. If P is a partition homogenous for d, then d is r-homogenous with $r = \min_{i \neq j} d(P_j, P_j) = \min_{i,j;\ x \in P_i,\ y \in P_j} d(x,y)$.

Conversely, if d is r-homogenous, it yields homogenous partitions by r-cliques.

III.2. Quasi-partitions

Definition 4. A distance d is k-ultrametric (k-UM) if for any k + 2 points $x_1 x_2 \dots x_{k+2}$ in E, the $\binom{k+2}{2}$ distances between them satisfy

$$d(x_i,x_j) \leq \bigvee_{(s,t),\{s,t\} \neq \{i,j\}} d(x_s,x_t) \quad \text{for all } i,j$$

The case k = 1 corresponds to d a UM; a k-UM is also a (k + 1) - UM. With our notations, the weakly k-ultrametricity [4] can be rephrased as:

Definition 5. A distance d is weakly k-ultrametric if for any r, two distinct r-cliques intersect in at most k-1 elements. Now we prove

Proposition 7. k-ultrametricity and weak k-ultrametricity are equivalent.

Proof. Suppose d k-ultrametric and consider two r-cliques C(r) and C'(r) intersecting in $t_1 \dots t_k$. Take any x in $C(r) - \{C(r) \cap C'(r)\}$, y in $C'(r) - \{C(r) \cap C'(r)\}$. Then the k + 2 points x, y, $t_1, \dots, t_k$ satisfy $d(x,y) \leq \vee d(t_i,t_j) \leq r$, $i, j \in \{1,2,\dots,k\}$. Hence $x \in C'(r)$, a contradiction and $|C(r) \cap C'(r)| \leq k - 1$. For the converse, suppose d is not k - UM; there exist k+2 points $x_1, x_2, \dots x_{k+2}$ with (say)

$$d(x_1,x_2) > \bigvee_{i,j,\{i,j\} \neq \{1,2\}} d(x_i,x_j) = d(x_p,x_q).$$

Then for $r = d(x_p,x_q)$, there exist two r-cliques C(r) and C'(r) with $x_1 \in C(r)$, $x_2 \notin C(r)$, $x_1 \notin C'(r)$, $x_2 \in C'(r)$ and $C(r) \cap C'(r) \supset \{x_3,\dots,x_{k+2}\}$. Hence $|C(r) \cap C'(r)| \geq k$ and d is not weakly k - UM. Let us consider the following generalization of partitions:

Definition 6. $P = \{P_i\}$ is a k-quasi-partition if $E = \bigcup_i P_i$ and $i \neq j \Rightarrow |P_i \cap P_j| \leq k - 1$. Of course, k = 1 gives the classical partitions and we have

Proposition 8. d is k-ultrametric iff for any r, the r-cliques form a k-quasi-partition of E.

### III.3. Fuzzy Partitions

For a distance d, we consider a family $P = \{P_1, P_2, \ldots, P_m\}$ of r-cliques covering E (i.e., $E \subset \cup P_i$), with $0 < r < 1$. P induces a fuzzy partition $\pi = \{A_1, A_2, \ldots, A_m\}$ of E, every fuzzy subset $A_i$ of E being defined by the following membership function:

$$\mu_i(x) = 0 \quad \text{if } x \notin P_i$$

$$\mu_i(x) = \Big( \sum_{y \in P_i} R(x,y) \Big) \Big/ \sum_{j \in \mathcal{C}(x)} \sum_{z \in P_j} R(x,z) \quad \text{if } x \in P_i ,$$

where $\mathcal{C}(x)$ is the set of cliques containing x. It comes:

$$\sum_{x \in E} \mu_i(x) > 0 \qquad \forall\, i \in \{1, \ldots, m\}$$

and

$$\sum_{i=1}^{m} \mu_i(x) = 1 \qquad x \in E$$

In the next section, we enumerate the fuzzy partitions.

## IV. COMBINATORIAL RESULTS

We propose to exhibit combinatorial results concerning the number of fuzzy subsets and partitions of E and we generalize well-known quantities used in classical combinatorics.

Let N be the number of elements in E. A fuzzy subset A of E is defined by a membership function $\mu_A : E \to I \subset [0,1]$. In the sequel we suppose that I is finite, with $|I| = k$, and we consider as a particular case the set

$$I_0 = \{0, \frac{1}{k-1}, \frac{2}{k-1}, \ldots, 1\}.$$

The cardinality of A is defined by: $p = \sum_{x \in E} \mu_A(x)$.

### IV.1. Enumeration of Fuzzy Subsets of E

Let $C_N^p(I)$ denote the number of fuzzy subsets of E of cardinality p. Clearly:

$$C_1^p(I) = 1 \qquad \forall p \in I$$

$$C_1^p(I) = 0 \qquad \forall p \notin I$$

$$C_N^0(I) = 1 \qquad \forall N \geq 1.$$

<u>Proposition 9</u>. For $k \geq 2$, $N \geq 2$, $p > 0$, we have:

$$C_N^p(I) = \sum_{i \in I} C_{N-1}^{p-i}(I) \tag{1}$$

and

$$\sum_{p=0}^{N} C_N^p(I) = k^N. \tag{2}$$

<u>Proof</u>. Any given fuzzy subset A of a set $E = \{e_1, \ldots, e_N\}$ with N elements is obtained from a fuzzy subset A' of $E' = \{e_1, \ldots, e_{n-1}\}$ by associating with $e_n$ the difference between the cardinalities of A and A', if this difference belongs to I, which proves equality (1). The left-hand side of equality (2) corresponds to the number of ways of associating every element of E with an element of I, which equals $k^N$. In the special case where $k = 2$, $C_N^p(I)$ equals $\binom{N}{p}$, with $p \in N$ and relation (1) is the equality defining Pascal's triangle. If we consider the set of values $I_0$, we can bind the coefficients $C_N^p(I_0)$ with the numbers $c(x, y, z)$ of compositions of z with exactly y parts, each $\leq x$, for three integers x, y, z [1]. It is easy to see that $\binom{N}{S}\, c(k-1, S, p(k-1))$ corresponds to the number of fuzzy subsets of E with cardinality p and number of elements S corresponding to a non-null value of the membership function. Consequently, by using the set $I_0$, we get:

$$C_N^p(I_0) = \sum_{S=1}^{p(k-1)} \binom{N}{S}\, c(k-1, S, p(k-1)). \tag{3}$$

<u>Example</u>. $I_0 = \{0, \frac{1}{2}, 1\}$, $k = 3$.

<u>Values of $C_N^p(I_0)$:</u>

| N \ P | 0 | 1/2 | 1 | 3/2 | 2 |
|---|---|---|---|---|---|
| 1 | 1 | 1 | 1 | 0 | 0 |
| 2 | 1 | 2 | 3 | 2 | 1 |
| 3 | 1 | 3 | 6 | 7 | 6 |
| 4 | 1 | 4 | 10 | 16 | 19 |
| 5 | 1 | 5 | 15 | 30 | 45 |

For N = 4, p = 2

$$C_4^2(I_0) = 19$$
$$= C_3^2(I_0) + C_3^{3/2}(I_0) + C_3^1(I_0)$$
$$= 6 + 7 + 6$$
$$= \sum_{S=1}^{4} \binom{4}{S} \quad (2, S, 4).$$

Let us study the particular case of fuzzy subsets of E corresponding to a non-null value of the membership function for s given elements of E. We denote by $D_{N,s}^p(I)$ the number of such subsets with cardinality p.

Proposition 10. For $n \geq 1$, $0 \leq j \leq N$, $p > 0$, we have

$$D_{N,s}^p(I) = \sum_{0<q\leq p} C_s^q(I^*)\, C_{N-s}^{p-q}(I) \tag{4}$$

and

$$D_{N,s}^p(I) = \sum_{r=o}^{N-s} \binom{N-s}{r} C_s^p(I^*) \tag{5}$$

with $I^* = I - \{0\}$.

Proof. Let $E_s = \{e_1, \ldots, e_s\}$ be the subset of E corresponding to non-null values of $\mu$ in such a subset A. The definition of $\mu$ is obtained by choosing a fuzzy subset $A_s$ of $E_s$ with values of $\mu$ in I*, and there exist $C_s^q(I^*)$ such possibilities for any given cardinality q of $A_s$, $0 < q \leq p$. Then, we have to define the values of $\mu$ for the elements of $E - E_s$; their sum equals p-q and they are taken in I. Equation (4) follows. Another way of evaluating $D_{N,s}^p(I)$ considers all the elements of E corresponding to non-null values of $\mu$. We have to choose r such elements in $E-E_s$, for any value r between 0 and N-s; we have $\binom{N-s}{r}$ such possibilities. Then, we obtain a set of s+r elements associated with value of I* which must have a sum equal to p. Equation (5) follows.

IV.2. Enumeration of Fuzzy Partitions

We consider a fuzzy partition $\pi = \{A_1, \ldots, A_m\}$ of E with m fuzzy classes. It can be represented by a matrix $A = (\mu_{ij})1 \leq i \leq m$, $1 \leq j \leq N$, where $\mu_{ij}$ is the value of the membership function defining the subset $A_i$ of E, for the element $e_j$; this matrix verifies:

$$\sum_i \mu_{ij} = 1 \qquad \forall j \; \varepsilon\{1, \ldots, N\}$$

$$\sum_i \mu_{ij} > 0 \qquad \forall i \; \varepsilon\{1, \ldots, m\}.$$

Let $S_N^m$ denote the number of fuzzy partitions of E with m classes, corresponding to the set of values I.

Proposition 11. For $k \geq 2$, $N \geq 2$, $m \geq 2$, we have:

$$S_N^m = S_{N-1}^m \, C_m^1(I) + \sum_{j=1}^{m-1} \binom{m}{j} S_{N-1}^j \, D_{m,m-j}^1(I). \qquad (6)$$

Proof. Any fuzzy partition $\pi$ of $E = \{e_1, \ldots, e_N\}$ can be obtained from a fuzzy partition $\pi'$ of $E' = \{e_1, \ldots, e_{N-1}\}$ by adding the element $e_N$, either in a class of $\pi$ which belongs to $\pi'$, or in a new class of $\pi$ which only contains $e_N$.

The first construction corresponds to $S_{N-1}^m$ partitions of E', each of them being associated with $C_m^1(I)$ definitions of $\{\mu_{iN}, 1 \leq i \leq m\}$. Thus, we get the first part of the right-hand side of (4).

The second construction corresponds to partitions $\pi'$ of E' with j fuzzy classes, for $1 \leq j \leq m-1$. Then, there exist $S_{N-1}^j$ such partitions, using j given classes of $\pi$, and $\binom{m}{j}$ choices of these classes are possible. For every given partition $\pi'$, we define $\pi$ by giving the values of $\{\mu_{iN}, 1 \leq i \leq m\}$; (m-j) such values are non-null and correspond to the classes of $\pi$ which do not belong to $\pi'$. Their sum equals q, for $0 < q \leq 1$, and $C_{m-j}^q(I^*)$ choices are possible.

For every determination of these (m-j) values, we have $C_j^{1-q}(I)$ possible choices of the other values $\mu_{iN}$. Then, for every given partition $\pi'$, we construct $D_{m,m-j}^1(I)$ partitions $\pi$. Thus, we obtain the second part of the right-hand side of equation (6).

In the case when k = 2, we get $C_m^1(I) = m$ and q equals 1 in the expression of $D_{m,m-j}^1(I)$ by (4). Consequently, $C_{m-j}^1(I^*) = 0$ if $m-j>1$. The second part of (6) corresponds to the unique value $j=m-1$ and equals $S_{N-1}^{m-1}$. Equation (6) is equivalent to the recursive relation defining Stirling numbers.

If we consider the set of values $I_0$, we can think of the numbers $c(X_1, \ldots, X_r ; y)$ of compositions of the vector $(X_1, \ldots, X_r)$ with y parts, for (r+1) integers $X_1, \ldots, X_r$, y. It is obvious that the number of fuzzy partitions of E with m classes equals $c(X_1, \ldots, X_N ; m)$, with $X_i = k-1$ for $1 \leq i \leq N$. Using the known value of $c(X_1, \ldots, X_N ; m)$ [1, 4.3.3], we obtain in the case of $I_0$:

$$S_N^m = \sum_{i=0}^{m} (-1)^i \binom{m}{i} \binom{k-2+m-i}{k-1}^N \tag{7}$$

Example. $I_0 = \{0, \frac{1}{2}, 1\}$ , $k = 3$.

Values of $S_N^m$ :

| N \ m | 1 | 2 | 3 |
|---|---|---|---|
| 1 | 1 | 1 | 0 |
| 2 | 1 | 7 | 12 |
| 3 | 1 | 25 | 138 |

## V. REFERENCES

[1] G. A. Andrews, "The Theory of Partitions," Addison-Wesley (1976).

[2] J. C. Bezdek and J. D. Harris, Fuzzy partitions and relations: An axiomatic basis for clustering, Fuzzy Sets and Systems 1, pp. 111-127 (1978).

[3] B. Bouchon, G. Cohen, and P. Frankl, Metrical properties of fuzzy partitions, Int. Colloqu. on Information Theory, Budapest (1981).

[4] H. K. Kim and F. W. Roush, Ultrametrics and matrix theory, J. of Math. Psychology , 18, pp. 195-203 (1978).

[5] L. A. Zadeh, Similarity relations and fuzzy orderings, Information Sciences, 3, pp. 177-200 (1971).

FUZZY SET STRUCTURE WITH STRONG IMPLICATION

Zhang Jinwen

Institute of Computing Technology
Academia Sinica
Beijing, People's Republic of China

Fuzzy set structure with strong implication has some very interesting logical characteristics. It is weaker than the normal fuzzy set structures, but a little stronger than some fuzzy set structures. We have the following results:

Theorem 1. Every axiom of weak logic calculus WL is true in this structure and every rule of inference of WL is also valid in it.

Theorem 2. Let G' be any complete quasi-Boolean sublattice. Then for any $\Sigma_0$-formula, $A(x_1,\dots,x_n)$ not including the negative $\neg$, and any $c_1, \dots, c_n \in F(G')$, we have

$$[\![A(C_1, \dots, C_n)]\!]^{G'} = [\![A(C_1, \dots, C_n)]\!]^{G}.$$

Theorem 3. Every axiom of a weak axiom system $ZF_1U$ of set theory is true in our structure; that is, our structure is a model of $ZF_1U$.

In this paper we apply the logic WL to generalize basic notions of fuzzy logic.

We shall show that there are some important interconnections between the system logic (including fuzzy logic) and the system methodology, and so the evolution of fuzzy logic enters into an important field of system methodology.

## 1. Fuzzy Sets and Higher Level Fuzzy Sets.

For any classic Set X, let F(X) be the set of all fuzzy

sets[15], that is,

$$F(X) := \{\underset{\sim}{A} \mid \underset{\sim}{A} \subseteq X\}$$

of course, F(X) is a classic Set, but its elements are fuzzy Sets.

Let M be a transitive model of the axiomatic Set theorem ZFU with urelements, S is the Set of urelements.

Definition 1. For the above M, S, by transfinite recursion on ordinal $\alpha$ we define the universe of fuzzy Sets as follows:

$$\mathcal{F}_0 := S \cup \{\phi\}$$

$$\mathcal{F}_{\alpha+I} := \mathcal{F}_\alpha \cup F(\mathcal{F}_\alpha)$$

$$\mathcal{J}_\lambda := \bigcup_{\alpha<\lambda} \mathcal{F}_\alpha, \text{ if } \lambda \text{ is a limit ordinal}$$

$$\mathcal{F} := \bigcup_{\alpha\varepsilon On} \mathcal{F}_\alpha,$$

Where On := $\{\alpha \mid \alpha\varepsilon M$ and $\alpha$ is a ordinal number$\}$.

Definition 2. For any $x\varepsilon\mathcal{F}$, we say that the least ordinal number $\alpha$ such that $x\varepsilon\mathcal{F}_\alpha$ is the rank of x and is denoted by $\rho(x)$.

Definition 3. For any $x\varepsilon\mathcal{F}$, we put

$$\text{dom}(x) := \{y \mid \mu_x(y) > 0 \text{ and } y\varepsilon\mathcal{F}_{\rho(x)}\},$$

where $\mu_x(y)$ is the membership grade of y in the fuzzy Set x. Certainly, $\mu_x(y)\varepsilon[0,1]$. Obviously, dom(x) is a classic Set, too.

For any $x\varepsilon\mathcal{F}$, if we know that the

I. Classic Set dom(x)

II. The value of $\mu_x(y)$, when $y\varepsilon$dom(x), then we know the fuzzy Set x. Therefore from now on, where we have defined I and II, we can say that we also defined fuzzy Set x.

From the above definition of $\mathcal{F}$ we observe that members of a fuzzy Set may be some fuzzy Sets. We call members of $\mathcal{F} - \mathcal{F}_1$ higher level fuzzy Sets. We had developed generalizations of the operations $\cup$ (union), $\cap$ (intersection), ' (relative complement) of fuzzy Sets under Zadeh's meaning to higher level fuzzy Sets. And for any ordinal $\alpha > 0$, $\mathcal{F}_\alpha - \mathcal{F}_0$ is a Zadeh's lattice, and also $\mathcal{F} - \mathcal{F}_0$ is one[5].

## 2. Fuzzy Logic with Strong Implication.

A quasi-Boolean lattice is a 6-tuple $\langle G, \oplus, \otimes, \circledast, 0_G, 1_G \rangle$; for an understanding of the definition and properties of quasi-Boolean lattice, see[4].

We will take the closed interval [0, 1] as G, and following Zadeh[15] we define as follows:

$$x \oplus y := \text{Max}\,\{x, y\},$$

$$x \otimes y := \text{Min}\,\{x, y\},$$

$$x \circledast := 1 - x.$$

Obviously, G is a complete lattice.

Now we define $\Rightarrow$ by

$$x \Rightarrow y := \begin{cases} 1 & \text{if } x \leq y \\ y & \text{otherwise}, \end{cases}$$

where $\leq$ is the natural order of real numbers, and $\Rightarrow$ is called strong implication. Dr. E. Sanchez called $\Rightarrow$ the α-operator. It is a very important operator in Sanchez[16].

And furthermore we define $\Leftrightarrow$ by

$$x \Leftrightarrow y := x \Rightarrow y \otimes y \Rightarrow x.$$

Lemma 1. If $a, b \in [0, 1]$, then $a \leq b$ iff $(a \Rightarrow b) = 1$.

Lemma 2. If $a, b, c \in [0, 1]$, then

$$a \otimes b \otimes (b \Rightarrow c) \leq a \otimes c.$$

Lemma 3. If $a, b, c \in [0, 1]$, then

$$a \Rightarrow (b \Rightarrow c) = a \otimes b \Rightarrow c.$$

Lemma 4. If $a, b, c \in [0, 1]$, then

$$a \otimes b \leq c \text{ iff } a \leq (b \Rightarrow c).$$

Lemma 5. If $a, b \in [0, 1]$, then

$$(a \Leftrightarrow b) = 1 \text{ iff } a = b.$$

Lemma 6. If $a, b, c \in [0, 1]$ and $b \leq c$, then

$a \Rightarrow b \leq a \Rightarrow c.$

Lemma 7. If a, b, c $\varepsilon$ [0, 1] and $b \leq c$, then

$(c \Rightarrow a) \leq (b \Rightarrow a).$

Lemma 8. There exists a number a $\varepsilon$ [0, 1], so that following statements:

$a \oplus a^{\circledast} < 1, \; a \otimes a^{\circledast} > 0.$

Lemma 9. There are a, b $\varepsilon$ [0, 1], such that:

$((a \Rightarrow b) \Rightarrow a) \Rightarrow a < 1.$

Lemma 10. There are a, b $\varepsilon$ [0, 1], such that

$(a \Rightarrow b) \Rightarrow ((a \Rightarrow b^{\circledast}) \Rightarrow a^{\circledast}) < 1.$

Lemma 11. There are a, b $\varepsilon$ [0, 1]), such that

$((a \Rightarrow b) \Rightarrow a) \Rightarrow b < 1.$

## 3. Definitions of Fuzzy Relations $\underset{\sim}{\varepsilon}$ and $\underset{\sim}{\equiv}$, and Fuzzy Set Structure $\langle F, \underset{\sim}{\varepsilon}, \underset{\sim}{\equiv} \rangle$.

In this paper we define fuzzy relations $\underset{\sim}{\varepsilon}$ and $\underset{\sim}{\equiv}$ which are different from [6], because their logical foundations are different; here is a fuzzy logic with strong implication, but that is a Boolean valued logic.

Definition 4. For any x, y$\varepsilon F$, the fuzzy relations $x \underset{\sim}{\varepsilon} y$, $x \underset{\sim}{\equiv} y$ are established by induction on $\rho(x)$, $\rho(y)$ as follows:

(1) When $\rho(x) = \rho(y) = 0$

$$x \underset{\sim}{\varepsilon} y := 0,$$

$$x \underset{\sim}{\equiv} y := \begin{cases} 1 & \text{if x and y are identical} \\ 0 & \text{otherwise.} \end{cases}$$

(2) When $\rho(x) > 0$ or $\rho(y) > 0$

$$x \varepsilon y := \begin{cases} 0 & \text{if } \rho(y) = 0 \\ \bigvee_{z \varepsilon dom(y)} & \mu y(z) \otimes z \underset{\sim}{\equiv} x, \text{ otherwise.} \end{cases}$$

$$
x \approx y := \begin{cases} 0 & \text{if } x\varepsilon s \text{ or } y\varepsilon s \\ \bigwedge_{z\varepsilon dom(x)} (\mu_x(z) \Longrightarrow 0) & \text{if } y = \phi, \\ \bigwedge_{z\varepsilon dom(y)} (\mu_y(z) \Longrightarrow 0) & \text{if } x = \phi, \\ \bigwedge_{z\varepsilon dom(y)} (\mu_y(z) \Longrightarrow z \underset{\sim}{\varepsilon} x) \otimes & \\ \bigwedge_{z\varepsilon dom(x)} (\mu_x(z) \Longrightarrow z \underset{\sim}{\varepsilon} y) & \text{otherwise.} \end{cases}
$$

Obviously, for any x, y$\varepsilon F$, relations $x \underset{\sim}{\varepsilon} y$ and $x \underset{\sim}{=} y$ take values in the interval [0,1].

Definition 5. By an ordered triple $\langle F, \underset{\sim}{\varepsilon}, \underset{\sim}{=} \rangle$, we mean a fuzzy Set Structure with Strong implication.

Basic Lemma 1: In our structure $\langle F, \underset{\sim}{\varepsilon}, \underset{\sim}{=} \rangle$, for any x, y, z$\varepsilon F$, the following propositions hold.

1) $x \underset{\sim}{=} x$,

2) $\mu_x(z) \le z \underset{\sim}{\varepsilon} x$, if $\rho(x) > 1$ and $z\varepsilon dom(x)$.

3) $x \underset{\sim}{=} y = y \underset{\sim}{=} x$.

4) $x \underset{\sim}{=} y \otimes y \underset{\sim}{=} z \le x \underset{\sim}{=} z$.

5) $x \underset{\sim}{=} y \otimes x \underset{\sim}{\varepsilon} z \le y \underset{\sim}{\varepsilon} z$.

6) $x \underset{\sim}{=} y \otimes z \underset{\sim}{\varepsilon} x \le z \underset{\sim}{\varepsilon} y$.

Proof: 1) is proved by induction on the well-founded relation $z\varepsilon dom(x)$. Assume that $(z \underset{\sim}{=} z) = 1$ for all $z\varepsilon dom(x)$. Then if $z\varepsilon dom(x)$, we have

$$z \underset{\sim}{\varepsilon} x = \bigvee_{y\varepsilon dom(x)} \mu_x(y) \otimes z \underset{\sim}{=} y \ge \mu_x(z) \otimes z \underset{\sim}{=} z = \mu_x(z). \quad (3.1)$$

Therefore $x \underset{\sim}{=} x = \bigvee_{z\varepsilon dom(x)} (\mu_x(z) \Longrightarrow z \underset{\sim}{\varepsilon} x) = 1.$

2) is proved as in (3.1) above, using 1).

3) holds by symmetry.

We show 4), 5), and 6), simultaneously by induction On $\alpha$ as follows, when $\alpha = \rho(x) = \rho(y) = \rho(z) = 0$, the three formula hold obviously.

I. When $\rho(x), \rho(y) < \alpha$ and $\rho(z) \leq \alpha$, we have

$$x \underset{\sim}{\equiv} y \otimes x\underset{\sim}{\varepsilon}y \leq y\underset{\sim}{\varepsilon}z.$$

II. When $\rho(z) < \alpha$ and $\rho(x), \rho(y) \leq \alpha$, we have

$$x \underset{\sim}{\equiv} y \otimes z\underset{\sim}{\varepsilon}x \leq z\underset{\sim}{\varepsilon}y.$$

III. When $\rho(x), \rho(y), \rho(z) \leq \alpha$, we have

$$x \underset{\sim}{\equiv} y \otimes y \underset{\sim}{\equiv} z \leq x \underset{\sim}{\equiv} z.$$

(1) We can obtain that

$$x \underset{\sim}{\equiv} y \otimes x\underset{\sim}{\varepsilon}z$$

$$= \bigvee_{t\varepsilon dom(z)} \mu_z(t) \otimes t \underset{\sim}{\equiv} x \otimes x \underset{\sim}{\equiv} y$$

$$\leq \bigvee_{t\varepsilon dom(z)} \mu_z(t) \otimes t \underset{\sim}{\equiv} y \quad \text{(by inductive hypothesis for III)}$$

$$= y\underset{\sim}{\varepsilon}z.$$

(2) If $t\varepsilon dom(y)$, then

$$x \underset{\sim}{\equiv} t \otimes \mu_y(t) \otimes y \underset{\sim}{\equiv} z$$

$$\leq x \underset{\sim}{\equiv} t \otimes \mu_y(t) \otimes (\mu_y(t) \Longrightarrow t\underset{\sim}{\varepsilon}z)$$

$$\leq x \underset{\sim}{\equiv} t \otimes t\underset{\sim}{\varepsilon}z \quad \text{(From Lemma 2)}$$

$$\leq x\underset{\sim}{\varepsilon}z. \quad \text{(by (1))}$$

Take the suprema over $t\varepsilon dom(y)$ in two sides of the unequality and we can obtain that $x = y \otimes y\varepsilon z \leq x\varepsilon z$.

(3) If $t\varepsilon dom(x)$, then

$$x \underset{\sim}{\equiv} y \otimes y \underset{\sim}{\equiv} z \otimes \mu_x(t)$$

$$\leq x \underset{\sim}{\equiv} y \otimes y \underset{\sim}{\equiv} z \otimes t\underset{\sim}{\varepsilon}x$$

$$\leq t\underset{\sim}{\varepsilon}y \otimes y \underset{\sim}{\equiv} z$$

$$\leq t\underset{\sim}{\varepsilon}z.$$

So we have

$$x \underset{\sim}{=} y \otimes y \underset{\sim}{=} z \le \bigwedge_{t \in dom(x)} (\mu_x(t) \Longrightarrow t \underset{\sim}{\varepsilon} x) \tag{3.2}$$

Besides, by symmetry, it is not difficult to obtain that

$$x \underset{\sim}{=} y \otimes y \underset{\sim}{=} z \le \bigwedge_{t \in dom(z)} (\mu_z(t) \Longrightarrow t \underset{\sim}{\varepsilon} z) \tag{3.3}$$

Hence, from (3.2) and (3.3) we get $x \underset{\sim}{=} y \otimes y \underset{\sim}{=} z \le x \underset{\sim}{=} z$.

The proof of our basic lemma 1 is now complete.

## 4. The Structure $\langle \mathcal{F}, \varepsilon, \underset{\sim}{=} \rangle$ is a Model of Weaker Predicate Calculus WL.

Following [5], for any formula A in first order language, we may define its valuation in this structure $\langle \mathcal{F}, \varepsilon, \underset{\sim}{=} \rangle$, and we also may determine whether a rule of inference of the first order logic is valid in $\langle \mathcal{F}, \underset{\sim}{\varepsilon}, \underset{\sim}{=} \rangle$.

Example 1. By this valuation, the formula $(A \to B) \to ((A \to \neg B) \to \neg A))$ is not true.

Example 2. Similarly, we can easily verify the following formula

$$((A \to B) \to A) \to A$$

also is not true in our structure $\langle \mathcal{F}, \underset{\sim}{\varepsilon}, \underset{\sim}{=} \rangle$.

From above two examples, we know that this structure is not a model of usual first order logic[10]. But we will show that it is a model of a weaker predicate logic following definited.

Now let WL be a weaker predicate logic which consists of the following axioms and rule of inference.

1 - 6, 8 - 12 as in Kleene[10].

7. $A \to \neg\neg A$;

13. $x = x$

14. $(x = y) \wedge \Phi(x) \to \Phi(y)$, where $\Phi(x)$, $\Phi(y)$ are formulas which do not contain negation $\neg$.

15. $\dfrac{A \to B}{\neg B \to \neg A}$ ,

16. $\neg(\neg A \wedge \neg B) \leftrightarrow A \vee B$,

17. $\forall x A(x) \leftrightarrow \neg \exists x \neg A(x)$,

18. $\exists x(A(x) \to B(x)) \to (\forall x A(x) \to \exists x B(x))$,

19. $(\exists x A(x) \to \forall x B(x)) \to \forall x(A(x) \to B(x))$.

Theorem 1. Every axiom of weaker predicate logic WL is true in this structure and every rule of inference of WL is also valid in it.

Proof: We will verify the axioms and rules in $< F, \underset{\sim}{\varepsilon}, \underset{\sim}{\equiv} >$ one by one.

1a. Since $[\![B \to A]\!] \geq [\![A]\!]$, we have $[\![A \to (B \to A)]\!] = 1$.

1b. We hope to prove that $[\![(A \to B) \to ((A \to (B \to C)) \to (A \to C))]\!] = 1$.

For this purpose, we will divide this verification into the following few steps:

1°. When $[\![A]\!] \leq [\![C]\!]$, obviously, our previous result holds.

2°. When $[\![C]\!] < [\![A]\!]$, we divide it further into two cases. First assume $[\![B]\!] \leq [\![C]\!]$, in which case we can easily verify desired result.

Otherwise $[\![C]\!] < [\![B]\!]$, because in this case we have $[\![A \to C]\!] = [\![C]\!]$, $[\![B \to C]\!] = [\![C]\!]$. Thus we get our desired result.

1b is proved.

2. We must prove that if $[\![A]\!] = 1$, $[\![A \to B]\!] = 1$, then $[\![B]\!] = 1$. Since $[\![A \to B]\!] = 1$, we have $[\![A]\!] \leq [\![B]\!]$. So obviously $[\![B]\!] = 1$.

3. It is asked to be proved that $[\![A \to (B \to A \wedge B)]\!] = 1$. It is not difficult that the desired result is obtained if we divide it into two cases: $[\![A]\!] \leq [\![B]\!]$ and $[\![B]\!] < [\![A]\!]$.

4a. $[\![A \wedge B \to A]\!] = 1$ and 4b. $[\![A \wedge B \to B]\!] = 1$. Both of them are evident.

5a. $[\![A \to A \vee B]\!] = 1$ and 5b. $[\![B \to A \vee B]\!] = 1$. They are also evident.

6. We must prove $[\![(A\to C)\to((B\to C)\to(A \vee B)\to C))]\!] = 1$. The proof is divided into four cases:

1°. When Max $\{ [\![A]\!], [\![B]\!] \} \leq [\![C]\!]$.

2°. When min $\{ [\![A]\!], [\![B]\!] \} > [\![C]\!]$.

Both the 1° and 2° are obvious.

3°. When $[\![A]\!] > [\![C]\!] \geq [\![B]\!]$, we have $( [\![A \vee B]\!] \Rightarrow [\![C]\!] ) = ( [\![A]\!] \Rightarrow [\![C]\!]) = [\![C]\!]$. So we get our desired result.

4°. When $[\![B]\!] > [\![C]\!] \geq [\![A]\!]$, $[\![A \vee B\to C]\!] = [\![C]\!]$ and $[\![B\to C]\!] = [\![C]\!]$, we can get the desired result. Also, we have the following two propositions.

7. $[\![A\to \neg\neg A]\!] = 1$ and 8. $[\![\neg\neg A\to A]\!] = 1$.

For the following, x is a variable, A(x) is a formula, C is a formula which does not contain x free, and t is a term which is free for x in A(x).

9. If $[\![C\to A(x)]\!] = 1$, then $[\![C\to \forall x\, A(x)]\!] = 1$. We have

$$
\begin{aligned}
[\![C\to \forall x A(x)]\!] &= [\![C]\!] \Rightarrow \bigwedge_{x\in F} [\![A(x)]\!] \\
&= \bigwedge_{x\in F} ( [\![C]\!] \Rightarrow [\![A(x)]\!] ) \\
&= \bigwedge_{x\in F} [\![C\to A(x)]\!] \\
&= [\![\forall x\, (C\to A(x))]\!] \\
&= [\![C\to A(x)]\!] \\
&= 1.
\end{aligned}
$$

10. Since $\bigwedge_{x\in F} A(x) \leq [\![A(t)]\!]$, we have $[\![\forall x A(x) \to A(t)]\!] = 1$.

11. Since $[\![A(t)]\!] \leq \bigvee_{x\in F} [\![A(x)]\!] = [\![\exists x A(x)]\!]$, we have $[\![A(t) \to \exists x A(x)]\!] = 1$.

12. By steps in analogy to 9, the desired result can be obtained.

By our basic lemma 1, 13 and the primitive cases in 14 are

complete. Because in order to prove

$$[\![(x = y) \wedge \Phi(x) \rightarrow \Phi(y)]\!] = 1 \tag{4.1}$$

we need to prove

$$(x \underset{\sim}{=} y) \otimes [\![\Phi(x)]\!] \leq [\![\Phi(y)]\!] \tag{4.2}$$

From basic lemma 1, we have

$$(x \underset{\sim}{=} y) \otimes x \underset{\sim}{\varepsilon} z \leq y \underset{\sim}{\varepsilon} z \tag{4.3}$$

$$(x \underset{\sim}{=} y) \otimes z \underset{\sim}{\varepsilon} x \leq z \underset{\sim}{\varepsilon} y \tag{4.4}$$

$$(x \underset{\sim}{=} y) \otimes y \underset{\sim}{=} z \leq x \underset{\sim}{=} z \tag{4.5}$$

Finally, 14 is proved by induction on the complexity of $\Phi$, in either cases of $\wedge$, $\vee$, $\exists x$, $\forall x$ it is easy to prove. So it suffices for us to prove it in case of $\rightarrow$, and from $\rightarrow$ and $\wedge$. The case of $\leftrightarrow$ can be proved naturally.

By letting that $\Phi(x) = \Phi_1(x) \rightarrow \Phi_2(x)$ and by the induction hypothesis, we have

$$(x \underset{\sim}{=} y) \otimes [\![\Phi_1(x)]\!] \leq [\![\Phi_1(y)]\!] \tag{4.6}$$

$$(x \underset{\sim}{=} y) \otimes [\![\Phi_2(x)]\!] \leq [\![\Phi_2(y)]\!] \tag{4.7}$$

From the symmetry of $x = y$, we have $y = x$; so we can obtain

$$(x \underset{\sim}{=} y) \otimes [\![\Phi_1(y)]\!] \leq [\![\Phi_1(x)]\!] \tag{4.8}$$

$$(x \underset{\sim}{=} y) \otimes [\![\Phi_2(y)]\!] \leq [\![\Phi_2(x)]\!] \tag{4.9}$$

Hence we have:

Either I

$$(x \underset{\sim}{=} y) \leq [\![\Phi_1(x)]\!] \tag{4.10}$$

$$(x \underset{\sim}{=} y) \leq [\![\Phi_1(y)]\!] \tag{4.11}$$

$$(x \underset{\sim}{=} y) \leq [\![\Phi_2(x)]\!] \tag{4.12}$$

$$(x \underset{\sim}{=} y) \leq [\![\Phi_2(y)]\!] \tag{4.13}$$

hold simultaneously, or II

$$[\![\Phi_1(x)]\!] = [\![\Phi_1(y)]\!] \tag{4.14}$$

$$[\![\Phi_2(x)]\!] = [\![\Phi_2(y)]\!] \tag{4.15}$$

hold simultaneously, or III

$$x \approxeq y \leq [\![\Phi_1(x)]\!] \tag{4.16}$$

$$x \approxeq y \leq [\![\Phi_1(y)]\!] \tag{4.17}$$

$$x \approxeq y > [\![\Phi_2(x)]\!] \tag{4.18}$$

$$x \approxeq y > [\![\Phi_2(y)]\!] \tag{4.19}$$

hold simultaneously, or IV

$$x \approxeq y \leq [\![\Phi_2(x)]\!] \tag{4.20}$$

$$x \approxeq y \leq [\![\Phi_2(y)]\!] \tag{4.21}$$

$$x \approxeq y > [\![\Phi_1(x)]\!] \tag{4.22}$$

$$x \approxeq y > [\![\Phi_1(y)]\!] \tag{4.23}$$

hold simultaneously.

In the four cases described above, we can obtain directly that

$$x \approxeq y \otimes [\![\Phi_1(x) \to \Phi_2(x)]\!] \leq [\![\Phi_1(y) \to \Phi_2(y)]\!]$$

$$x \approxeq y \otimes [\![\Phi_1(y) \to \Phi_2(y)]\!] \leq [\![\Phi_1(x) \to \Phi_2(x)]\!]$$

Thus we have proved the rule 14.

We shall omit the proof of rules 15-19 since they are quite obvious.

Thus we have proved the theorem 1.

Remark 1. When the formula $\Phi(x)$ contains negation $\neg$ , the the equation (4.2) cannot be obtained. For example, when $\Phi(x) = \neg\Phi_1(x)$, for $\Phi_1(x)$ we have

$$x \approxeq y \otimes \Phi_1(x) \leq \Phi_1(y) \tag{4.24}$$

By invoking the property of symmetry, we also have

$$x \approxeq y \otimes [\![\Phi_1(y)]\!] \leq [\![\Phi_1(x)]\!] \tag{4.25}$$

However, we can not obtain the following two inequalities generally:

$$x \underset{\sim}{\equiv} y \otimes [\![ \neg \Phi_1(x) ]\!] \leq [\![ \neg \Phi_1(y) ]\!] \tag{4.26}$$

$$x \underset{\sim}{\equiv} y \otimes [\![ \neg \Phi_1(y) ]\!] \leq [\![ \neg \Phi_1(x) ]\!] \tag{4.27}$$

Now, we shall give an example which satisfies (4.24) and (4.25), but not the equations (4.26) and (4.27) simultaneously.

Let

$$\text{dom}(x) = S(\subseteq F_0)$$

$$\mu_x(t) = \frac{7}{10} \qquad \text{if } t \varepsilon \text{dom}(x)$$

$$\text{dom}(y) = S(\subseteq F_0)$$

$$\mu_y(t) = 1 \qquad \text{if } t \varepsilon \text{dom}(y).$$

It is not difficult to calculate that

$$x \underset{\sim}{\equiv} y = \frac{7}{10}.$$

Let us further define the fuzzy set u, such that

$$\text{dom}(u) = \{x, y\},$$

$$\mu_u(x) = \frac{9}{10}, \ \mu_u(y) = \frac{8}{10}.$$

Hence we have:

$$\begin{aligned} x \underset{\sim}{\varepsilon} u &= \bigvee_{z \varepsilon \text{dom}(u)} \mu_u(z) \otimes z \underset{\sim}{\equiv} x \\ &= \mu_u(x) \otimes x \underset{\sim}{\equiv} x \oplus \mu_u(y) \otimes y \underset{\sim}{\equiv} x \\ &= \frac{9}{10} \otimes 1 \oplus \frac{8}{10} \otimes \frac{7}{10} \\ &= \frac{9}{10}. \end{aligned}$$

Similarly, we have $y \underset{\sim}{\varepsilon} u = \frac{8}{10}$. It is quite obvious that (4.24), (4.25) and (4.26) do hold where equation (4.27) does not hold.

## 5. The Technique of $\Sigma_0$ Formulas.

$\Sigma_0$ formulas[11] are quite important and they possess good properties. Besides, they are a strong tool in either forcing method or the method of Boolean-Valued model. in the fuzzy set structures, we are also using this technique to our advantage.

A formula $\Phi$ is a $\Sigma_0$-formula if it has only restricted quantifiers $\forall x \varepsilon y$ and $\exists x \varepsilon y$ (i.e., if $\Phi$ is a restricted form $x \varepsilon y A(x)$ or $x \varepsilon y A(x)$) where

$$\exists x \varepsilon y A(x) := \exists x(x \varepsilon y \wedge A(x)),$$

$$\forall x \varepsilon y A(x) := \forall x(x \varepsilon y \rightarrow A(x)).$$

This class of formulas shall be used in the next section.

Basic Lemma 2. $\Phi(x)$ is any formula which does not contain negation $\neg$ . For any fuzzy set $u \varepsilon F$, we can establish that:

$$(1) \quad [\![\forall x \varepsilon u A(x)]\!] = \bigwedge_{x \varepsilon dom(u)} (\mu_u(x) \Longrightarrow [\![\Phi(x)]\!]),$$

$$(2) \quad [\![\exists x \varepsilon u A(x)]\!] = \bigvee_{x \varepsilon dom(u)} (\mu_u(x) \otimes [\![\Phi(x)]\!]).$$

Proof: We shall prove (1) first;

$$[\![\forall x \varepsilon u \Phi(x)]\!]$$

$$= [\![\forall x(x \varepsilon u \rightarrow \Phi(x))]\!]$$

$$= \bigwedge_{x \varepsilon F} [\![x \varepsilon u \rightarrow \Phi(x)]\!]$$

$$= \bigwedge_{x \varepsilon F} ([\![x \varepsilon u]\!] \Rightarrow [\![\Phi(x)]\!])$$

$$= \bigwedge_{x \varepsilon F} (\bigvee_{z \varepsilon dom(u)} \mu_u(z) \otimes z \underset{\sim}{=} x \Rightarrow [\![\Phi(x)]\!])$$

$$= \bigwedge_{z \varepsilon dom(u)} \bigwedge_{x \varepsilon F} (\mu_u(z) \Rightarrow (z \underset{\sim}{=} x \Rightarrow [\![\Phi(x)]\!])$$

$$= \bigwedge_{z \varepsilon dom(u)} (\mu_u(z) \Rightarrow \bigwedge_{x \varepsilon F} (x \underset{\sim}{=} z \Rightarrow [\![\Phi(x)]\!])$$

$$= \bigwedge_{z \varepsilon dom(u)} (\mu_u(z) \Rightarrow [\![\forall x(z \underset{\sim}{=} x \rightarrow \Phi(x)]\!])$$

$$= \bigwedge_{z \varepsilon dom(u)} (\mu_u(x) \Rightarrow [\![\Phi(x)]\!]) .$$

Similarly, we can prove (2).

## 6. Sublattices and Substructures.

Let G' be a complete quasi-Boolean sublattice of G (of course it also possesses the strong implication $\Rightarrow$). For the sake of simplicity, we will write the structure $\langle F, \varepsilon, \approx \rangle$ as defined in 3, $F^{(G)}$, and the structure $F^{(G')}$ henceforth defined in parallel with G'. Now we intend to show that $F^{(G')}$ is a substructure of $F^{(G)}$.

Basic Lemma 3. Let G' be any complete quasi-Boolean sublattice of G, then

(1) $F^{(G')} \leq F^{(G)}$

(2) For any u, $v \varepsilon F^{(G')}$, we have

$$[\![u \varepsilon v]\!]^{G'} = [\![u \varepsilon v]\!]^{G} \tag{6.1}$$

$$[\![u = v]\!]^{G'} = [\![u = v]\!]^{G} \tag{6.2}$$

Proof: Making use of the induction principle on the well-founded relation $y \varepsilon dom(v)$ and for any $u \varepsilon F^{(G')}$, we shall assume that

$$[\![y \varepsilon v]\!]^{G'} = [\![y \varepsilon v]\!]^{G} \tag{6.3}$$

$$[\![u = y]\!]^{G'} = [\![u = y]\!]^{G} \tag{6.4}$$

$$[\![y \varepsilon u]\!]^{G'} = [\![y \varepsilon u]\!]^{G} \tag{6.5}$$

First, we shall prove (6.1)

$$[\![u \varepsilon v]\!]^{G'} = \bigvee_{y \varepsilon dom(v)} \mu_v(y) \otimes [\![y = u]\!]^{G'}$$

$$= \bigvee_{y \varepsilon dom(u)} \mu_v(y) \otimes [\![y = u]\!]^{G}$$

$$= [\![u \varepsilon v]\!]^{G}.$$

Secondly, we prove (6.2). When $\rho(u) = \rho(v) = 0$, the equation (6.2) obviously holds. Generally,

$$[\![u = v]\!]^{G'} = \bigwedge_{y \varepsilon dom(u)} (\mu_u(y) \Rightarrow [\![y \varepsilon v]\!]^{G'}) \otimes \bigwedge_{z \varepsilon dom(v)} (\mu_v(z) \Rightarrow [\![z \varepsilon u]\!]^{G'})$$

$$= \bigwedge_{y \varepsilon dom(u)} (\mu_u(y) \Rightarrow [\![y \varepsilon v]\!]^{G}) \otimes \bigwedge_{z \varepsilon dom(v)} (\mu_v(z) \Rightarrow [\![z \varepsilon u]\!]^{G})$$

$= [\![u = v]\!]^G.$

Thus we have proved the basic lemma 3.

Theorem 2. Let G' be any complete quasi-Boolean sublattice. Then for any $\Sigma_0$ formula $\Phi(x_1,\ldots,x_n)$, not including the negation and any $c_1,\ldots,c_n \in F^{(G')}$, we have

$$[\![\Phi(c_1,\ldots,c_n)]\!]^{G'} = [\![\Phi(c_1,\ldots,c_n)]\!]^G .$$

Proof: By performing the induction on the complexity of formula $\Phi(x_1,\ldots,x_n)$, the proof can be obtained directly as in the primitive cases by invoking the basic lemma 3. So far as the inductive steps are concerned, it suffices to show only two steps; namely when $\Phi(x_1,\ldots,x_n)$ is $\exists x \varepsilon u \psi(x,x_1,\ldots,x_n)$ or $\forall x \varepsilon u \psi(x,x_1,\ldots,x_n)$ it holds. The details of the proof are outlines as follows:

If $c,c_1,\ldots,c_n$ are in $F^{(G')}$, and $\Phi(x_1,\ldots,x_n)$ is $\exists x \varepsilon u \psi(x,x_1,\ldots,x_n)$,

$$[\![\Phi(c,c_1,\ldots,c_n)]\!] = [\![\exists x \varepsilon c \psi(x,c_1,\ldots,c_n)]\!]^G$$

$$= \bigvee_{x \varepsilon F^{(G')}} (\mu_c(x) \otimes [\![\psi(x,c_1,\ldots,c_n)]\!]^{G'})$$

$$= \bigvee_{= x \varepsilon F^{(G)}} (\mu_c(x) \otimes [\![\psi(x,c_1,\ldots,c)]\!]^G)$$

$$= [\![\Phi(c,c_1,\ldots,c_n)]\!]^G.$$

Similarly, we have $\forall x \varepsilon u \psi(x,x_1,\ldots,x_n)$.

## 7. The Standard Substructure.

We should note that the binary algebra $Z = \{0,1\}$ is a complete sub-algebra of any complete quasi-Boolean lattice. That is to say, the law of excluded middle does hold. So $F^{(2)}$ is a substructure of $F^{(G)}$ and is isomorphic to the model M.

Definition 6. For any element x in M, we will define the corresponding element $\check{x}$ in $F^{(G)}$ as follows:

(1) When $x = \phi$ or $x \varepsilon s$, $\check{x} := x$.

(2) when $\rho(x) > 0$, put

$$\check{x} = \{\langle \check{y}, 1 \rangle \mid y \varepsilon x\}.$$

We will call the element which is form $\check{x}$ the standard element of

$F^{(G)}$. This set of elements is also called the standard set. Now let us point out some basic properties on the standard sets.

By invoking the basic lemma 3, we can establish the following formulae for any x, y$\varepsilon$M.

$$\check{x} \varepsilon F^{(2)} \subseteq F^{(G)} \quad (7.1)$$

$$[\![ \check{x} \varepsilon \check{y} ]\!]^G = [\![ \check{x} \varepsilon \check{y} ]\!]^2 \varepsilon 2 \quad (7.2)$$

$$[\![ \check{x} = \check{y} ]\!]^G = [\![ \check{x} \varepsilon \check{y} ]\!]^2 \varepsilon 2 \quad (7.3)$$

$$x \varepsilon y \text{ iff } \check{x} \underset{\sim}{\varepsilon} \check{y} \quad (7.4)$$

$$x - y \text{ iff } \check{x} \underset{\sim}{=} \check{y} \quad (7.5)$$

Therefore, the mapping from x to $\check{x}$ is one-to-one injection from M to $F^{(G)}$.

Lemma 12. For any x$\varepsilon$M, u$\varepsilon F^{(G)}$, we have

$$[\![ u \varepsilon \check{x} ]\!]^G = \bigvee_{y \varepsilon x} [\![ u = \check{y} ]\!]^G.$$

Proof:
$$[\![ u \varepsilon \check{x} ]\!] = \bigvee_{z \varepsilon dom(\check{x})} \mu_{\check{x}}(z) \otimes [\![ u = z ]\!]^G$$
$$= \bigvee_{y \varepsilon x} \mu_{\check{x}}(\check{y}) \otimes [\![ u = \check{y} ]\!]^G$$
$$= \bigvee_{y \varepsilon x} [\![ u = \check{y} ]\!]^G.$$

Lemma 13. For any u$\varepsilon F^{(2)}$, there is a unique $\check{x}$ in M such that $F^{(2)} \models u = \check{x}$.

Proof: The uniqueness property can be obtained directly by making use of (7.4) and (7.5). The existence of x is proved by induction on the well-founded relation x$\varepsilon$dom(u). Assuming that u$\varepsilon F^{(2)}$ and $\forall$x$\varepsilon$dom(u)$\exists$y$\varepsilon$M($[\![ x = \check{y} ]\!]^2$), we intend to prove that there is a set t in M, such that

$$[\![ u = \check{t} ]\!]^2 = 1 \quad (7.6)$$

Since $[\![ u = \check{t} ]\!]^2 = \bigwedge_{x \varepsilon dom(u)} (\mu_u(x) \Rightarrow [\![ u \varepsilon \hat{t} ]\!]^2) \otimes \bigwedge_{y \varepsilon t} [\![ \check{y} \varepsilon u ]\!]^2.$

By invoking the equation (7.6), there is t$\varepsilon$M iff:

1) $x \varepsilon dom(u)$ implies that

$$\mu_u(x) \leq [\![ x \varepsilon \check{t} ]\!]^2 = \bigvee_{y \varepsilon t} [\![ x = \check{y} ]\!]$$

2) $y \varepsilon t$ implies that

$$1 = [\![ \check{y} \varepsilon u ]\!]^2 = \bigvee_{x \varepsilon dom(u)} (\mu_u(x) \otimes [\![ x = \check{y} ]\!]^2).$$

Obviously, in order to satisfy 2), we must take

$$t = \{y \varepsilon M | \exists x \varepsilon dom(u) (\mu_u(x) = 1 \text{ and } [\![ x = \check{y} ]\!] = 1)\}.$$

From (7.4) and (7.5), making use of the inductive hypothesis, we shall be able to obtain 1).

Thus we have proved the lemma 13.

Basic lemma 4. For any formula $\Phi(x_1,\ldots,x_n)$ and any $c_1,\ldots, c_n \varepsilon M$, we have

1) $\Phi(c_1,\ldots,c_n)$ is true iff $F^{(2)} \models \Phi(\check{c}_1,\ldots,\check{c}_n)$. When $\Phi$ is a $\Sigma_0$ formular, we have

2) $\Phi(c_1,\ldots,c_n)$ is true iff $F^{(G)} \models \Phi(\check{c}_1,\ldots,\check{c}_n)$.

Proof: It can be proved by induction on the complexity of $\Phi$. In primitive cases it is not difficult to obtain the desired result by invoking (7.4) and (7.5). In inductive steps, we shall omit the proof of 1) because it is both tedious and obvious. So far as 2) is concerned, let the $\Phi(x_1,\ldots,x_n)$ be $\Phi_1(x_1,\ldots,x_n) \vee \Phi_2(x_1,\ldots,x_n)$. We further assume that for $\Phi_1$ and $\Phi_2$, the result of 2) has held because we have the following:

$\Phi_1(c_1,\ldots,c_n)$ is true iff $F^{(G)} \models \Phi_1(\check{c}_1,\ldots,\check{c}_n)$,

$\Phi_2(c_1,\ldots,c_n)$ is true iff $F^{(G)} \models \Phi_2(\check{c}_1,\ldots,\check{c}_n)$.

Obviously we have established the desired result. Similarly, for the connectives $\neg$, $\wedge$, $\rightarrow$, $\leftrightarrow$, it is also quite evident.

Let $\Phi(x_0,x_2,\ldots,x_n)$ be $\exists x_1 \varepsilon x_0 \psi(x_1,x_2,\ldots,x_3)$. Via the inductive hypothesis, we have

$\psi(c_1,\ldots,c_n)$ is true iff $F^{(G)} \models \psi(\check{c}_1,\ldots,\check{c}_n)$.

Now, let us prove the desired result. Suppose $\Phi(c_0,c_2,\ldots,c_n)$ is true, that is,

$\exists x_1 \varepsilon c_0 \; \psi(x_1, c_2, \ldots, c_n)$ is true. (7.7)

So $c_1 \varepsilon c_0$ is true and $\psi(c_1, c_2, \ldots, c_n)$ is also true. Therefore, we have

$$F^{(G)} \models \check{c}_1 \varepsilon \check{c}_0 \wedge \psi(\check{c}_1, \check{c}_2, \ldots, \check{c}_n) \tag{7.8}$$

$$F^{(G)} \models \exists x_1 \varepsilon \check{c}_0 \; \psi(x_1, \check{c}_2, \ldots, \check{c}_n) \tag{7.9}$$

$$F^{(G)} \models \Phi(\check{c}_0, \check{c}_2, \ldots, \check{c}_n) \tag{7.10}$$

Whereas, we will assume that (7.10) is true. Furthermore, equations (7.9) and (7.8) are true, and hence we have that $c_1 \varepsilon c_0 \wedge \psi(c_1, c_2, \ldots, c_n)$ is true and equation (6.7) is true. Hence we have established the desired result.

Similarly, we may prove the case in which $\Phi(x_0, x_2, \ldots, x_n)$ is $\forall x_1 \varepsilon x_0 \psi(x_1, x_2, \ldots, x_3)$.

Thus the proof for basic lemma 4 is completed.

## 8. The Structure $\langle F, \underset{\sim}{\varepsilon}, \underset{\sim}{=} \rangle$ is a Nonstandard Model of Axiomatic Set Theory.

Lemma 14. The empty set axiom is true in $F^{(G)}$, namely, $[\![\exists x \forall y(\neg y \varepsilon x \wedge \neg x \varepsilon s)]\!] = 1$.

Proof: Since the empty set $\phi \varepsilon F$, it is quite obvious that the above statement is true.

Lemma 15. The weak axiom of the urelement set is true in our structure, namely

(1) $[\![\forall x(x\varepsilon S \to \neg(x = \phi) \wedge \neg \exists z(z\varepsilon x))]\!] = 1.$

Proof: $[\![\forall x(x\varepsilon S \to \neg(x = \phi) \wedge \neg\exists z(z\varepsilon x))]\!]$

$$= \bigwedge_{x\varepsilon F} [\![x\varepsilon S \to \neg(x = \phi) \wedge \exists z(z\varepsilon x)]\!]$$

$$= \bigwedge_{x\varepsilon S} ([\![x\varepsilon S]\!] \Rightarrow [\![\neg(x = \phi) \wedge \neg\exists z(z\varepsilon x)]\!])$$

$$\otimes \bigwedge_{x\varepsilon F-S} ([\![x\varepsilon S]\!] \Rightarrow [\![\neg(x = \phi) \wedge \neg\exists z(z\varepsilon x)]\!])$$

$$= \bigwedge_{x\varepsilon S} ([\![x\varepsilon S]\!] \Rightarrow [\![\neg(x = \phi) \wedge \neg\exists z(z\varepsilon x)]\!]).$$

When $x \varepsilon S$, it is obvious that $x \neq \phi$. Invoking the definition of 4 (1), we have $[\![ x = \phi ]\!] = 0$, and $[\![ \exists z(z \varepsilon x) ]\!] = 0$, Hence $[\![ x = \phi ]\!]^* = 1$ and $[\![ \exists z(z \varepsilon x) ]\!]^{\circledast} = 1$. Therefore, we have, when $x \varepsilon S$,

$$[\![ x = \phi ]\!]^* \otimes [\![ \exists z(z \varepsilon x) ]\!]^{\circledast} = 1,$$

$$[\![ \neg(x = \phi) \vee \neg \exists z(z \varepsilon x) ]\!] = 1.$$

Hence

$$\bigwedge_{x \varepsilon S} [\![ (x \varepsilon S \rightarrow \neg(x=\phi) \wedge \neg \exists z(z \varepsilon x) ]\!] = 1.$$

Thus lemma 15 is proved.

Remark 2. The axiom of the urelement set

$$\forall x(x \varepsilon S \leftrightarrow \neg(x = \phi) \wedge \neg \exists z(z \varepsilon x))$$

does not hold in this structure. We illustrate a counterexample as shown in the following: Let

$$\mathrm{dom}(x_1) = S$$

$$\mu_{x_1}(t) = \frac{1}{2}, \quad \text{if } t \varepsilon \mathrm{dom}(x_1).$$

Under this hypothesis we have

$$[\![ x_1 = \phi ]\!] = \bigwedge_{z \varepsilon \mathrm{dom}(x_1)} (\mu_{x_1}(z) \Rightarrow 0)$$

$$= \bigwedge_{z \varepsilon S} (\tfrac{1}{2} \Rightarrow 0)$$

$$= 0,$$

$$[\![ \exists z(z \varepsilon x_1) ]\!] = \bigvee_{z \varepsilon F} [\![ z \varepsilon x_1 ]\!].$$

We shall note that when $z \varepsilon S$,

$$[\![ z \varepsilon x_1 ]\!] = \bigvee_{t \varepsilon \mathrm{dom}(x_1)} \mu_{x_1}(t) \otimes [\![ t = z ]\!]$$

$$= \bigvee_{t \varepsilon S} \mu_{x_1}(t) \otimes [\![ t = z ]\!]$$

$$= \mu_{x_1}(z) \otimes [\![ z = z ]\!] = \frac{1}{2}$$

while $z\varepsilon S$

$$[\![ z\varepsilon x_1 ]\!] = \bigvee_{t\varepsilon S} \mu_{x_1}(t) \otimes [\![ t = z ]\!]$$

$$= 0 \quad \text{(because in the case } [\![ t = z ]\!] \text{ is always o).}$$

Hence $[\![ \exists z(z\varepsilon x_1) ]\!] = \bigvee_{z\varepsilon F} [\![ z\varepsilon x_1 ]\!] = \frac{1}{2}$. Therefore

$$[\![ x_1 = \phi ]\!]^* \otimes [\![ \exists z(z\varepsilon x_1) ]\!]^* = \frac{1}{2}.$$

But

$$[\![ x_1 \varepsilon S ]\!] = \bigvee_{z\varepsilon dom(S)} \mu_S(z) \otimes [\![ z = x_1 ]\!]$$

$$= \bigvee_{z\varepsilon S} \mu_S(z) \otimes [\![ z = x_1 ]\!]$$

$$= 0 \quad \text{(Since } \rho(x_1) = 1, \text{ z is an urelement).}$$

So

$$([\![ \neg(x = \phi) \wedge \neg \exists z(z\varepsilon x_1) ]\!] \Rightarrow [\![ x_1 \varepsilon S ]\!]) = 0.$$

Hence

$$\bigwedge_{x\varepsilon F} ([\![ \neg(x = \phi) \wedge \neg \exists z(z\varepsilon x_1) \rightarrow x\varepsilon S ]\!]) = 0.$$

That is

$$[\![ \forall x(\neg(x = \phi) \wedge \neg \exists z(z\varepsilon x) \rightarrow x \varepsilon S) ]\!] = 0.$$

So we have just obtained the following result:

$$F^{(G)} \not\models \forall x(x \varepsilon S \leftrightarrow \neg(x=\phi) \wedge \neg \exists z(z\varepsilon x)).$$

Lemma 16. The extensionality axiom is true in our structure, namely

(2) $$[\![ \forall x \forall y(\neg x \varepsilon S \wedge \neg y \varepsilon S) \rightarrow (\forall z(z \varepsilon x \leftrightarrow z \varepsilon y) \rightarrow x = y) ]\!]$$

$$= 1.$$

Proof: For any $x, y\varepsilon F$, we assume that $\neg x\varepsilon S$, $\neg y\varepsilon S$, namely $[\![ \neg x\varepsilon S \wedge \neg y\varepsilon S ]\!] = 1$. Then, since the negation is absent from the formulas $z\varepsilon x$ and $z\varepsilon y$, we have the following result by invoking the basic lemma 2:

$$[\![\forall z(z\varepsilon x \leftrightarrow z\varepsilon y)]\!] = [\![\forall z\varepsilon x(z\varepsilon y) \wedge \forall z\varepsilon y(z\varepsilon x)]\!]$$

$$= \bigwedge_{z\varepsilon dom(x)} (\mu_x(z) \Rightarrow z\hat{\varepsilon}y) \otimes \bigwedge_{z\varepsilon dom(y)} (\mu_y(z) \rightarrow z\tilde{\varepsilon}x)$$

$$= [\![x = y]\!] .$$

Hence we can obtain that

$$[\![\forall z(z\varepsilon x \leftrightarrow z\varepsilon y) \rightarrow x = y]\!] = 1.$$

Therefore, we get the expected result. Secondly, when

$$[\![\neg x\varepsilon S \wedge \neg y\varepsilon S]\!] < 1,$$

and since the set S is nonfuzzy, we have $[\![\neg x\varepsilon S \wedge \neg y\varepsilon S]\!] = 0$. Thus our argument was justified.

Lemma 17. The pairing axiom is true in our structure, namely

(3) $[\![\forall x \forall y \exists z \forall u(u\varepsilon z \leftrightarrow u = x \vee u = y)]\!] = 1.$

Proof: For x, $y\varepsilon F$, we define z as follows:

$dom(z) = \{x, y\}$

$\mu_z(x) = \mu_z(y) = 1.$

Hence

$$[\![u\varepsilon z]\!] = \bigvee_{t\varepsilon dom(z)} (\mu_z(t) \otimes [\![t = u]\!])$$

$$= \mu_z(x) \otimes [\![x = u]\!] \oplus \mu_z(y) \otimes [\![y = u]\!]$$

$$= [\![x = u]\!] \oplus [\![y = u]\!]$$

$$= [\![x = u \vee y = u]\!] .$$

Therefore $[\![u\varepsilon z \leftrightarrow u = x \vee u = y]\!] = 1$. Hence it is rather obvious that we can get (3).

Lemma 18. The union axiom is true in our structure, namely

(4) $[\![\forall x \exists y \forall z(z\varepsilon y \leftrightarrow \exists t(t\varepsilon x \wedge z\varepsilon t))]\!] = 1.$

Proof: For any $x\varepsilon F$, when $x = \phi$ or $x\varepsilon S$, if we put $y = \phi$, then (4) holds. When $\rho(x) > 1$, we will define fuzzy set y as shown in the following:

$$\mathrm{dom}(y) = \cup \{\mathrm{dom}(t) \mid t \varepsilon \mathrm{dom}(x)\}$$

$$\mu_y(z) = [\![ \exists t \varepsilon x (z \varepsilon t) ]\!] \,, \quad \text{if } z \varepsilon \mathrm{dom}(y).$$

In order to obtain (4), it is sufficient to show:

$$[\![ \forall z \varepsilon y \exists t \varepsilon x (z \varepsilon t) ]\!] = 1 \tag{8.1}$$

$$[\![ \forall z (\exists t \varepsilon x (z \varepsilon t) \to z \varepsilon y) ]\!] = 1 \tag{8.2}$$

For (8.1), since $\exists t \varepsilon x(z \varepsilon t)$ satisfies the condition of the basic lemma 2, we have:

$$[\![ \forall z \varepsilon y \exists t \varepsilon x (z \varepsilon t) ]\!] = \bigwedge_{z \varepsilon \mathrm{dom}(y)} (\mu_y(z) \Rightarrow [\![ \exists t \varepsilon x (z \varepsilon t) ]\!])$$

$$= \bigwedge_{z \varepsilon \mathrm{dom}(y)} (\mu_y(z) \Rightarrow \mu_y(z))$$

$$= 1.$$

For (8.2), we notice the fact that;

$$\mu_x(t) \otimes \mu_t(z) \leq \mu_x(t) \otimes [\![ (z \varepsilon t) ]\!]$$

$$\leq \bigvee_{t \varepsilon \mathrm{dom}(x)} \mu_x(t) \otimes [\![ z \varepsilon t ]\!]$$

$$= [\![ \exists t \varepsilon x (z \varepsilon t) ]\!]$$

$$= \mu_y(z)$$

and that

$$\mathrm{WL} \vdash \forall z (\exists t \varepsilon x (z \varepsilon t) \to z \varepsilon y) \leftrightarrow \forall t \varepsilon x \forall z \varepsilon t (z \varepsilon y).$$

So we have

$$[\![ \forall z (\exists t \varepsilon x (z \varepsilon t) \to z \varepsilon y ]\!]$$

$$= [\![ \forall t \varepsilon x \forall z \varepsilon t \ (z \varepsilon x) ]\!]$$

$$= \bigwedge_{t \varepsilon \mathrm{dom}(x)} \bigwedge_{z \varepsilon \mathrm{dom}(t)} (\mu_x(t) \otimes \mu_t(z) \Rightarrow [\![ z \varepsilon y ]\!])$$

$$\geq \bigwedge_{t \varepsilon \mathrm{dom}(x)} \bigwedge_{z \varepsilon \mathrm{dom}(t)} (\mu_x(t) \otimes \mu_t(z) \Rightarrow \mu_y(z))$$

$$= 1.$$

Thus the lemma is proved.

Lemma 19. The power set axiom is true in our structure, namely,

(5) $[\![\forall x(\neg x \varepsilon S \rightarrow \exists y \forall z(z\varepsilon y \leftrightarrow \forall t\varepsilon z(t\varepsilon x)))]\!] = 1.$

Proof: We shall prove it by delineating the proof into two cases:

(i) when $x = \phi$, we define fuzzy set y as follows

$dom(y) = \{\phi\} \qquad (\subseteq F_0)$

$\mu_y(z) = 1$, if $z\varepsilon dom(y)$, namely $\mu_y(\phi) = 1$.

In this instance, we need to show that

$[\![\forall z(z\varepsilon y \leftrightarrow z \subseteq \phi)]\!] = 1.$

To this end, it suffices to show:

$[\![\forall z\varepsilon y(z \subseteq \phi)]\!] = 1$ (8i,1)

$[\![\forall z(z \subseteq \phi \rightarrow z\varepsilon y)]\!] = 1.$ (8i,2)

First, we will prove (8i,2).

For any $z\varepsilon F$, we shall show that

$[\![ z \subseteq \phi \rightarrow z\varepsilon y ]\!] = 1$ (8i,3)

It suffices to show the following two equalities:

$[\![ z \subseteq \phi \rightarrow z = \phi]\!] = 1$ (8i,4)

$[\![ z \subseteq \phi \rightarrow \phi\varepsilon y ]\!] = 1$ (8i,5)

Furthermore, in order to prove (8i,4), we shall note that $[\![\phi \subseteq z ]\!] = 1$. From this very fact that the extensionality axiom is true in our structure, we may write the following expression:

$$[\![z \subseteq \phi]\!] \leq [\![z \subseteq \phi \wedge \phi \subseteq z]\!] \leq [\![ z = \phi]\!].$$

Hence we have proved (8i,4). To prove (8i,5), we note that $[\![\phi\varepsilon y]\!] \geq \mu_y(\phi) = 1$, and we may say $([\![z \subseteq \phi ]\!] \Rightarrow [\![ \phi\varepsilon y ]\!]) = 1$. Consequently (8i,5) holds. Finally, we say (8i,3 holds, too. Due to the fact that $z\varepsilon F$ is arbitrary, we get the proof of (8i,2).

Now, we are in a position to prove (8i,1):

$$[\![\forall z\varepsilon y(z \subseteq \phi)]\!] = \bigwedge_{z\varepsilon \mathrm{dom}(y)} (\mu_y(z) \Rightarrow [\![z \subseteq \phi]\!])$$

$$= \bigwedge_{z\varepsilon\{\phi\}} (\mu_y(z) \Rightarrow [\![z \subseteq \phi]\!])$$

$$= (\mu_y(\phi) \Rightarrow [\![\phi \subseteq \phi]\!])$$

$$= 1.$$

It follows quite obviously that

$$\bigwedge_{t\varepsilon F} [\![t \; \varepsilon \; \phi \rightarrow t \; \varepsilon \; \phi]\!] = 1.$$

So the proof of (i) has been accomplished.

(ii) For any $x\varepsilon F - F_0$, we designate the fuzzy set y as follows:

$$\mathrm{dom}(y) = F(\mathrm{dom}(x))$$

$$\mu_y(z) = [\![z \subseteq x]\!] = [\![\forall t\varepsilon z(t\varepsilon x)]\!], \text{ if } t\varepsilon \mathrm{dom}(y).$$

First for any $z\varepsilon F$, we have

$$[\![z\varepsilon y]\!] = \bigvee_{t\varepsilon \mathrm{dom}(y)} (\mu_y(t) \otimes [\![t = z]\!])$$

$$= \bigvee_{t\varepsilon \mathrm{dom}(y)} ([\![t \subseteq x]\!] \otimes [\![t = z]\!])$$

$$\leq [\![z \subseteq x]\!] \tag{8.3}$$

So for any $z\varepsilon F$, we have

$$[\![z\varepsilon y \rightarrow (z \subseteq x)]\!] = 1$$

Therefore

$$[\![\forall z(z\varepsilon y \rightarrow z \subseteq x)]\!] = 1 \tag{8.4}$$

Secondly, we need to prove that

$$[\![\forall z(z \subseteq x \rightarrow z\varepsilon y)]\!] = 1 \tag{8.5}$$

To this end, for any given fuzzy set $z\varepsilon F$, we designate the fuzzy set $z'\varepsilon F$ as follows:

$$\mathrm{dom}(z') = \mathrm{dom}(x)$$

$\mu_{z'}(t) = [\![ t \varepsilon z ]\!]$, if $t \varepsilon dom(z')$.

So when $z' \varepsilon dom(y)$, it is desirable to establish the following two equalities:

$$[\![ z \subseteq x \rightarrow z = z' ]\!] = 1 \tag{8.6}$$

$$[\![ z \subseteq x \rightarrow z' \varepsilon y ]\!] = 1 \tag{8.7}$$

From these two equalities, it is obvious that

$$[\![ z \subseteq x \rightarrow z \varepsilon y ]\!] = 1.$$

Hence it suffices to prove that (8.6) and (8.7) are true. First we will prove (8.6), since

$$\begin{aligned} [\![ t \varepsilon z' ]\!] &= \bigvee_{u \varepsilon dom(z')} (\mu_{z'}(u) \otimes [\![ u = t ]\!] ) \\ &= \bigvee_{t \varepsilon dom(z')} ( [\![ u \varepsilon z ]\!] \otimes [\![ u = t ]\!] ) \\ &= [\![ t \varepsilon z ]\!] . \end{aligned}$$

Therefore we have

$$[\![ z' \subseteq z ]\!] = [\![ \forall t \varepsilon z' (t \varepsilon z) ]\!] = 1 \tag{8.8}$$

For any $t \varepsilon F$, we have

$$\begin{aligned} [\![ t \varepsilon x \wedge t \varepsilon z ]\!] &= \bigvee_{u \varepsilon dom(x)} (\mu_x(u) \otimes [\![ u = t ]\!] \otimes [\![ t \varepsilon z ]\!] \\ &\leq \bigvee_{u \varepsilon dom(x)} ( [\![ u = t ]\!] \otimes [\![ u \varepsilon z ]\!] ) \\ &= \bigvee_{u \varepsilon dom(x)} ( [\![ u = t ]\!] \otimes \mu_{z'}(u)) \\ &= t \varepsilon z' . \end{aligned}$$

Hence, it can be established that $[\![ x \cap z \subseteq z' ]\!] = 1$. Based upon this and (8.8) we can obtain that:

$$[\![ z \subseteq x ]\!] \leq [\![ z \cap x \subseteq z' \wedge z' \subseteq z \wedge z \subseteq x ]\!] \leq [\![ z = z' ]\!] .$$

So equation (8.6) has been obtained. Now, we shall proceed to prove (8.7). Since

$$
\begin{aligned}
[\![ z \subseteq x ]\!] &= [\![ \forall t(t \in z \to t \in x) ]\!] \\
&= \bigwedge_{t \in F} ( [\![ t\in z ]\!] \Rightarrow [\![ t\in x ]\!] ) \\
&\leq \bigwedge_{t \in dom(z')} ( [\![ t\in z ]\!] \Rightarrow [\![ t\in x ]\!] ) \\
&= \bigwedge_{t \in dom(z')} ( \mu_{z'}(t) \Rightarrow [\![ t\in x ]\!] ) \\
&= [\![ \forall t\in z' (t\in x) ]\!] \\
&= [\![ z' \in x ]\!] \\
&= \mu_y(z') \\
&\leq [\![ z' \in y ]\!] .
\end{aligned}
$$

Hence (8.7) holds.

Thus the proof of lemma 19 is completed.

Lemma 20. The infinity axiom is true in our structure, namely

(6) $[\![ \exists x(\phi\in x \wedge \neg x\in S \wedge \forall y\in x \exists z\in x(y\in z) ]\!] = 1.$

Proof: Since $(\phi\in x \wedge \neg x\in S \wedge \forall y\in x \exists z\in x(y\in z))$ is a $\Sigma_0$ formula which may be written as $\Phi(x)$, from basic lemma 4, it can obviously be obtained in our structure as

$$[\![ \Phi(\overset{\vee}{\omega}) ]\!] = 1.$$

Thus, lemma 20 is also proved.

Lemma 21. For any formula $\Phi(x,y)$ not including the negation , we have

(7) $[\![ \forall t(\forall x\in t \exists y\Phi(x,y) \to \exists u \forall x\in t \exists y\in u\Phi(x,y)) ]\!] = 1.$

That is to say, a weaker form of the replacement axiom (the formula $\Phi(x,y)$ not including the negation $\neg$ ) holds in our structure.

Proof: For any $t\in F$, we have

$$[\![ \forall x\in t \exists y\Phi(x,y) ]\!] = \bigwedge_{x\in dom(t)} (\mu_t(x) \Rightarrow \bigvee_{y\in F} [\![ \Phi(x,y) ]\!] ). \qquad (8.9)$$

Since the interval [0, 1] is a set in M, we can use the axiom of replacement in M to obtain a map from x to $\alpha_x$ with the domain dom(t) and the range being a set of ordinals such that, for each $x \varepsilon dom(t)$:

$$\bigvee_{y\varepsilon F} [\![\Phi(x,y)]\!] = \bigvee_{y\varepsilon F_{\alpha_x}} [\![\Phi(x,y)]\!] . \tag{8.10}$$

Let $\alpha = \bigcup \{\alpha_x | x\varepsilon dom(t)\}$, then from (8.10), we have

$$\bigwedge_{x\varepsilon dom(t)} (\mu_t(x) \Rightarrow \bigvee_{y\varepsilon F} [\![\Phi(x,y)]\!]$$

$$= \bigwedge_{x\varepsilon dom(t)} (\mu_t(x) \Rightarrow \bigvee_{x\varepsilon F_{\alpha_x}} [\![\Phi(x,y)]\!]$$

$$\leq \bigwedge_{x\varepsilon dom(t)} (\mu_t(x) \Rightarrow \bigvee_{y\varepsilon F_\alpha} [\![\Phi(x,y)]\!] . \tag{8.11}$$

Now, we designate $u = F_\alpha \times \{1\}$, and $u\varepsilon F$.

$$\bigvee_{y\varepsilon F_\alpha} [\![\Phi(x,y)]\!] = [\![\exists y\varepsilon u\Phi(x,y)]\!] .$$

So from (8.9) and (8.11), we have

$$[\![\forall x\varepsilon t \exists y\Phi(x,y)]\!] \leq \bigwedge_{x\varepsilon dom(t)} (\mu_t(t) \Rightarrow [\![\exists y\varepsilon u\Phi(x,y)]\!] )$$

$$= [\![\forall x\varepsilon t \exists y\varepsilon u\Phi(x,y)]\!] .$$

Hence we can establish (7); that is to say, lemma 21 is proved.

Lemma 22. The regularity axiom is true in our structure, namely

(8) $[\![\forall x(\neg x = \phi \wedge \neg x\varepsilon S \rightarrow \exists y\varepsilon x \wedge \neg \exists z(z\varepsilon y \wedge z\varepsilon x))]\!] = 1.$

Proof: It is sufficient to show that for any $x\varepsilon F$, $\rho(x) > 0$, we have

$$[\![\forall y\varepsilon x \exists z\varepsilon y(z\varepsilon x)]\!] = 0. \tag{8.12}$$

This is not difficult to show and hence we shall omit the proof.

Lemma 23. The axiom of choice is true in our structure, namely

(9) $[\![\forall u(\neg u = \phi \wedge \neg u\varepsilon S \rightarrow \exists \alpha \exists f (Fun(f) \wedge dom(f) = \alpha \wedge u \subseteq ran(f)))]\!] = 1.$

Proof: Taking any fuzzy set $u \varepsilon F - F_0$, from the axiom of choice in M, we may establish that there exists an ordinal $\alpha$ and a function g of $\alpha$ onto dom(u). By defining $f \varepsilon F$, we may write:

$$f = \{ < \check{\beta}, g(\beta) > | \ \beta < \alpha \} \times \{1\}.$$

So,

$$\text{dom}(f) = \{ < \check{\beta}, g(\beta) > | \ \beta < \alpha \}$$

$$\mu_f(x) = 1, \quad \text{if } x \varepsilon \text{dom}(f).$$

Since $\alpha$ is an ordinal number, namely On($\alpha$), and On(x) is a $\Sigma_0$ formula, so $F^{(G)} \models \text{On}(\check{\alpha})$. Hence it suffices to prove that

I. $[\![ \forall z \varepsilon f \exists x \exists y (z = < x,y > ) ]\!] = 1.$

Suppose $z \varepsilon F$, then

$$[\![ z \varepsilon f ]\!] = \bigvee_{\beta<\alpha} [\![ z = < \check{\beta}, g(\beta) > ]\!] \leq [\![ \exists x \exists y (z = < x,y > ) ]\!] .$$

II. $[\![ \text{Fun}(f) ]\!] = 1.$

For any x, y, $z \varepsilon F$, we have

$$[\![ < x, y > \varepsilon f \wedge < x,z > \varepsilon f ]\!]$$

$$= \bigvee_{\beta<\alpha} \bigvee_{\gamma<\alpha} [\![ < x,y > = < \check{\beta}, g(\beta) > \wedge < x,z > = < \check{\gamma}, g(\gamma) > ]\!]$$

$$\leq \bigvee_{\beta<\alpha} \bigvee_{\gamma<\alpha} ( [\![ \check{\beta} = \check{\gamma} ]\!] \otimes [\![ y = g(\beta) ]\!] \otimes [\![ z = g(\gamma) ]\!] )$$

$$= \bigvee_{\beta<\alpha} [\![ \check{\beta} = \check{\beta} ]\!] \otimes [\![ y = g(\beta) ]\!] \otimes [\![ z = g(\beta) ]\!]$$

$$\leq [\![ y = z ]\!] .$$

In the course of deriving the steps as shown above, Eq. (7.5) has been employed. When $\beta \neq \gamma$, we have $[\![ \check{\beta} = \check{\gamma} ]\!] = 0$.

III. $[\![ \text{dom}(f) = \alpha ]\!] = 1.$

For any $x \varepsilon F$, we have

$$[\![ \exists y < x,y > \varepsilon f ]\!] = \bigvee_{z \varepsilon F} [\![ < x,z > \varepsilon f ]\!]$$

$$= \bigvee_{z\varepsilon F} \bigvee_{\beta<\alpha} ( [\![ \check{\beta}=x ]\!] \otimes [\![ g(\beta)=z ]\!] )$$

$$= \bigvee_{\beta<\alpha} [\![ \check{\beta}=x ]\!]$$

$$= [\![ x\varepsilon\check{\alpha} ]\!] .$$

In the process of the derivation presented above, we have used the fact that when u is nonempty and $\beta<\alpha$, the following is always true:

$$\bigvee_{z\varepsilon F} [\![ g(\beta)=z ]\!] = 1.$$

This proves III.

IV. $[\![ u \subseteq ran(f) ]\!] = 1.$

Since $[\![ \exists x \langle x,y\rangle \varepsilon f ]\!]$

$$= \bigvee_{x\varepsilon F} [\![ \langle x,y\rangle \ \varepsilon f ]\!]$$

$$= \bigvee_{\beta<\alpha} ( [\![ g(\beta) = y ]\!] \otimes \bigvee_{x\varepsilon F} [\![ \beta = x ]\!] )$$

$$= \bigvee_{\beta<\alpha} [\![ g(\beta) = y ]\!]$$

$$= \bigvee_{z\varepsilon dom(u)} [\![ z = y ]\!]$$

$$\geq \bigvee_{z\varepsilon dom(u)} ( [\![ z = y ]\!] \otimes \mu_u(z))$$

$$= [\![ y\varepsilon u ]\!] .$$

Therefore $[\![ y\varepsilon u \rightarrow \exists x \langle x,y\rangle \varepsilon f ]\!] = 1$, which proves IV. Hence the proof of lemma 23 is now complete.

From lemmas 14 to 23 as presented above we have shown that the Zermelo - Fraenkel set theory system $ZF_1U$ in which the replacement axioms take a weaker form is true in $F$, that is:

Theorem 6. Every axiom of $ZF_1U$ is always true in our structure; namely our structure $\langle F, \underset{\sim}{\varepsilon}, \underset{\sim}{=} \rangle$ is a model of $ZF_1U$.

Based upon the corollary of theorem 1 and the theorem itself, we have obtained that our structure $\langle F, \underset{\sim}{\varepsilon}, \underset{\sim}{=} \rangle$ is a nonstandard model of WL + $ZF_1U$. This fact, in a sense, may be viewed as the main conclusion of this article.

REFERENCES


1. Zhang Jinwen, The Normal Fuzzy Set Structures and the Boolean-valued Models, J. of Huazhong Institute of Technology, vol. VII, no. 2, 1979 (in Chinese with an English abstract); English Edition, vol. 2, no. 1, 1980.
2. Zhang Jinwen, Some Basic Properties of Normal Fuzzy Set Structures, J. of Huazhong Institute of Technology, vol. VII, no. 3, 1979 (in Chinese with an English abstract). There is a separate edition (in English).
3. Zhang Jinwen, Fuzzy Set Structures and Normal Fuzzy Set Structures, submitted for formal publication, to appear, Symposium on Fuzzy Sets and Possibility Theory, December 12-16, 1980 at Acapulco, Mexico.
4. Zhang Jinwen, A Unified Treatment of Fuzzy Set Theory and Boolean-valued Set Theory - Fuzzy Set Structures and Normal Fuzzy Set Structures, J. of Mathematical Analysis and Applications, vol. 76, no. 1, July 1980.
5. Zhang Jinwen, A Kind of Nonstandard Models of the Axiomatic Set Theory with Urelements - the Normal Fuzzy Set Structure, Selected papers on Mathematics at 1979's Annual Meeting of Beijing Society for Mathematics.
6. Zhang Jinwen, Normal Fuzzy Set Structures, Fuzzy Sets and Systems, to appear.
7. Pan Xuehai and Zhang Jinwen, Fuzzy Set Theory, Applications of the Computers and Applied Mathematics (in Chinese) September 1976.
8. P. J. Cohen, "Set Theory and the Continuum Hypothesis," W. A. Benjamin, Inc., New York, 1966, Amsterdam.
9. B. R. Gaines, Foundations of Fuzzy Reasoning, in: "Fuzzy Automata and Decision Processes", M. M. Gupta and G. N. Saridis, eds., pp. 19-76, North-Holland/Elsevier, Amsterdam, 1977.
10. S. C. Kleene, "Introduction to Metamathematics", Amsterdam and Groningen, New York and Toronto, 1952.
11. A. Levy, Formula Hierarchy in Set Theory, 1974.
12. D. Scott, Boolean-valued Models for Set Theory. Mimeographed notes for the 1967 American Math. Soc. Symposium on axiomatic set theory.
13. J. R. Shoenfield, "Mathematical Logic", Addison-Wesley Publishing Co., 1967.
14. M. Sugeno, Fuzzy Measures and Fuzzy Integrats: A Survey, Ibid, [9], pp. 89-101.
15. A. Zadeh, Fuzzy Set Theory: A Perspective, Ibid, pp. 3-4.
16. E. Sanchez, Compositions of Fuzzy Relations, in: "Advances in Fuzzy Set Theory and Applications", M. M. Gupta, R. K. Racade, and R. R. Yager, eds., pp. 421-461, North-Holland, 1979.

# INFERENCE REGIONS FOR FUZZY PROPOSITIONS*

I. B. Türksen

Department of Industrial Engineering
University of Toronto
Toronto, Ontario M5S 1A4 Canada

## INTRODUCTION

Ever since L. A. Zadeh's introduction of fuzzy sets [1], a concerned body of literature has appeared on the appropriate interpretations of (i) the logical fuzzy connectives, (Bellman and Gertz [2], Hamacher [3], Gaines [4], Bellman and Zadeh [5], Rödder [6], Zimmerman, et. al. [7], Zimmerman [8], Oden [9], Hersh and Caramazza [10], Goguen [11], Yager [12]) and (ii) the structural semantics of fuzzy logic and fuzzy conditional inference, (Zadeh [13,14], Gottwald [15], Mamdani [16], Fukami-Mizumoto-Tanaka [18], Mizumoto-Fukami-Tanaka [19], Hisdal [20], Türksen [21], and Türksen and Yao [22]). Naturally, these two avenues of concern are not independent of each other.

Various approaches may be considered to explain some of these interactions. In this paper, we shall examine the semantic implications of the structures known as the normal forms of fuzzy relational propositions. We shall identify regions for fuzzy inference via these normal forms. We shall show that these regions of inference are different from the regions identified by Yager [12] in a general class of fuzzy connectives. But we shall also show that these regions envelop some models of fuzzy conditional inference suggested by Zadeh [13, 14], Fukami-Mizumoto-Tanaka [18, 19].

---

*Supported by the Natural Science and Engineering Council of Canada.

Various details of the arguments developed here may be found in Türksen [21], and Türksen and Yao [22].

## FUZZY RELATIONAL PROPOSITIONS

In two valued logic, we can construct sixteen propositions by logical composition of two primitive statements (Table 1). In an anologuous manner, we can express sixteen fuzzy relational propositions, $P_k$, k=1,...,16, for possible relationships between fuzzy subsets $A \subseteq U_1$, and $B \subseteq U_2$, where $U_1$ and $U_2$ are two universes of discourse.

Table 1. The Sixteen Propositions

| k | $P_k$ (Name) |
|---|---|
| 1 | Complete Affirmation |
| 2 | Complete Negation |
| 3 | Disjunction |
| 4 | Conjunctive Negation |
| 5 | Incompatibility |
| 6 | Conjunction |
| 7 | Implication |
| 8 | Non-Implication |
| 9 | Inverse Implication |
| 10 | Non-Inverse Implication |
| 11 | Equivalence |
| 12 | Exclusion |
| 13 | Affirmation (1) |
| 14 | Negation (1) |
| 15 | Affirmation (2) |
| 16 | Negation (2) |

For example for two fuzzy statements $S_1$-x is A, $S_2$-y is B, we have:

$P_3$ - fuzzy disjunction: $S_1$ OR $S_2$

$P_6$ - fuzzy conjunction: $S_1$ AND $S_2$

$P_7$ - fuzzy implication: IF $S_1$ THEN $S_2$ .

Initially, Zadeh [1,13,14] have stated various normative interpretations for most of these sixteen propositions. For example, $P_3$ was expressed as

$A \times U_2 \cup U_1 \times B$, $P_6$ as $A \times U_2 \cap U_1 \times B$, and

$P_7$ as $\bar{A} \times U_2 \cup U_1 \times B$,

where in the membership domain x and $\cap$ were interpreted as "min", and $\cup$ was "max". Later, theoretical and empirical studies [3,4,5, 6,7,8,12] raised the concern for appropriate interpretations of the connectives. More recently, we have observed various structural interpretations of the fuzzy conditional inference in current literature [13,14,16,18,19].

In theory, one could argue therefore, each fuzzy relational proposition could be expressed in infinitely many ways depending on the interpretation of the user and user context. Any yet, there is a natural concern for the appropriateness of these expressions. For example, Fukami-Mizumoto-Tanaka [18] state several criteria to assess the methods for fuzzy conditional inference.

In order to provide another perspective, we propose to identify inference regions for fuzzy relational propositions in general and for propositions $P_3$, $P_6$, $P_7$, and $P_{10}$ in particular.

For the development of this proposal, fuzzy normal forms are chosen as models of fuzzy relational propositions. For illustration, the fuzzy disjunctive normal forms, FDNK(k), and the fuzzy conjunctive normal forms FCNF(k) k=3,6,7,10 are shown in Table 2.

We observe that the initial interpretations for $P_3$ and $P_7$ were expressed in the fuzzy conjunctive normal forms, FCNF(3) and

Table 2. FDNF(k) and FCNF(k), k=3,6,7,10

| | |
|---|---|
| $P_3$: | $FDNF(3) = (AxU_2 \cap U_1xB) \cup (AxU_2 \cap U_1x\bar{B}) \cup (\bar{A}xU_2 \cap U_1xB)$ |
| | $FCNF(3) = AxU_2 \cup U_1xB$ |
| $P_6$: | $FDNF(6) = AxU_2 \cap U_1xB$ |
| | $FCNF(6) = (AxU_2 \cup U_1xB) \cap (AxU_2 \cup U_1x\bar{B}) \cap (\bar{A}xU_2 \cup U_1xB)$ |
| $P_7$: | $FDNF(7) = (AxU_2 \cap U_1xB) \cup (\bar{A}xU_2 \cap U_1B) \cup (\bar{A}xU_2 \cap U_1x\bar{B})$ |
| | $FCNF(7) = \bar{A}xU_2 \cup U_1B$ |
| $P_{10}$: | $FDNF(10) = \bar{A} \times U_2 \cap U_1B$ |
| | $FCNF(10) = (AxU_2 \cup U_1xB) \cap (\bar{A}xU_2 \cup U_1B) \cap (\bar{A}xU_2 \cup U_1x\bar{B})$ |

FCNF(7), respectively, whereas for $P_6$ and $P_{10}$, they were expressed as the fuzzy disjunctive normal forms, FDNF(6) and FDNF(10), respectively.

The disjunctive and conjunctive normal forms are equivalent in the two valued propositional calculus, whereas the fuzzy disjunctive and conjunctive normal forms are not equivalent. We identify this non-equivalence and define an inference region based on this non-equivalence of the fuzzy normal forms. In order to develop this, we next review certain definitions and lemmas.

Definitions:

(1) Suppose A and B are two fuzzy subsets in two universes of discourse $U_1$ and $U_2$ respectively, then we define A and B as:

$$A \triangleq \sum_{U_1} \mu_i^{(1)} | u_i^{(1)}, \quad B \triangleq \sum_{U_2} \mu_j^{(2)} | u_j^{(2)}$$

(2) A fuzzy relation R between A and B is a fuzzy subset of the product space $U_1 \times U_2$ and is defined as: $R \triangleq A \times B \subseteq U_1 \times U_2$. This can be expressed as:

$$r_{ij} \triangleq (R)_{ij} = \min(\mu_i^A, \mu_j^B) = \mu_i^{(1)} \wedge \mu_j^{(2)}$$

where $r_{ij}$ denotes the (i, j)th element of the fuzzy relation (matrix) R.

(3) Since a fuzzy relation is a fuzzy subset, the "implication" of two fuzzy relations is defined as:

$$R_1 \subseteq R_2 \quad \text{iff} \quad r_{ij}^{(1)} \leq r_{ij}^{(2)}, \ \forall\, i,j.$$

where $r_{ij}^{(1)} \triangleq (R_1)_{ij}$, $r_{ij}^{(2)} \triangleq (R_2)_{ij}$.

(4) Suppose $R_1 \subseteq U_1 \times U_2$ and $R_2 \subseteq U_2 \times U_3$ are two fuzzy relations, then the composition of $R_1$ and $R_2$ is a fuzzy subset of the product space $U_1 \times U_3$:

$$(R)_{ij} \triangleq (R_1 \circ R_2)_{ij} = \bigvee_k (r_{ik}^{(1)} \wedge r_{kj}^{(2)})$$

where $R \subseteq U_1 \times U_3$, "o" is the min-max product $\vee$ and $\wedge$ are the max and min operators, respectively.

(5) The compliment of fuzzy relation R is defined as

$$(\overline{R})_{ij} \triangleq 1 - (R)_{ij} .$$

For this paper, we set aside our concern that this normative definition needs to be reassessed in the light of empirical measurement results [23, 24].

Lemmas:

(1) If $R_1 \subseteq R_2$ then $\overline{R}_1 \supseteq \overline{R}_2$

(2) If $R_1 \subseteq R_1'$, $R_2 \subseteq R_2'$ then $R_1 \circ R_2 \subseteq R_1' \circ R_2'$

For all the sixteen fuzzy relational propositions, it can be shown that the fuzzy disjunctive normal form is contained in the fuzzy conjunctive normal form. Hence, the following theorem.

Theorem 1. $\text{FDNF}(k) \subseteq \text{FCNF}(k)$

$k=1,\ldots,16$

Recalling the definitions and the lemmas stated earlier, the proof of Theorem 1 may be discussed by showing that

$$r(\text{FDNF}(k))_{ij} \leq r(\text{FDNF}(k))_{ij}, \ \forall\ i,j, \text{ and } k.$$

For illustration the cases for k=3,6,7 and 10 are shown here, the rest can be shown to hold in a similar manner.

$$P_3: \quad r(\text{FDNF}(3))_{ij} = (\mu_i^{(1)} \wedge \mu_j^{(2)}) \vee (\mu_i^{(1)} \wedge (1-\mu_j^{(2)})) \vee ((1-\mu_i^{(1)}) \wedge \mu_j^{(2)})$$

$$r(\text{FCNF}(3))_{ij} = (\mu_i^{(1)} \vee \mu_j^{(2)}).$$

Since $\mu_i^{(1)} \wedge \mu_j^{(2)} \leq \mu_i^{(1)}$, $\mu_i^{(1)} \wedge (1-\mu_j^{(2)}) \leq \mu_i^{(1)}$, and $(1-\mu_i^{(1)}) \wedge \mu_j^{(2)} \leq \mu_j^{(2)}$, $\forall$ i,j, we have

$$r(\text{FDNF}(3))_{ij} \leq r(\text{FCNF}(3))_{ij}, \ \forall\ i,j$$

$$P_6: \quad r(\text{FDNF}(6))_{ij} = (\mu_i^{(1)} \wedge \mu_j^{(2)})$$

$$r(\text{FCNF}(6))_{ij} = (\mu_i^{(1)} \vee \mu_j^{(2)}) \wedge (\mu_i^{(1)} \vee (1-\mu_j^{(2)}) \wedge ((1-\mu_i^{(1)}) \vee \mu_j^{(2)}) .$$

Since $\mu_i^{(1)} \wedge \mu_j^{(2)} \leq \mu_i^{(1)} \vee \mu_j^{(2)}$

$\mu_i^{(1)} \leq \mu_i^{(1)} \vee (1-\mu_j^{(2)})$ and $\mu_j^{(2)} \leq (1-\mu_i^{(1)}) \vee \mu_j^{(2)}$ .
We have $r(FDNF(6))_{ij} \leq r(FCNF(6))_{ij}$, $\forall$ i,j.

$$P_7: \quad r(FDNF(6))_{ij} = (\mu_i^{(1)} \wedge \mu_j^{(2)}) \vee ((1-\mu_i^{(1)}) \wedge \mu_j^{(2)}) \vee ((1-\mu_i^{(1)}) \wedge (1-\mu_j^{(2)})$$

$$r(FCNF(6))_{ij} = (1-\mu_i^{(1)}) \vee \mu_j^{(2)},$$

Since $\mu_i^{(1)} \wedge \mu_j^{(2)} \leq \mu_j^{(2)}$, and

$$(1-\mu_i^{(1)}) \wedge (1-\mu_j^{(2)}) \leq (1-\mu_i^{(1)}), \forall\ i,j$$

we have $r(FDNF(7))_{ij} \leq r(FCNF(7))_{ij}$, $\forall$ i,j.

$$P_{10}: \quad r(FDNF(3)) = (1-\mu_i^{(1)}) \wedge \mu_j^{(2)}$$

$$r(FDNF(3)) = (\mu_i^{(1)} \vee \mu_j^{(2)}) \wedge (1-\mu_i^{(1)} \vee \mu_j^{(2)}) \wedge (1-\mu_i^{(1)}) \vee (1-\mu_j^{(2)})$$

Since $\mu_j^{(2)} \leq \mu_i^{(1)} \vee \mu_j^{(2)}$,

$(1-\mu_i^{(1)}) \leq (1-\mu_i^{(1)}) \vee \mu_j^{(2)}$, and $(1-\mu_i^{(1)}) \leq (1-\mu_i^{(1)}) \vee (1-\mu_j^{(2)})$
we have $r(FDNF(10))_{ij} \leq r(FCNF(10))_{ij}$, $\forall$ i,j.

Definition: Based on Theorem 1, we define Fuzzy Inference Regions, FIR(k), k=1,...,16, as the regions enveloped by the fuzzy disjunctive and conjunctive normal forms; i.e.

$$FIR(k) = [FDNF(k), FCNF(k)]$$

such that

$$\mu(FIR(k))_{ij} \in [r(FCNF(k))_{ij}, r(FDNF(k))_{ij}], \forall\ i,j,k.$$

This is an admissible operation on the interval scale [23, 24].

Corollary 1. If $A \subseteq U$ and $B \subseteq U$, then $FDNF(k) \subseteq FCNF(k)$, $\forall k$, still holds. We give a graphical display for $P_3$ in Figure 1.

Corollary 2. If A and B are crisp subsets of $U_1$ and $U_2$ respectively,

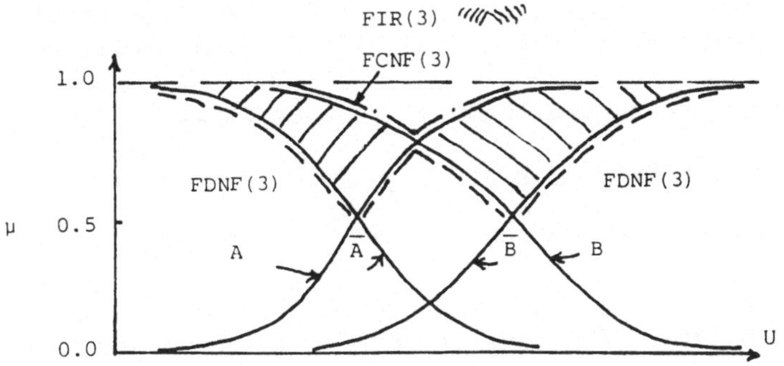


Fig. 1. FDNF(4) $\subseteq$ FCN(3)

then FDNF(k) = FCNF(k), ∀k.

Corollary 3. $\overline{\text{FDNF(k)}} \supseteq \overline{\text{FCNF(k)}}$.

Corollary 4. $\text{FDNF}_k \circ \text{FDNF}_{k'} \subseteq \text{FCNF}_k \circ \text{FCNF}_{k'}$

k, k' = 1,...,16 where

$\text{FDNF}_k$ and $\text{FCNF}_k$ are defined on $U_1 \times U_2$ and

$\text{FDNF}_{k'}$ and $\text{FCNF}_{k'}$ are defined on $U_2 \times U_3$.

We shall now compare, the fuzzy inference regions defined in the previous section with
(i) the regions identified by Yager [12] and
(ii) the models of the fuzzy conditional inference proposed by Zadeh [13, 14] and Fukami-Mizumoto-Tanaka [18, 19].

ON FUZZY CONNECTIVES

Yager [12] defined, for $p \geq 1$,

(1) $D_p = A \cup_p B$ for fuzzy disjunction, $P_3$,

and (2) $C_p = A \cap_p B$ for fuzzy conjunction, $P_6$.
It is shown that

(1)' $\mu_{D_p} \geq \mu_A \vee \mu_B$ = FCNF(3) and

(2)' $\mu_{C_p} \leq \mu_A \wedge \mu_B$ = FDNF(6), $p \geq 1$.

Thus, Yager's [12] general class of fuzzy connectives defines a region for fuzzy propositions, $P_3$ and $P_6$ that is entirely outside of the fuzzy inference region identified by the fuzzy normal forms.

Observing these two regions, side by side, we find that

$$FDNF(3) \subseteq FCNF(3) \subseteq D_p,$$

and $C_p \subseteq FDNF(6) \subseteq FCNF(6)$, $p \geq 1$.

Our interpretation is that "AND" connective loses its "strength" from $C_p$ to FDNF(6) as p tends toward $\infty$, in the sense of simultaneous satisfaction of both $S_1$ and $S_2$. But then the structural semantics of min-max take over from FDNF(6) toward FCNF(6) for various interpretations of the fuzzy relational proposition $P_6$. On the other hand, for "OR" connective, the "interchangeability" or collective satisfaction of $S_1$ and $S_2$ are lost as p tends toward $\infty$. But then again the structural semantics take over from FCNF(3) toward FDNF(3) for various interpretations of the fuzzy relational proposition $P_3$. It is natural to ask what would happen to these two non-overlapping regions if $p \neq \infty$ were used in the definition of the fuzzy inference regions via normal forms. This we shall explore in a future paper.

## ON FUZZY REASONING

The importance we place on fuzzy inference regions defined for fuzzy relational propositions appear more clearly when we examine various methods of fuzzy reasoning.

First, let us consider the generalized modus ponens:

Ant 1: IF $S_1$ THEN $S_2$

Ant 2: $S_3$

---

Cons : $S_4$

where $S_1$ and $S_2$ as defined previously and $S_3$-x is A' and $S_4$-y is B', x and y are the names of objects, and A, A', B, B' are fuzzy concepts represented by fuzzy sets in universes of discourse $U_1$ and $U_2$, A, A' $\subseteq U_1$ and B, B' $\subseteq U_2$. In the previous section we have seen that Ant 1 is expressed as $P_7$, thus from a structural point of view we can define $P_7$ to be a whole set of fuzzy relational propositions:

$$P_7 \triangleq \{P \mid FDNF(7) \subseteq P \subseteq FCNF(7)\}$$

and hence the consequence of the "generalized modus ponens" a whole set of fuzzy relational propositions (c.f. corollary 4):

$$B' \triangleq \{B^* | A' \text{ o } FDNF(7) \subseteq B^* \subseteq A' \text{ o } FCNF(7)\}.$$

Next let us consider the generalized modus tollens:

Ant 1: IF $S_1$ THEN $S_2$

Ant 2: $S_4$

Cons : $S_3$.

Under our definition of $P_7$, the consequence is

$$A' = \{A^* | FDNF(7) \text{ o } B' \subseteq A^* \subseteq FCNF(7) \text{ o } B'\}$$

The methods suggested by Zadeh [13, 14] and Fukami-Mizumoto-Tanaka [18, 19] usually extend a certain inference rule in logic, such as the "Lukasiewic's logic" in [13, 14], the "standard sequence" and the "Godelian sequence" in [18, 19]. Due to a natural concern for appropriateness of these extensions, the authors also suggest the assessment of the consequences with a set of subjective criteria. Some of these reasoning methods are unable to satisfy such basic requirements as syllogism and contrapositive [18, 19]. However, Corollaries 3 and 4 of the previous section show that our definitions satisfy both syllogism and contrapositive. We shall now reassess some of the current fuzzy reasoning methods.

Example 1: A reasoning method due to Zadeh [13, 14] is $P(Z_1(7)) = A \times B \cup \bar{A} \times U_2$ which yields $r(Z_1(7))_{ij} = (\mu_i^{(1)} \wedge \mu_j^{(2)}) \vee (1-\mu_i^{(1)})$, $\forall$, i,j. It can be shown that

$$r(FDNF(7))_{ij} \leq r(Z_1(7))_{ij} \leq r(FCNF(7))_{ij}, \; \forall \; i, j.$$

Hence we have $FDNF(7) \subseteq P(Z_1(7)) \subseteq FCNF(7)$. That is Zadeh's first interpretation of the fuzzy implication, identified here as $P(Z_1(7))$, falls within the fuzzy inference region defined by the fuzzy disjunctive and conjunctive normal forms.

Example 2: Another method due to Zadeh [13, 14] is $P(Z_2(7)) = (\bar{A} \times U_2) \oplus (U_1 \times B)$ which yields $r(Z_2(7))_{ij} = 1 \wedge (1-\mu_i^{(1)} + \mu_j^{(2)})$, $\forall$ i,j. It can be shown that

$$r(FCNF(7))_{ij} \leq r(Z_2(7))_{ij}, \; \forall \; i,j.$$

Hence, we have $FCNF(7) \subseteq P(Z_2(7))$. That is Zadeh's second interpretation of fuzzy implication, identified here as $P(Z_2(7))$, falls outside the fuzzy inference region defined in this paper.

In our discussions of the measurement of fuzziness [23, 24], we have suggested that arithmetical addition of the membership values are not admissible because the membership values appear to be on interval scale. We ask, therefore, that Zadeh's second interpretation be reassessed by other researchers.

Example 3: A reasoning method due to Fukami, et. al., [18, 19] is

$$P(F_1) = A \times U_2 \rightarrow U_1 \times B \text{ which yields}$$

$$r(F_1) = \begin{cases} 1 & \text{if } \mu_i^{(1)} \leq \mu_j^{(2)} \\ 0 & \text{if } \mu_i^{(1)} > \mu_j^{(2)} \end{cases} .$$

Obviously, we always have

$$FDNF(2) \equiv \phi \subseteq P(F_1) \subseteq FCNF(1) \equiv U_1 \times U_2 .$$

Example 4: Another reasoning method due to Fukami, et al., [18, 19] is $P(F_2) = A \times U_2 \rightarrow U_1 \times B$ which now yields

$$r(F_2)_{ij} = \begin{cases} 1 & \text{if } \mu_i^{(1)} \leq \mu_j^{(2)} \\ \mu_j^{(2)} & \text{if } \mu_i^{(1)} > \mu_j^{(2)} \end{cases} .$$

(Noting that the structure is the same but the interpretation of the membership values are different?)

If we have the condition

$$\mu_i^{(1)} > \mu_j^{(2)} \text{ and } 1-\mu_j^{(2)} > \mu_j^{(2)}$$

then it can be shown that

$$r(FDNF(10))_i \leq r(F_2)_{ij} \leq r(FCNF(10))_{ij}, \ \forall\ i,j.$$

Hence we have

$$FDNF(10) \subseteq P(F_2) \subseteq FCNF(10)$$

under the condition specified above.

## CONCLUSION

We have defined fuzzy inference regions via the normal forms of the fuzzy relational propositions. We have shown firstly that these regions do not overlap with Yager's [12] regions that contain infinitely many interpretations of the fuzzy connectives. Secondly, we have demonstrated that one of Zadeh's [13, 14] interpretations of the fuzzy implications is inside but the other is outside of the fuzzy inference region defined for the fuzzy implication. Finally, we have shown that, under special conditions, an interpretation of Fukami, et. al., falls within the fuzzy inference region identified for the fuzzy non-inverse implication.

Above and beyond these observations, the perspective of the fuzzy inference regions may also be useful in certain applications where there is a need to satisfy syllogism and contrapositive as basic logical requirements.

## REFERENCES


[1] L. A. Zadeh, Fuzzy sets, Information and Control, 8 (1965), 338-353.

[2] R. E. Bellman and M. Gertz, On the analytic formalism of the theory of fuzzy sets, Information Science, 5 (1973), 149-156.

[3] H. Hammacher, On logical connectives of fuzzy statements and their affiliated truth functions, Third Europ. Meeting on Cybernetics and Systems Research, Vienna (April, 1976).

[4] B. R. Gaines, Foundations of fuzzy reasoning, Int. J. Man-Machine Studies, 8 (1976), 623-668.

[5] R. E. Bellman and L. A. Zadeh, Local and fuzzy logics, in: J. M. Dunn and D. Epstein, eds., "Modern Uses of Multiple Valued Logic" (D. Reidel, Dordrecht, 1977), 103-165.

[6] R. Rödder, On "and" and "or" connectives in fuzzy logic, EURO I Brussels (January, 1975).

[7] H. J. Zimmerman, Results of empirical studies in fuzzy set theory, Proc. Int. Conference on Applied General Systems Theory, Binghamton, N.Y. (1977).

[8] H. J. Zimmerman, Theory and applications of fuzzy sets, in: K. B. Haley, ed., OR '78, North Holland Publishing Company (1978), N18m, 1-17.

[9] G. Oden, Integration of fuzzy logical information, J. Exp. Psychol. 3 (1977), 565-575.

[10] H. M. Hersh and A. Caramazza, A fuzzy set approach to modifiers and vagueness in natural language, J. Exp. Psychol. (1975), 254-276.

[11] J. A. Goguen, The logic of inexact concepts, Synthese 19 (1969), 325-375.

[12] R. R. Yager, On a general class of fuzzy connectives, Fuzzy Sets and Systems 4, (1980) 235-242.

[13] L. A. Zadeh, Calculus of fuzzy restrictions, in: L. A. Zadeh and K. Tanaka, et. al., eds., "Fuzzy Sets and Their Applications to Cognitive and Decision Processes," Academic Press, N.Y. (1975), 1-39.

[14] L. A. Zadeh, The concept of linguistic variable and its application to approximate reasoning, I, II, III, Information Sci. 8 (1975), 199-251; 8 (1975), 301-357; 9 (1975), 43-80.

[15] S. Gottwald, Fuzzy propositional logics, Working Paper, Sktion Mathematik, Karl Marx Universitat, Leipzig, G.D.R. (1976).

[16] E. H. Mamdani, Application of fuzzy logic to approximate reasoning using linguistic systems, IEEE Trans. Comp. C-26 (1977), 1182-1191.

[17] N. Rescher, Many Valued Logic, McGraw Hill, N.Y., (1969).

[18] S. Fukami, M. Mizumoto, K. Tanaka, Some considerations on fuzzy conditional inference, Fuzzy Sets and Systems, 4 (1980), 243-273.

[19] M. Mizumoto, S. Fukami, K. Tanaka, Some methods of fuzzy reasoning, in: M. M. Gupta, et. al., eds., "Advances in Fuzzy Set Theory and Applications," North-Holland, Amsterdam (1979), 117-136.

[20] E. Hisdal, Generalized fuzzy sets systems and particularization, Fuzzy Sets and Systems 4 (1980), 275-291.

[21] I. B. Türksen, Lattices and groups of fuzzy propositions, Working Paper #80-006, Department of Industrial Engineering, University of Toronto, Toronto, Ontario, M5S 1A4, Canada.

[22] I. B. Türksen and D. D. Yao, Normal forms of fuzzy relational propositions, their lattice structures and applications, Working Paper #81-022, Department of Industrial Engineering, University of Toronto, Toronto, Ontario, M5S 1A4, Canada.

[23] A. M. Norwich and I. B. Türksen, Measurement and Scaling of Membership Functions, Proc. of Int. Conf. on Applied Systems Research and Cybernetics, Acapulco, Mexico, (Dec. 1980) (to appear).

[24] I. B. Türksen and A. M. Norwich, Measurement of Fuzziness, Proc. of Int. Conf. on Policy Analysis and Information Systems, Tapei, Taiwan, (August, 1981), 745-754.

# FUZZY TREE GRAMMAR AND FUZZY FOREST GRAMMAR

Chu Shang-Yong

Shanghai Jiao Tong University

Shanghai, People's Republic of China

## I. INTRODUCTION

In 1956, Chomsky put forward a kind of mathematical model about grammar to approximately describe natural languages. Thus, formal language became one of the important research fields of computer science.

In order to narrow the gaps between formal languages and natural languages, the concepts of randomness (Fu, Huang, 1972) and fuzziness (Lee, Zadeh, 1969) were introduced to extend the utility value of the formal languages.

Lee and Zadeh initiated the discussion on fuzzy string grammar and fuzzy string language. Mizumoto (1973) defined the concept of n-fold fuzzy grammar and n-fold fuzzy language to extract context-sensitive language from fuzzy context-free language by setting an appropriate threshold. Santos (1975), Depalma and Yau (1975) studied fuzzy string grammar and fuzzy string language, however, from still different viewpoints respectively. As a generalization of string grammar as defined by Chomsky, Brainerd (1969) put forward the tree grammar and Takahashi (1975) extended it to forest grammar.

The present paper consists of three parts. Begin with the concept of the fuzzy directed graph, a kind of fuzzy tree (Def. 1.4) is designated and its properties are discussed. Secondly, the fuzzy tree grammar (Def. 2.1) and fuzzy tree language (Def. 2.4) are studied, as well as the genealized concatenated operation (Def. 2.6). Two significant results are obtained: the relations between fuzzy tree language and fuzzy context-free language as defined by Santos are established (Th. 2.7, Th. 2.9) via leaves drawing operation (Def. 2.7); the relations between fuzzy tree language and tree language by Brainerd are established (Corollary to Th. 2.12) via $\lambda$-cut concept (Def. 2.11). Furthermore, fuzzy forest grammar (Def. 3.6) and fuzzy forest language (Def. 3.10) are discussed. Similar results corresponded with fuzzy tree language are obtained. An "isomorphism" between fuzzy tree language and fuzzy forest language is established. (Th. 3.1). As a conclusion, it is worthwhile to point out that the type 2 n-fold fuzzy language by Mizumoto may be replaced by a fuzzy forest language via leave drawing operation.

## II. FUZZY TREE

Definition 1.1: (1) a fuzzy directed graph (or abbreviating as a f-digraph) $\underset{\sim}{D}$ is an ordered triplet $(D,\sigma,\mu)$, there $D=(v,E,\psi)$ is a finite digraph, $\sigma$ is a map from finite set V into interval (0,1], and $\mu$ is a map from finite set E into interval (0,1] satisfying

$$\mu(e) \leq \sigma(x) \wedge \sigma(y), \quad \forall e \in E$$

where $\psi(e) = xy$, $x,y \in V$, i.e., e has head y and tail x. Also, e is called out-edge about node x, or in-edge about node y. Sometimes, we call e is directed edge joining x and y. If $\psi(e)=xy$, there is only one directed edge joining x and y, as denoted by e=xy. Throughout this paper, if directed edge xy exists, we always assume that it is the only one. We further denote the number of x's out-edges by r(x).

A node $(x,\sigma(x))$ in fuzzy digraph D is an element x, with membership function $\sigma(x)$, or set V. We denote it by $(x,\sigma(x)) \in \underset{\sim}{D}$.

An edge $(e,\mu(e))$ (or $(xy,\mu(xy))$ in fuzzy digraph $\underset{\sim}{D}$ is an element e, with membership function $\mu(xy)$, of the set E. We denote it by $(xy,\mu(xy)) \in \underset{\sim}{D}$. We suppose that, if $(xy,\mu(xy)) \in \underset{\sim}{D}$, then $(x,\sigma(x))$, $(y,\sigma(y)) \in \underset{\sim}{D}$.

(2) A directed walk (or a walk) from the node $W(x_0,\sigma(x_0))$ to $(x_k,\sigma(x_k))$ in $\underset{\sim}{D}$ with the length k $(k>0)$ is a finite non-hull sequence $W(x_0,x_k)=\langle(x_0,\sigma(x_0)), (x_0x_1,\mu(x_0x_1)),\ldots,(I_k,\sigma(I_k))\rangle$ where its terms are alternately nodes and edges, such that, for $i=1,\ldots,k$, the edge $(x_{i-1}x_i,\mu(x_{i-1}x_i))$ has head $(x_i,\sigma(x_i))$ and tail $(x_{i-1},\sigma(x_{i-1}))$, and $(x_{i-1}x_i,\mu(x_{i-1}x_i)) \in \underset{\sim}{D}$.

Let $W(x_j,x_r)=\langle(x_j,\sigma(x_j)), (x_jx_{j+1},\mu(x_jx_{j+1})),\ldots,(x_{r-1}x_r, \mu(x_{r-1}x_r)), (x_r,\sigma(x_r))\rangle$ $(0 \leq j < r \leq k)$ be a subsequence in $W(x_0,x_k)$. The walk $W(x_j,x_r)$ is called a subdirected walk in $W(x_0,x_k)$. It is often represented simply by $W(x_j,x_r) \subset W(x_0,x_k)$.

If $W(x,y)$ is a walk in $\underset{\sim}{D}$, then designating it by $W(x,y) \subset \underset{\sim}{D}$. If $(z,\sigma(z))$ is a node and $(z_1 z_2, \mu(z_1 z_2))$ is a edge in $W(x,y)$, then writing them by $(z,\sigma(z)) \in W(x,y)$ and $(z_1 z_2, \mu(z_1 z_2)) \in W(x,y)$.

If $W(x,y) = \langle (x,\sigma(x)), (xx_1, \mu(xx_1)), \ldots, (y,\sigma(y)) \rangle$, $W(y,z) = \langle (y,\sigma(y)), (yy_1, \mu(yy_1)), \ldots, (z,\sigma(z)) \rangle$, then the union of $W(x,y)$ and $W(y,z)$ is defined as follow:

$$W(x,y) \cup W(y,z) = \langle (x,\sigma(xx)), (xx_1, \mu(xx_1)), \ldots, (y,\sigma(y)),$$

$$(yy_1, \mu(yy_1)), \ldots, (z,\sigma(z)) \rangle.$$

It is obvious, we have $W(x,z) = W(x,y) \cup W(y,z)$, $\forall W(x,y), W(y,z) \subset \underset{\sim}{D}$, if it satisfies the condition that the walk is only one when it exists for any two nodes, and its converse is true, i.e., for any $W(x,z) \subset \underset{\sim}{D}$, $(y,\sigma(y)) \in W(x,z)$, there exist $W(x,y)$, $W(y,z) \subset \underset{\sim}{D}$, such that $W(x,z) = W(x,y) \cup W(y,z)$, or another represents $W(x,y) = W(x,z) \dot{-} W(y,z)$, $W(y,z) = W(x,z) \dot{-} W(x,y)$.

Generally, above results may be written as follow: $\forall (x_{ij}, \sigma(x_{ij})) \in W(x_0,x_k) = \langle (x_0,\sigma(x_0)), \ldots, (x_{i1}, \sigma(x_{i1})), \ldots, (x_{ij}, \sigma(x_{ij})), \ldots, (x_{i_{\ell-1}}, \sigma(x_{i_{\ell-1}})), \ldots, (x_k, \sigma(x_k)) \rangle$, we have $W(x_0,x_k) = \bigcup_{\substack{0 \le j \le \ell-1 \\ i_0=0,\, i_\ell=k}} W(x_{i_j}, x_{i_{j+1}})$.

(3) Define connected strength $S[W(x_0,x_k)]$ of a walk $W(x_0,x_k)$ as follow:

$$S[W(x_0,x_k)] = \bigwedge_{0 \le i \le k-1} \mu(x_i, x_{i+1}) \text{ where "}\Lambda\text{" means minimum,}$$

for all i.

(4) The root $(r,\sigma(r))$ of a fuzzy digraph $\underset{\sim}{D}$ is a node $(r,\sigma(r))$ satisfying $(r,\sigma(r)) \in \underset{\sim}{D}$, and for $\forall (t,\sigma(t)) \in \underset{\sim}{D}$, we have $W(r,t) \subset \underset{\sim}{D}$. We denote the root $(r,\sigma(r))$ of a fuzzy digraph $\underset{\sim}{D}$ by $R(\underset{\sim}{D})$.

(5) A close walk is a walk $W(x_0,x_k) = \langle (x_0,\sigma(x_0)), ((x_0,x_1), \mu(x_0,x_1)), \ldots, (x_k,\sigma(x_k)) \rangle$, where $x_0 = x_k$.

If $W(x_0,x_k) = \langle (x_0,\sigma(x_0)), (x_0x_1, \mu(x_0x_1), \ldots, (x_k,\sigma(x_k)) \rangle$ is a close walk and $x_i \ne x_j$ for $i \ne j$ $(0 < i, j < k)$, then $W(x_0,x_k)$ is called a fuzzy cycle.

A no-cyc-digraph is a f-digraph without any fuzzy cycle.

When the same letter $x \in V$ represents the different nodes in f-digraph $\underset{\sim}{D}$, we have to label to letters in alphabet V.

Definition 1.2: Let $J=\{1,2,\ldots,k\}$, a relation V' over set $V\times J$, where $V'\subset V\times J = \{(x,n)|x\in V,\ n\in J\}$, is called a finite labeled set over set V.

Usually, we shall not distinguish the difference between a set V and a finite labeled set over set V, so we also name set V is a finite labeled set. We represent the element of a finite labeled set V by (x:n). x is called the first label of a node in f-digraph $\underset{\sim}{D}$, and n is called the second. If there is no need to distinguish the difference of second labels or it is not vague in every sense, we may use $(x,\sigma(x))$ directly to represent the node in $\underset{\sim}{D}$, where x is an element of finite labeled set V.

Definition 1.3: A fuzzy directed tree $\underset{\sim}{T}$ is a f-digraph satisfying the condition there is a root of $\underset{\sim}{T}$, $\underset{\sim}{T}$ is a no-cyc-digraph and corresponding V is a finite labeled set.

Definition 1.4: A fuzzy directed tree in which there is only one walk from its root to any node (non-root) is called a strong fuzzy directed tree.

A simple fuzzy directed tree T is a fuzzy directed tree, if satisfying $\sigma(x)=1$ for all node $(x,\sigma(x))$ in $\underset{\sim}{T}$. In this case, a node (x:n,1) may be denoted by (x:n), or (x,1) being denoted by x.

A fuzzy tree is a strong and simple fuzzy directed tree.

The set of all fuzzy trees in which the first labels of nodes are over B is denoted by $\mathcal{J}$(B).

$\forall\underset{\sim}{T}\in\mathcal{J}(B)$, $\underset{\sim}{T}$ is called a fuzzy tree over B. Defining that the single tree {x} and the null tree $\Lambda$ are fuzzy trees, i.e., {x}, $\Lambda\in$ (B). However, for the convenience in the following discussion, we do not consider the properties about null tree $\Lambda$ except a special announcement.

Definition 1.5: Let the tree T be a common tree as same nodes as fuzzy tree $\underset{\sim}{T}$, satisfying $xy\in T$ iff $(xy,\mu(xy)\in\underset{\sim}{T}$, then tree T is called a basic tree of fuzzy tree $\underset{\sim}{T}$. We know that all trees over B is denoted by $B^{\#}$.

Definition 1.6: A full fuzzy subtree $\underset{\sim}{T}_1$ of fuzzy tree $\underset{\sim}{T}$ is a fuzzy tree satisfying

$$R(\underset{\sim}{T}_1)=(x:n)\in\underset{\sim}{T} \text{ and } W(x:n,y)\subset\underset{\sim}{T}_1 \text{ iff } W(x:n,y)\subset\underset{\sim}{T}.$$

Also, above $\underset{\sim}{T}_1$ may be called a full fuzzy subtree of $\underset{\sim}{T}$ at (x:n), being written $\underset{\sim}{T}_1=\underset{\sim}{T}/(x:n)$. Obviously, $\underset{\sim}{T}/(s:1)=\underset{\sim}{T}$, if $\tilde{R}(\underset{\sim}{T})=(s:1)$.

A fuzzy subtree $\underset{\sim}{T}_1$ of fuzzy tree $\underset{\sim}{T}$ is a fuzzy tree, where $R(\underset{\sim}{T}_1)=(x:n)\in\underset{\sim}{T}$, and if $W(x:n,y)\subset\underset{\sim}{T}_1$, then $W(x:n,y)\subset\underset{\sim}{T}$. Being written $\underset{\sim}{T}_1=\underset{\sim}{T}//(x:n)$ or $\underset{\sim}{T}_1\subset\underset{\sim}{T}$.

Definition 1.7: A fuzzy tree $\underset{\sim}{T}_1$ equals a fuzzy tree $\underset{\sim}{T}_2$, if $\underset{\sim}{T}_1$ is a fuzzy subtree of $\underset{\sim}{T}_2$ and $\underset{\sim}{T}_2$ is a fuzzy subtree of $\underset{\sim}{T}_1$.

According to the above definition, we have

Proposition 1.1: Let $\underset{\sim}{T}_1,\underset{\sim}{T}_2$ be fuzzy trees, then $\underset{\sim}{T}_1=\underset{\sim}{T}_2$ iff $\forall W(x,y)\subset\underset{\sim}{T}_1$, we have $W(x,y)\subset\underset{\sim}{T}_2$, and $\forall W(x,y)\subset\underset{\sim}{T}_2$, we have $W(x,y)\subset\underset{\sim}{T}_1$.

Corollary: Let $\underset{\sim}{T}_1,\underset{\sim}{T}_2$ be fuzzy trees, then $\underset{\sim}{T}_1=\underset{\sim}{T}_2$ iff $\forall W(R(\underset{\sim}{T}_1),y)\subset\underset{\sim}{T}_1$, we have $W(R(\underset{\sim}{T}_1),y)\subset\underset{\sim}{T}_2$, and $\forall W(R(\underset{\sim}{T}_2),y)\subset\underset{\sim}{T}_2$, we have $W(R(\underset{\sim}{T}_2),y)\subset\underset{\sim}{T}_1$.

Definition 1.8: A node (x:n) of fuzzy tree $\underset{\sim}{T}$ is called an ancestor of a node (y:m), or a node (y:m) of fuzzy tree $\underset{\sim}{T}$ is called a descendent of a node (x:n), if there exists walk $W(x:n,y:m)\subset\underset{\sim}{T}$.

A node (x:n) of fuzzy tree $\underset{\sim}{T}$ is called a direct ancestor of a node (y:m), or a node (y:m) of fuzzy tree $\underset{\sim}{T}$ is called a direct descendent of a node (x:n), if there exists edge $((x:n)(y:m),\mu((x:n)(y:m)))\in\underset{\sim}{T}$.

The number of direct descendents of a node (x:n) is denoted by r(x:n).

A leaf is a node (z:1) in $\underset{\sim}{T}$, where no a node in $\underset{\sim}{T}$ can be a descendent of (z:1). The set of all leaves in $\underset{\sim}{T}$ is written as $F_r(\underset{\sim}{T})$.

Defining $(x:n)<(y:m)$ in $\underset{\sim}{T}$, if (x:n) is an ancestor of (y:m) in $\underset{\sim}{T}$, else $(x:n)\nless(y:m)$.

If (x:n), (y:m) do not belong to the same fuzzy tree, then there are no relation "$<$" and relation "$\nless$" between nodes (x:n) and (y:m), also is itself of (x:n).

The following conclusions are evident.

Proposition 1.2: In fuzzy tree T

(1) If $(x_1:n_1)\neq(x_2:n_2)$, then one and only one relation is true between $(x_1:n_1)<(x_2:n_2)$ and $(x_1:n_1)\nless(x_2:n_2)$;

(2) If $(x_1:n_1)<(x_2:n_2)$, then $(x_2:n_2)\nless(x_1:n_1)$;

(3)   If $(x_1:n_1) < (x_2:n_2)$, $(x_2:n_2) < (x_3:n_3)$, then $(x_1:n_1) < (x_3:n_3)$;

(4)   If $(x_1:n_1) < (x_2:n_2)$, $(x_1:n_1) \nless (x_3:n_3)$, then $(x_2:n_2) \nless (x_3:n_3)$;
      if adding to $(x_3:n_3) \nless (x_1:n_1)$, $(x_1:n_1) < (x:n)$, then $(x:n) \nless$
      $(x_3:n_3)$ and $(x_3:n_3) \nless (x:n)$;

(5)   If $(y:m) \in \underset{\sim}{T}_1 = \underset{\sim}{T}//(x_1:n_1)$, $(y:m) \in \underset{\sim}{T}_2 = \underset{\sim}{T}//(x_2:n_2)$, then
      $(x_1:n_1) \leq (x_2:n_2)$ in $\underset{\sim}{T}_1$, or $(x_2:n_2) < (x_1:n_1)$ in $\underset{\sim}{T}_2$.

Definition 1.9:   Let $\underset{\sim}{T}_1 = \underset{\sim}{T}//(x_1:n_1)$, $\underset{\sim}{T}_2 = \underset{\sim}{T}//(x_2:n_2)$,

(1)   If $W(x_1:n_1, x_2:n_2) \subset \underset{\sim}{T}_1$, then the fuzzy subtree $\underset{\sim}{T}_1 \cup \underset{\sim}{T}_2$ is called
      a union of fuzzy subtrees $\underset{\sim}{T}_1$ and $\underset{\sim}{T}_2$, where

$$\underset{\sim}{T}_1 \cup \underset{\sim}{T}_2 = \{W(x_1:n_1, y) \mid W(x_1:n_1, y) \subset \underset{\sim}{T}_1\} \cup \{W(x_1:n_1, z) =$$

$$W(x_1:n_1, x_2:n_2) \cup W(x_2:n_2, z) \mid W(x_2:n_2, z) \subset \underset{\sim}{T}_2\};$$

(2)   If $(x:n) = \min\{(y:m) \mid (y:m) \in \underset{\sim}{T}_1, (y:m) \in \underset{\sim}{T}_2\}$, then the fuzzy sub-
      tree $\underset{\sim}{T}_1 \cap \underset{\sim}{T}_2$ is called an intersection of fuzzy subtrees $\underset{\sim}{T}_1$
      and $\underset{\sim}{T}_2$, where $\underset{\sim}{T}_1 \cap \underset{\sim}{T}_2 = \{W(x:n, z) \mid W(x:n, z) \subset \underset{\sim}{T}_1, W(x:n, z) \subset \underset{\sim}{T}_2\}$
      $\underset{\sim}{T}_1 \cap \underset{\sim}{T}_2 = \wedge$ iff there is not any node in $T_1$ as in $T_2$.

Definition 1.10:   A set $\Sigma$ of fuzzy subtrees of fuzzy tree
T is called a co-family set of fuzzy subtrees of $\underset{\sim}{T}$, if $\forall \underset{\sim}{T}_1, \underset{\sim}{T}_2 \in \Sigma$,
we have $\underset{\sim}{T}_1 \cap \underset{\sim}{T}_2 \neq \wedge$, and if for any $\underset{\sim}{T}_3$ satisfying $\underset{\sim}{T}_3 \cap \underset{\sim}{T} \neq \wedge$, where
$\underset{\sim}{T} \in \Sigma$, we have $\underset{\sim}{T}_3 \in \Sigma$.

Theorem 1.3:   A co-family set $\Sigma$ of any fuzzy subtrees of fuzzy
tree $\underset{\sim}{T}$ constitutes a distributive lattice under operations "$\cup$" and
"$\cap$".

Proof: $\forall \underset{\sim}{T}_1, \underset{\sim}{T}_2 \in \Sigma$, $\underset{\sim}{T}_1 \cap \underset{\sim}{T}_2 \neq \wedge$, by proposition 1.2/(5), we
get $R(\underset{\sim}{T}_1) \leq R(\underset{\sim}{T}_2)$ in $\underset{\sim}{T}_1$, or $R(\underset{\sim}{T}_2) < R(\underset{\sim}{T}_1)$ in $\underset{\sim}{T}_2$. We may suppose
$R(\underset{\sim}{T}_1) \leq R(\underset{\sim}{T}_2)$, then $R(\underset{\sim}{T}_1 \cup \underset{\sim}{T}_2) = R(\underset{\sim}{T}_1)$, $R(\underset{\sim}{T}_1 \cap \underset{\sim}{T}_2) = R(\underset{\sim}{T}_2)$.

$\forall \underset{\sim}{T}' \in \Sigma$, according to the definition, $\underset{\sim}{T}_1 \cup \underset{\sim}{T}_2 \supset \underset{\sim}{T}_2$. But
$\underset{\sim}{T}_2 \cap \underset{\sim}{T}' \neq \wedge$, thus $(\underset{\sim}{T}_1 \cup \underset{\sim}{T}_2) \cap \underset{\sim}{T}' \neq \wedge$. Then $\underset{\sim}{T}_1 \cup \underset{\sim}{T}_2 \in \Sigma$. Since
$R(\underset{\sim}{T}_1 \cap \underset{\sim}{T}_2) = R(\underset{\sim}{T}_2)$, $\underset{\sim}{T}_2 \cap \underset{\sim}{T}' \neq \wedge$, by proposition 1.2/(5) $R(\underset{\sim}{T}_2) \leq R(\underset{\sim}{T}')$
in $\underset{\sim}{T}_2$, or $R(\underset{\sim}{T}') < R(\underset{\sim}{T}_2)$ in $\underset{\sim}{T}'$.

If $R(\underset{\sim}{T}') < R(\underset{\sim}{T}_2)$, then $R(\underset{\sim}{T}' \cap \underset{\sim}{T}_2) = R(\underset{\sim}{T}_2)$, i.e., $R(\underset{\sim}{T}_2) \in (\underset{\sim}{T}_1 \cap \underset{\sim}{T}_2)$
$\cap \underset{\sim}{T}'$, thus $(\underset{\sim}{T}_1 \cap \underset{\sim}{T}_2) \cap \underset{\sim}{T}' \neq \wedge$. If $R(\underset{\sim}{T}_2) \leq R(\underset{\sim}{T}')$ in $\underset{\sim}{T}_2$, by proposition
1.2/(3) and $R(\underset{\sim}{T}_1) \leq R(\underset{\sim}{T}_2)$, $R(\underset{\sim}{T}_1) \leq R(\underset{\sim}{T}')$ in $\underset{\sim}{T}$. Now, by the definition
of co-family set and proposition 1.2/(5), $R(\underset{\sim}{T}_2) \leq R(\underset{\sim}{T}')$ in $\underset{\sim}{T}_1$, i.e.,
$R(\underset{\sim}{T}') \in \underset{\sim}{T}_1 \cap \underset{\sim}{T}_2$ therefore $(\underset{\sim}{T}_1 \cap \underset{\sim}{T}_2) \cap \underset{\sim}{T}' \neq \wedge$, thus $\underset{\sim}{T}_1 \cap R_2 \in \Sigma$. And
obviously $\underset{\sim}{T}_1 \cup \underset{\sim}{T}_2 = \underset{\sim}{T}_2 \cup \underset{\sim}{T}_1$, $\underset{\sim}{T}_1 \cap \underset{\sim}{T}_2 = \underset{\sim}{T}_2 \cap \underset{\sim}{T}_1$, $(\underset{\sim}{T}_1 \cup \underset{\sim}{T}_2) \cup \underset{\sim}{T}_3 = \underset{\sim}{T}_1 \cup$
$(\underset{\sim}{T}_2 \cup \underset{\sim}{T}_3)$, $(\underset{\sim}{T}_1 \cap \underset{\sim}{T}_2) \cap \underset{\sim}{T}_3 = \underset{\sim}{T}_1 \cap (\underset{\sim}{T}_2 \cap \underset{\sim}{T}_3)$. By the definition of the

operation $\cup$ and $\cap$, we may prove following easily,

$$\underset{\sim}{T}_1 \cap (\underset{\sim}{T}_2 \cup \underset{\sim}{T}_3) = (\underset{\sim}{T}_1 \cap \underset{\sim}{T}_2) \cup (\underset{\sim}{T}_1 \cap \underset{\sim}{T}_3)$$

$$\underset{\sim}{T}_1 \cup (\underset{\sim}{T}_2 \cap \underset{\sim}{T}_3) = (\underset{\sim}{T}_1 \cup \underset{\sim}{T}_2) \cap (\underset{\sim}{T}_1 \cup \underset{\sim}{T}_3). \quad ¤$$

Definition 1.11: Let $\underset{\sim}{T}$, $\underset{\sim}{T}_1$ be fuzzy trees, $\underset{\sim}{T} \cap \underset{\sim}{T}_1 = \Lambda$, $(z:1)$, $(x:n) \in \underset{\sim}{T}$, $(z:1)$ is a direct ancestor of $(x:n)$, $R(\underset{\sim}{T}) = (s:1)$, and $R(\underset{\sim}{T}_1) = (r:1)$, then fuzzy tree $\underset{\sim}{T}[(x:n) \leftarrow \underset{\sim}{T}_1]$ is called a replacement of $\underset{\sim}{T}_1$ in $\underset{\sim}{T}$ at $(x:n)$, where $\underset{\sim}{T}[(x:n) \leftarrow \underset{\sim}{T}_1] = \{W(s,y) \mid W(s,y) \subset \underset{\sim}{T}, (x:n) \nleq (y:m)\} \cup \{W(x,z) \cup W(z,r) \cup W(r,t) \mid W(z,r) = \langle (z:1), ((z:1)(r:1), \mu((z:1)(x:n))), (r:1) \rangle, W(r,t) \subset \underset{\sim}{T}_1\}$, and assuming $(x:n) \neq (s:1)$. We have $\underset{\sim}{T}[(x:n) \leftarrow \underset{\sim}{T}_1]/(r:1) = \underset{\sim}{T}_1$.

Proposition 1.4: If T is a fuzzy tree, $(x:n)$, $(y:m) \in \underset{\sim}{T}$ and $(x:n) < (y:m)$, then $[\underset{\sim}{T}/(x:n)]/(y:m) = \underset{\sim}{T}/(y:m)$.

Proof: $(x:n) < (y:m)$, $\exists W(x,y) \subset \underset{\sim}{T}$, $W(x,y) \subset \underset{\sim}{T}/(x:n)$, $(y:m) \in \underset{\sim}{T}/(x:n)$.

Let $W(y,z) \subset \underset{\sim}{T}$, thus $W(x,z) = W(x,y) \cup W(y,z) \subset \underset{\sim}{T}$, then $W(x,z) \subset \underset{\sim}{T}/(x:n)$, $W(y,z) = W(x,z) \dot{-} W(x,y) \subset \underset{\sim}{T}/(x:n)$.

Inversely, if $W(y,z) \subset \underset{\sim}{T}/(x:n)$, we get $W(y,z) \subset \underset{\sim}{T}$, thus $W(y,z) \subset \underset{\sim}{T}/(y:m)$, i.e., $[\underset{\sim}{T}/(x:n)]/(y:m) = \underset{\sim}{T}/(y:m)$. ¤

Corollary: Let $\underset{\sim}{T}$, $\underset{\sim}{T}_1$ be fuzzy trees, $\underset{\sim}{T} \cap \underset{\sim}{T}_1 = \Lambda$, $(x:n) \in \underset{\sim}{T}$, $(y:m) \in \underset{\sim}{T}_1$, then $\underset{\sim}{T}[(x:n) \leftarrow \underset{\sim}{T}_1]/(y:m) = \underset{\sim}{T}_1/(y:m)$.

Proposition 1.5: If $\underset{\sim}{T}$, $\underset{\sim}{T}_1$ are fuzzy trees, $\underset{\sim}{T} \cap \underset{\sim}{T}_1 = \Lambda$, $(x:n)$, $(y:m) \in \underset{\sim}{T}$ and $(x:n) < (y:m)$, then $\underset{\sim}{T}[(y:m) \leftarrow \underset{\sim}{T}_1]/(x:n) = [\underset{\sim}{T}/(x:n)][(y:m) \leftarrow \underset{\sim}{T}_1]$.

Proposition 1.6: If $\underset{\sim}{T}$, $\underset{\sim}{T}_1$ are fuzzy trees, $\underset{\sim}{T} \cap \underset{\sim}{T}_1 = \Lambda$,
(1) Let $(x_1:n_1)$, $(x:n)$, $(x_2:n_2) \in \underset{\sim}{T}$, and $(x:n) \nleq (x_2:n_2)$, $(x:n) < (x_1:n_1)$, then $(x_2:n_2) \in \underset{\sim}{T}[(x:n) \leftarrow \underset{\sim}{T}_1]$, and $(x_1:n_1)$ does not.
(2) Let $(x:n) \in \underset{\sim}{T}$, $W(y,z) \subset \underset{\sim}{T}_1$, then $W(y,z) \subset \underset{\sim}{T}[(x:n) \leftarrow \underset{\sim}{T}_1]$.

Proposition 1.7: If $\underset{\sim}{T}$, $\underset{\sim}{T}_1$, $\underset{\sim}{T}_2$ are fuzzy trees, $\underset{\sim}{T} \cap \underset{\sim}{T}_1 = \underset{\sim}{T} \cap \underset{\sim}{T}_2 = \underset{\sim}{T}_1 \cap \underset{\sim}{T}_2 = \Lambda$, $(x:n) \in \underset{\sim}{T}$, $(y:m) \in \underset{\sim}{T}_1$, then $\underset{\sim}{T}[(x:n) \leftarrow \underset{\sim}{T}_1[(y:m) \leftarrow \underset{\sim}{T}_2]] = \underset{\sim}{T}[(x:n) \leftarrow \underset{\sim}{T}_1][(y:m) \leftarrow \underset{\sim}{T}_2]$.

Proof. Let $R(T) = (s:1)$, $R(T_1) = (s_1;1)$, $R(T) = (s_2:1)$. $\forall W(s,z) \subset \underset{\sim}{T}[(x:n) \leftarrow \underset{\sim}{T}_1[(y:m) \leftarrow \underset{\sim}{T}_2]]$. If $(s_1:1) \nleq (z:1)$, then $W(s,t) \subset \underset{\sim}{T}$ and $(x:n) \nleq (z:1)$ in $\underset{\sim}{T}$. By proposition 1.6/(1), $W(s,z) \subset \underset{\sim}{T}[(x:n) \leftarrow \underset{\sim}{T}_1]$, but $(s_1:1) \nleq (z:1)$ in $\underset{\sim}{T}[(x:n) \leftarrow \underset{\sim}{T}_1]$, $(s_1:1) < (y:m)$, by proposition 1.2/(4), $(y:m) \nleq (z:1)$, thus $W(s,z) \subset \underset{\sim}{T}[(x:n) \leftarrow \underset{\sim}{T}_1][(y:m) \leftarrow \underset{\sim}{T}_2]$.

If $(s_1:1) < (z:1)$ and $(s_2:1) \nless (z:1)$, supposing $(x_1:n_1)$ is an ancestor of $(S:1)$ in $\underset{\sim}{T}[(x:n) \leftarrow \underset{\sim}{T}_1[(y:m) \leftarrow \underset{\sim}{T}_2]]$, then $W(s,z) = W(s,x_1) \cup W(x_1:s_1) \cup W(s_1,z)$, where $W(s,x_1) \subset \underset{\sim}{T}$, $W(x_1,s_1) = <(x_1:n_1), ((x_1:n_1)(s_1:1), \mu((x_1:n_1)(s_1:1))>$, $(s_1:1)$ $W(s_1,z) \subset \underset{\sim}{T}_1$. Thus, $W(s,z) \subset \underset{\sim}{T}[(x:n) \leftarrow \underset{\sim}{T}_1]$.

Since $(s_2:1) \nless (z:1)$ in $\underset{\sim}{T}[(x:n) \leftarrow \underset{\sim}{T}_1[(y:m) \leftarrow \underset{\sim}{T}_2]]$, and $(y:m) \nless (z:1)$ in $\underset{\sim}{T}[(x:n) \leftarrow \underset{\sim}{T}_1]$, thus $W(s,z) \subset \underset{\sim}{T}[(x:n) \leftarrow \underset{\sim}{T}_1][(y:m) \leftarrow \underset{\sim}{T}_2]$.

If $(s_2:1) < (z:1)$, let $(x_2:n_2)$ be an ancestor of $(s_2:1)$ in $\underset{\sim}{T}[(x:n) \leftarrow \underset{\sim}{T}_1[(y:m) \leftarrow \underset{\sim}{T}_2]]$, $W(s,z) = W(s,x_2) \cup W(x_2,s_2) \cup W(s_2,z)$, where $W(s_1,z) \subset \underset{\sim}{T}_2$, $W(s,x_2) \subset \underset{\sim}{T}[(x:n) \leftarrow \underset{\sim}{T}_1]$, $W(x_2,s_2) = <(x_2:n_2), ((x_2:n_2)(s_2:1), \mu((x_2:n_2)(s_2:1))), (s_2:1)>$, then $W(s,z) \subset \underset{\sim}{T}[(x:n) \leftarrow \underset{\sim}{T}_1][(y:m) \leftarrow \underset{\sim}{T}_2]$.

Its proof is as same as above. For any $W(s,z) \subset \underset{\sim}{T}[(x:n) \leftarrow \underset{\sim}{T}_1][(y:m) \leftarrow \underset{\sim}{T}_2]$, we have $W(s,z) \subset \underset{\sim}{T}[(x:n) \leftarrow \underset{\sim}{T}_1[(y:m) \leftarrow \underset{\sim}{T}_2]]$. ¤

Proposition 1.8: If $\underset{\sim}{T}$, $\underset{\sim}{T}_1$ are fuzzy trees, $\underset{\sim}{T} \cap \underset{\sim}{T}_1 = \Lambda$, $(x:n)$, $(x_1:n_1)$, $(x_2:n_2) \in \underset{\sim}{T}$ satisfying $(x_1:n_1) < (x:n)$, $(x:n) \nless (x_2:n_2)$ and $(x_2:n_2) \nless (x:n)$, then for any $(z:1) \in \underset{\sim}{T}_1$, $(x_1:n_1) < (z:1)$, $(z:1) \nless (x_2:n_2)$, $(x_2:n_2) \nless (z:1)$ in $\underset{\sim}{T}[(x:n) \leftarrow \underset{\sim}{T}_1]$.

Proof: Let $R(\underset{\sim}{T}) = (s:1)$, $R(\underset{\sim}{T}_1) = (s_1:1)$. Since $(x_1:n_1) < (x:n)$, there exists $W(x_1,x) \subset \underset{\sim}{T}$, thus $W(x_1,s_1) \subset \underset{\sim}{T}[(x:n) \leftarrow \underset{\sim}{T}_1]$. Since $\exists W(s_1,z) \subset \underset{\sim}{T}_1$, thus $W(x_1,z) = W(x_1,s_1) \cup W(s_1,z) \subset \underset{\sim}{T}[(x:n) \leftarrow \underset{\sim}{T}_1]$, i.e., $(x_1:n_1) < (z:1)$.

Since $(x_2:n_2) \nless (x:n)$ in $\underset{\sim}{T}$, then there does not exist a walk $W(x_2,x)$ in $\underset{\sim}{T}$, i.e., there does not exist a walk $W(x_2,s_1)$ in $\underset{\sim}{T}[(x:n) \leftarrow \underset{\sim}{T}_1]$.

Since $(z:1) \in \underset{\sim}{T}_1$, there does not exist a walk $W(x_2,z)$ in $\underset{\sim}{T}[(x:n) \leftarrow \underset{\sim}{T}_1]$, i.e., $(x_2:n_2) \nless (z:1)$.

Since $(x:n) \nless (x_2:n_2)$ in $\underset{\sim}{T}$, thus $(s_1:1) \nless (x_2:n_2)$ in $\underset{\sim}{T}[(x:n) \leftarrow \underset{\sim}{T}_1]$.

But $(s_1:1) < (z:1)$, by proposition 1.2/(4), $(z:1) \nless (x_2:n_2)$. ¤

Proposition 1.9: If $\underset{\sim}{T}$, $\underset{\sim}{T}_1$, $\underset{\sim}{T}_2$ are fuzzy trees, $\underset{\sim}{T} \cap \underset{\sim}{T}_1 = \underset{\sim}{T} \cap \underset{\sim}{T}_2 = \underset{\sim}{T}_1 \cap \underset{\sim}{T}_2 = \Lambda$, $(x_1:n_1)$, $(x_2:n_2) \in \underset{\sim}{T}$, and $(x_1:n_1) \nless (x_2:n_2)$, $(x_2:n_2) \nless (x_1:n_1)$, then $\underset{\sim}{T}[(x_1:n_1) \leftarrow \underset{\sim}{T}_1][(x_2:n_2) \leftarrow \underset{\sim}{T}_2] = \underset{\sim}{T}[(x_2:n_2) \leftarrow \underset{\sim}{T}_2][(x_1:n_1) \leftarrow \underset{\sim}{T}_1]$.

Proof: Let $R(\underset{\sim}{T}) = (s:1)$, $R(\underset{\sim}{T}_1) = (s_1:1)$, $R(\underset{\sim}{T}_2) = (s_2:1)$. We take $\forall W(s,y) \subset \underset{\sim}{T}[(x_2:n_2) \leftarrow \underset{\sim}{T}_2][(x_1:n_1) \leftarrow \underset{\sim}{T}_1]$. If $(s_1:1) \nless (y:m)$, then $W(s,y) \subset \underset{\sim}{T}[(x_2:n_2) \leftarrow \underset{\sim}{T}_2]$ and $(x_1:n_1) \nless (y:m)$. When $(s_2:1) \nless (y:m)$ in $\underset{\sim}{T}[(x_2:n_2) \leftarrow \underset{\sim}{T}_2]$, since $W(s,y) \subset \underset{\sim}{T}$, $(x_2:n_2) \nless (y:m)$ in $\underset{\sim}{T}$, we have $W(s,y) \subset \underset{\sim}{T}[(x_1:n_1) \leftarrow \underset{\sim}{T}_1][(x_2:n_2) \leftarrow \underset{\sim}{T}_2]$.

When $(s_2:1) < (y:m)$ in $\tilde{T}[(x_2:n_2) \leftarrow \tilde{T}_2]$, let $(y_2:m_2)$ is an ancestor of $(s_2:1)$ in $\tilde{T}[(x_2:n_2) \leftarrow \tilde{T}_2]$, $W(s,y) = W(s,y_2) \cup W(y_2,s_2) \cup W(s_2,y)$, where $W(s,y_2) \subset \tilde{T}$, $W(y_2,s_2) = <(y_2:m_2), ((y_2:m_2)(s_2:1), \mu((y_2:m_2)(s_2:1))), (s_2:1)>$, $W(s_2,y) \subset \tilde{T}_2$.

Now, by proposition 1.6/(1) and $(x_1:n_1) \nless (x_2:n_2)$ in $\tilde{T}$, we get $W(s,x_2) \subset \tilde{T}[(x_1:n_1) \leftarrow \tilde{T}_1]$. Thus, $W(s,s_2) = W(s,y_2) \cup W(y_2,s_2) \subset \tilde{T}[(x_1:n_1) \leftarrow \tilde{T}_1][(x_2:n_2) \leftarrow \tilde{T}_2]$, $W(s,y) = W(s,y_2) \cup W(y_2,s_2) \cup W(s_2,y) \subset \tilde{T}[(x_1:n_1) \leftarrow \tilde{T}_1][(x_2:n_2) \leftarrow \tilde{T}_2]$.

If $(s_1:1) < (y:m)$, then $(y:m) \varepsilon \tilde{T}_1$. By $(x_1:n_1) \nless (x_2:n_2)$, $(x_2:n_2) \nless (x_1:n_1)$ in $\tilde{T}$, and proposition 1.8, thus $(x_2:n_2) \nless (y:m)$ in $\tilde{T}[(x_1:n_1) \leftarrow \tilde{T}_1]$.

By proposition 1.6/(1), we get $W(s,y) \subset \tilde{T}[(x_1:n_1) \leftarrow \tilde{T}_1][(x_2:n_2) \leftarrow \tilde{T}_2]$.

Due to the symmetry property, we also get $\tilde{T}[(x_1:n_1) \leftarrow \tilde{T}_1][(x_2:n_2) \leftarrow \tilde{T}_2]$ is a fuzzy subtree of $\tilde{T}[(x_2:n_2) \leftarrow \tilde{T}_2][(x_1:n_1) \leftarrow \tilde{T}_1]$. ¤

Proposition 1.10: If $\tilde{T}$, $\tilde{T}_1$ are fuzzy trees, $\tilde{T} \cap \tilde{T}_1 = \Lambda$, $(x:n)$, $(y:m) \varepsilon \tilde{T}$, and $(x:n) \nless (y:m)$, $(y:m) \nless (x:n)$, then $\tilde{T}[(x:n) \leftarrow \tilde{T}_1]/(y:m) = \tilde{T}/(y:m)$.

Proof: Since $(x:n) \nless (y:m)$, then $(y:m) \varepsilon \tilde{T}[(x:n) \leftarrow \tilde{T}_1]$. Thus $\tilde{T}[(x:n) \leftarrow \tilde{T}_1]/(y:m)$ may be defined.

Let $R(\tilde{T}_1) = (s_1:1)$, let us consider $\forall W(y,z) \subset \tilde{T}[(x:n)$ $\tilde{T}_1]/(y:m)$. Since $(x:n) \nless (y:m)$, and $(y:m) \nless (x:n)$ in $\tilde{T}$, then it follows $(s_1:1) \nless (y:m)$ and $(y:m) \nless (s_1:1)$ in $\tilde{T}[(x:n) \leftarrow \tilde{T}_1]$.

Since $\forall (z_1:1_1) \varepsilon \tilde{T}_1$, we have $(s_1:1) < (z_1:1_1)$ in $\tilde{T}_1$, as same as in $\tilde{T}[(x:n) \leftarrow \tilde{T}_1]$. Now, by proposition 1.2/(4), $(y:m) \nless (z_1:1_1)$ in $\tilde{T}[(x:n) \leftarrow \tilde{T}_1]$, i.e., there does not exist any walk $W(y,z_1)$ in $\tilde{T}[(x:n) \leftarrow \tilde{T}_1]$. Thus $W(y,z) \subset \tilde{T}$, i.e., $W(y,z) \subset \tilde{T}/(y:m)$.

Inversely, if $W(y,z) \subset \tilde{T}/(y:m)$, then $(y:m) < (z:1)$ in $\tilde{T}$. By proposition 1.2/(4) and $(y:m) \nless (x:n)$, $(x:n) \nless (y:m)$, thus $(z:1) \nless (x:m)$, $(x:n) \nless (z:1)$.

Now, by proposition 1.6/(1), $W(y,z) \subset \tilde{T}[(x:n) \leftarrow \tilde{T}_1]$, i.e., $W(y,z) \subset \tilde{T}[(x:n) \leftarrow \tilde{T}_1]/(y:m)$. ¤

We extend definition 1.11 as follow.

Definition 1.12: Let $\tilde{T}$, $\tilde{T}_1$ be fuzzy trees, $\tilde{T} \cap \tilde{T}_1 = \Lambda$, $(z:1)$, $(x:n) \varepsilon \tilde{T}$, $(z:1)$ is a direct ancestor of $(x:n)$, $R(\tilde{T}) = (s:1)$, $R(\tilde{T}_1) = (s_1:1)$, $f \varepsilon (0,1]$, then fuzzy tree $\tilde{T}[(x:n) \leftarrow \tilde{T}_1]$ is called a replacement of $\tilde{T}_1$ with membership f in $\tilde{T}$ at $(x:n)$, where

$\underset{\sim}{T}[(x:n) \overset{f}{\leftarrow} \underset{\sim}{T}] = \{W(s,y) \mid W(x,y) \subset \underset{\sim}{T}, (x:n) \not< (y:m)\} \cup \{W(s,z) \cup W(z,s_1) \cup W(s_1,t) \mid W(z,s_1) = <(z:1), ((z:1)\ (s_1:1)), \mu((z:1)\ (s_1:1)) \wedge f),$
$(s_1:1)>, W(s_1,t) \subset \underset{\sim}{T}_1\}$, and assuming $(x:n) \neq (s:1)$.

Repeating the foregoing discussions, we have

Proposition 1.4': If $\underset{\sim}{T}, \underset{\sim}{T}_1$ are fuzzy trees, $\underset{\sim}{T} \cap \underset{\sim}{T}_1 = \wedge$, $(x:n) \varepsilon \underset{\sim}{T}$, $(y:m) \varepsilon \underset{\sim}{T}_1$, $f \varepsilon (0,1]$, then $\underset{\sim}{T}[(x:n) \overset{f}{\leftarrow} \underset{\sim}{T}_1]/(y:m) = \underset{\sim}{T}_1/(y:m)$.

Proposition 1.5': If $\underset{\sim}{T}, \underset{\sim}{T}_1$ are fuzzy trees, $\underset{\sim}{T} \cap \underset{\sim}{T}_1 = \wedge$, $(x:n)$, $(y:m) \varepsilon \underset{\sim}{T}$, $(x:n) < (y:m)$, $f \varepsilon (0,1]$, then $\underset{\sim}{T}[(y:m) \overset{f}{\leftarrow} \underset{\sim}{T}_1]/(x:n) = [\underset{\sim}{T}/(x:n)][(y:m) \overset{f}{\leftarrow} \underset{\sim}{T}_1]$.

Proposition 1.6': If $\underset{\sim}{T}, \underset{\sim}{T}_1$ are fuzzy trees, $\underset{\sim}{T} \cap \underset{\sim}{T}_1 = \wedge$, $f \varepsilon (0,1]$.
(1) Let $(x_1:n_1)$, $(x_2:n_2)$, $(x:n) \varepsilon \underset{\sim}{T}$, $(x:n) \not< (x_2:n_2)$, $(x:n) < (x_1:n_1)$, then $(x_2:n_2) \varepsilon \underset{\sim}{T}[(x:n) \overset{f}{\leftarrow} \underset{\sim}{T}_1]$ and $(x_1:n_1) \bar{\varepsilon} \underset{\sim}{T}[(x:n) \overset{f}{\leftarrow} \underset{\sim}{T}_1]$;
(2) Let $(x:n) \varepsilon \underset{\sim}{T}$, $W(y,z) \subset \underset{\sim}{T}_1$, then $W(y,z) \subset \underset{\sim}{T}[(x:n) \overset{f}{\leftarrow} \underset{\sim}{T}_1]$.

Proposition 1.7': If $\underset{\sim}{T}, \underset{\sim}{T}_1, \underset{\sim}{T}_2$ are fuzzy trees, $\underset{\sim}{T} \cap \underset{\sim}{T}_1 = \underset{\sim}{T} \cap \underset{\sim}{T}_2 = \underset{\sim}{T}_1 \cap \underset{\sim}{T}_2 = \wedge$, $(x:n) \varepsilon \underset{\sim}{T}$, $(y:m) \varepsilon \underset{\sim}{T}_1$, $f_1, f_2 \varepsilon (0,1]$, then $\underset{\sim}{T}[(x:n) \overset{f_1}{\leftarrow} \underset{\sim}{T}_1[(y:m) \overset{f_2}{\leftarrow} \underset{\sim}{T}_2] = \underset{\sim}{T}[(x:n) \overset{f_1}{\leftarrow} \underset{\sim}{T}_1][(y:m) \overset{f_2}{\leftarrow} \underset{\sim}{T}_2]$.

Proposition 1.8': If $\underset{\sim}{T}, \underset{\sim}{T}_1$ are fuzzy trees, $\underset{\sim}{T} \cap \underset{\sim}{T}_1 = \wedge$, $(x:n)$, $(x_1:n_1)$, $(x_2:n_2) \varepsilon \underset{\sim}{T}$, $(x_1:n_1) < (x:n)$, $(x:n) \not< (x_2:n_2)$, $(x_2:n_2) \not< (x:n)$, $f \varepsilon (0,1]$, then for any $(z:1) \varepsilon \underset{\sim}{T}_1$ we have $(x_1:n_1) < (z:1)$, $(z:1) \not< (x_2:n_2)$, $(x_2:n_2) \not< (z:1)$ in $\underset{\sim}{T}[(x:n) \overset{f}{\leftarrow} \underset{\sim}{T}_1]$.

Proposition 1.9': If $\underset{\sim}{T}, \underset{\sim}{T}_1, \underset{\sim}{T}_2$ are fuzzy trees, $\underset{\sim}{T} \cap \underset{\sim}{T}_1 = \underset{\sim}{T} \cap \underset{\sim}{T}_2 = \underset{\sim}{T}_1 \cap \underset{\sim}{T}_2 = \wedge$, $(x_1:n_1)$, $(x_2:n_2) \bar{\varepsilon} \underset{\sim}{T}$ and $(x_1:n_1) \not< (x_2:n_2)$, $(x_2:n_2) \not< (x_1:n_1)$, $f_1, f_2 \varepsilon (0,1]$, then $\underset{\sim}{T}[(x_1:n_1) \overset{f_1}{\leftarrow} \underset{\sim}{T}_1][(x_2:n_2) \overset{f_2}{\leftarrow} \underset{\sim}{T}_2] = \underset{\sim}{T}[(x_2:n_2) \overset{f_2}{\leftarrow} \underset{\sim}{T}_2][(x_1:n_1) \overset{f_1}{\leftarrow} \underset{\sim}{T}_1]$.

Proposition 1.10': If $\underset{\sim}{T}, \underset{\sim}{T}_1$ are fuzzy trees, $\underset{\sim}{T} \cap \underset{\sim}{T}_1 = \wedge$ $(x:n)$, $(y:m) \varepsilon \underset{\sim}{T}$, $(x:n) \not< (y:m)$, $(y:m) \not< (x:n)$, $f \varepsilon (0,1]$, then $\underset{\sim}{T}[(x:n) \overset{f}{\leftarrow} \underset{\sim}{T}_1]/(y:m) = \underset{\sim}{T}/(y:m)$.

Definition 1.13: The strength $\mu(\underset{\sim}{T})$ of a fuzzy tree $\underset{\sim}{T}$ is defined as follow:
$$\mu(\underset{\sim}{T}) = \underset{\forall W \subset \underset{\sim}{T}}{\wedge} S(W),$$ where $W$ may be all walks in $\underset{\sim}{T}$.
The following propositions are proven easily.

Proposition 1.11: For fuzzy tree $\underset{\sim}{T}$ and $\forall (x:n) \varepsilon \underset{\sim}{T}$, we have $\mu(\underset{\sim}{T}) \leq \mu(\underset{\sim}{T}//(x:n))$.

Corollary 1: If $\underset{\sim}{T}, \underset{\sim}{T}_1$ are fuzzy trees, $\underset{\sim}{T} \cap \underset{\sim}{T}_1 = \wedge$, $(x:n) \varepsilon \underset{\sim}{T}$, $f \varepsilon (0,1]$, then $\mu(\underset{\sim}{T}[(x:n) \overset{f}{\leftarrow} \underset{\sim}{T}_1]) \not< \mu(\underset{\sim}{T}_1) \wedge f$.

Corollary 2:  If T is a fuzzy tree, $(x_1:n_1)$, $(x_2:n_2) \varepsilon \underset{\sim}{T}$, $(x_1:n_1) < (x_2:n_2)$, then $\mu(\underset{\sim}{T}/(x_1:n_1)) \nleq \mu(\underset{\sim}{T}/(\bar{x}_2:\bar{n}_2))$.

Proposition 1.12:  If $\underset{\sim}{T}$, $\underset{\sim}{T}_1$ are fuzzy trees, $\underset{\sim}{T} \cap \underset{\sim}{T}_1 = \wedge$, $(x:n) \varepsilon \underset{\sim}{T}$, $f \varepsilon (0,1]$, then  $\mu(\underset{\sim}{T}_1) \wedge f \wedge \mu(\underset{\sim}{T}) \leq \mu(\underset{\sim}{T}[(x:n) \leftarrow \underset{\sim}{T}_1])$.

Corollary:  If $\underset{\sim}{T}$, $\underset{\sim}{T}_1$, $\underset{\sim}{T}_2$ are fuzzy trees, $\underset{\sim}{T} \cap \underset{\sim}{T}_1 = \underset{\sim}{T} \cap \underset{\sim}{T}_2 = \wedge$, $(x:n) \varepsilon \underset{\sim}{T}$, $f \varepsilon (0,1]$, $\mu(\underset{\sim}{T}_1^+) \leq \mu(\underset{\sim}{T}_2)$, then  $\mu(\underset{\sim}{T}[(x:n) \overset{f}{\leftarrow} \underset{\sim}{T}_1]) \leq \mu(\underset{\sim}{T}[(x:n) \overset{f}{\leftarrow} \underset{\sim}{T}_2])$.

Proposition 1.13:  If $\underset{\sim}{T}$, $\underset{\sim}{T}_1$, $\underset{\sim}{T}_2$ are fuzzy trees, $\underset{\sim}{T} \cap \underset{\sim}{T}_1 = \underset{\sim}{T} \cap \underset{\sim}{T}_2 = \underset{\sim}{T}_1 \cap \underset{\sim}{T}_2 = \wedge$, $f_1$, $f_2 \varepsilon (0,1]$, $(x:n) \varepsilon \underset{\sim}{T}$, $(y:m) \varepsilon \underset{\sim}{T}_1$, then $\mu(\underset{\sim}{T}) \wedge \mu(\underset{\sim}{T}_1) \wedge \mu(\underset{\sim}{T}_2) \wedge f_1 \wedge f_2 \leq \mu(\underset{\sim}{T}[(x:n) \overset{f_1}{\leftarrow} \underset{\sim}{T}_1[(y:m) \overset{f_2}{\leftarrow} \underset{\sim}{T}_2]]) \leq \mu(\underset{\sim}{T}_2) \wedge f_2$.

## III.  FUZZY TREE GRAMMAR

Definition 2.1:  A fuzzy tree grammar (FTG) is a system G, $G = (V_N, V_T, P, \Gamma, J, f)$, satisfying
(1)  $V_N$ is a finite set of nonterminal symbols;
(2)  $V_T$ is a finite set of terminal symbols, and $V_N \cap V_T = \phi$ ;
(3)  $\Gamma$ is a finite set of axioms, where $\Gamma \subset \mathcal{J}(V_N \cup V_T)$;
(4)  P is a finite set of productions of the form $(r)\Phi \to \Psi$, $f(r)$, where  $\Phi, \Psi \varepsilon (V_N \cup V_T)$.
     A production may be a family in which we consider that the second label n of a node (x:n) in $\Phi, \Psi$ is a set;
(5)  $J = \{r\}$ is a finite labeled set of productions in P;
(6)  f is a map from J into (0,1] and is called a membership of production r.

Definition 2.2:  Let G be a FTG, fuzzy trees $\underset{\sim}{T}_1$, $\underset{\sim}{T}_2 \varepsilon \mathcal{J}(V_N \cup V_T)$, $(x:n) \varepsilon \underset{\sim}{T}_1$, then a fuzzy direct deduction $\underset{\sim}{T}_1 \xrightarrow[(x:n), \sigma]{G, f(r)} \underset{\sim}{T}_2$ at node $(x:n)$ from $\underset{\sim}{T}_1$ to $\underset{\sim}{T}_2$ under G, or an fd-deduction $\underset{\sim}{T}_1 \xrightarrow[(r)]{f(r)} \underset{\sim}{T}_2$ from $\underset{\sim}{T}_1$ to $\underset{\sim}{T}_2$ briefly, means $\exists [(r)\Phi \to \Psi f(r)] \varepsilon P, \ni \underset{\sim}{T}_1/(x:n) = \Phi$, $\underset{\sim}{T}_1[(x:n) \overset{f(r)}{\longleftarrow} \Psi] = \underset{\sim}{T}_2$, where (x:n) is a node of this deduction and $\Psi \cap \underset{\sim}{T} = \wedge$ - Don't point out this in next arguments, as we always suppose it.

A fuzzy deduction (or f-deduction) $\underset{\sim}{T} \xrightarrow[(r_1, \ldots, r_k)]{G, (f(r_1), \ldots, f(r_k))} \underset{\sim}{Q}$, or $\underset{\sim}{T} \xrightarrow[(r_1, \ldots, r_k)]{f(r_1), \ldots, f(r_k)} \underset{\sim}{Q}$, or $\underset{\sim}{T} \overset{\#}{\Longrightarrow} \underset{\sim}{Q}$, from $\underset{\sim}{T}$ to $\underset{\sim}{Q}$ means $\exists \underset{\sim}{T}_1, \ldots, \underset{\sim}{T}_{k-1} \varepsilon$

$$\mathcal{J}(V_N \cup V_T), \ni \underset{\sim}{T} \xrightarrow[(r_1)]{f(r_1)} \underset{\sim}{T}_1 \xrightarrow[(r_2)]{f(r_2)} \cdots \xrightarrow[(r_{k-1})]{f(r_{k-1})} \underset{\sim}{T}_{k-1} \xrightarrow[(r_k)]{f(r_k)} \underset{\sim}{Q}.^3$$

This expression is called a $\underset{\sim}{T}$-$\underset{\sim}{Q}$ f-deduction chain. Also, $\underset{\sim}{T}$ is called a start tree, $\underset{\sim}{Q}$ is an end tree, $\underset{\sim}{T}_1,\ldots,\underset{\sim}{T}_{k-1}$ are middle trees of f-deduction chain, and all of the productions $\{(r_i)\Phi_i \to \Psi_i\ f(r_i),\ i=1,\ldots,k\}$ used in each fd-deduction are called a set of productions of an f-deduction chain.

There are several f-deduction chains with start tree $\underset{\sim}{T}$ and end tree $\underset{\sim}{Q}$ under G.

In $\underset{\sim}{T}$-$\underset{\sim}{Q}$ f-deduction chain, $\underset{\sim}{Q}$ is called an end tree of fuzzy tree grammar G and this f-deduction chain is called a $\underset{\sim}{Q}$ f-deduction chain of G, if its start tree $\underset{\sim}{T} \in \Gamma$.

Definition 2.3: The grade of the generation of an end tree of a fuzzy tree grammar G, which is denoted as $\mu_G(\underset{\sim}{Q})$ is given as follow:

$$\mu_G(\underset{\sim}{Q}) = \bigvee_{\substack{T \Rightarrow Q \\ T \in \Gamma}} \bigwedge_{i=1}^{k} \left[f(r_i) \wedge \mu(\Phi_i) \wedge \mu(\Psi_i) \wedge \mu(\underset{\sim}{T})\right],$$

where $\{(r_i)\ \Phi_i \to \Psi_i\ f(r_i),\ i=1,\ldots,k$ is a set of productions in $\underset{\sim}{T} \xrightarrow[(r_1)]{f(r_1)} \underset{\sim}{T}_1 \xrightarrow[(r_2)]{f(r_2)} \cdots \underset{\sim}{T}_{k-1} \xrightarrow[(r_k)]{f(r_k)} \underset{\sim}{Q}$ and "V" means a maximum which is taken over all the f-deduction chains from $T \in \Gamma$ to $\underset{\sim}{Q}$.

If $\forall \underset{\sim}{T} \in \Gamma$, $\bar{\exists}\ \underset{\sim}{T}$-$\underset{\sim}{Q}$ f-deduction chain, we shall write $\mu_G(\underset{\sim}{Q})=0$.

The grade $\mu_\beta(\underset{\sim}{Q})$ of the generation of an end tree of an f-deduction chain $\beta$ is given

$$\mu_\beta(\underset{\sim}{Q}) = \bigwedge_{i=1}^{k} \left[f(r_i) \wedge \mu(\Phi_i) \wedge \mu(\Psi_i) \wedge \mu(\underset{\sim}{T})\right]$$

where $\{(r_i)\Phi_i \to \Psi_i\ f(r_i),\ i=1,\ldots,k\}$ is a set of productions in $\beta$ :

$$T \xrightarrow[(r_1)]{f(r_1)} T_1 \xrightarrow[(r_2)]{f(r_2)} \cdots T_{k-1} \xrightarrow[(r_k)]{f(r_k)} Q.$$

Clearly, $\exists \underset{\sim}{Q}$ f-deduction chain $\beta$, $\ni \mu_G(\underset{\sim}{Q}) = \mu_\beta(\underset{\sim}{Q})$, if $\underset{\sim}{Q}$ is an end tree of G.

Proposition 2.1: $\underset{\sim}{T}$, $\underset{\sim}{Q} \in \mathcal{J}(V_N \cup V_T)$, $(x{:}n)$, $(y{:}m) \in \underset{\sim}{T}$, and $(x{:}n) < (y{:}m)$, $\underset{\sim}{T} \xrightarrow[(y:m)(r)]{f(r)} \underset{\sim}{Q}$, then $\underset{\sim}{T}/(x{:}n) \xrightarrow[(y:m)(r)]{f(r)} \underset{\sim}{Q}/(x{:}n)$.

Proof: $\underset{\sim}{T} \xrightarrow[(y:m)(r)]{f(r)} \underset{\sim}{Q}$, $\exists [(r)\Phi \to \Psi\ f(r)] \in P$, $\ni \underset{\sim}{T}/(y{:}m) = \Phi$,

$\underset{\sim}{T}[(y:m) \xleftarrow{f(r)} \Psi] = \underset{\sim}{Q}$.

Since $(x:n) < (y:m)$, by proposition 1.4, $[\underset{\sim}{T}/(x:n)]/(y:m) = \underset{\sim}{T}/(y:m) = \Phi$, and by proposition 1.5',

$$[\underset{\sim}{T}/(x:n)][(y:m) \xleftarrow{f(r)} \Psi] = \underset{\sim}{T}[(y:m) \xleftarrow{f(r)} \Psi]/(x:n) = \underset{\sim}{Q}/(x:n),$$

then, $\underset{\sim}{T}/(x:n) \underset{(y:m),(r)}{\overset{f(r)}{\Longrightarrow}} \underset{\sim}{Q}/(x:n)$. ¤

Proposition 2.2: $\underset{\sim}{T}, \underset{\sim}{Q}, \underset{\sim}{U} \in \mathcal{J}(V_N \cup V_T)$, $(x:n) \in \underset{\sim}{U}$, $f_1 \in (0,1]$, $T \underset{(r)}{\overset{f(r)}{\Longrightarrow}} Q$, then $\underset{\sim}{U}[(x:n) \xleftarrow{f_1} \underset{\sim}{T}] \underset{(r)}{\overset{f(r)}{\Longrightarrow}} \underset{\sim}{U}[(x:n) \xleftarrow{f_1} \underset{\sim}{Q}]$.

Proof: $\underset{\sim}{T} \underset{(r)}{\overset{f(r)}{\Longrightarrow}} \underset{\sim}{Q}$, $\exists (y:m) \in \underset{\sim}{T}$ and $[(r)\Phi \to \Psi \; f(r)] \in P$, $\ni \underset{\sim}{T}/(y:m) = \Phi$ and $\underset{\sim}{T}[(y:m) \xleftarrow{f(r)} \Psi] = \underset{\sim}{Q}$.

Since $\underset{\sim}{U}[(x:n) \xleftarrow{f_1} \underset{\sim}{T}]/(y:m) = \underset{\sim}{T}/(y:m) = \Phi$ , by proposition 1.7'

$$\underset{\sim}{U}[(x:n) \xleftarrow{f_1} \underset{\sim}{T}][(y:m) \xleftarrow{f(r)} \Psi] = \underset{\sim}{U}[(x:n) \xleftarrow{f_1} \underset{\sim}{T}[(y:m) \xleftarrow{f(r)} \Psi]]$$
$$= \underset{\sim}{U}[(x:n) \xleftarrow{f_1} \underset{\sim}{Q}],$$

thus, $\underset{\sim}{U}[(x:n) \xleftarrow{f_1} \underset{\sim}{T}] \underset{(r)}{\overset{f(r)}{\Longrightarrow}} \underset{\sim}{U}[(x:n) \xleftarrow{f_1} \underset{\sim}{Q}]$. ¤

Proposition 2.3: $\underset{\sim}{T}, \underset{\sim}{Q}, \underset{\sim}{U} \in \mathcal{J}(V_N \cup V_T)$, $(x:n)$, $(y:m) \in \underset{\sim}{T}$ and $(y:m) \nless (x:n)$, $(x:n) \nless (y:m)$, $f_1 \in (0,1]$, $\underset{\sim}{T} \underset{(y:m),(r)}{\overset{f(r)}{\Longrightarrow}} \underset{\sim}{Q}$, then $\underset{\sim}{T}\left[(x:n) \xleftarrow{f_1} \underset{\sim}{U}\right] \underset{(y:m),(r)}{\overset{f(r)}{\Longrightarrow}} \underset{\sim}{Q}\left[(x:n) \xleftarrow{f_1} \underset{\sim}{U}\right]$.

Proof: $\underset{\sim}{T} \underset{(y:m),(r)}{\overset{f(r)}{\Longrightarrow}} \underset{\sim}{Q}$, $\exists [(r)\Phi \to \Psi \; f(r)] \in P$, $\ni \underset{\sim}{T}/(y:m) = \Phi$ ,

$\underset{\sim}{T}[(y:m) \xleftarrow{f(r)} \Psi] = \underset{\sim}{Q}$.

Since $(y:m) \nless (x:n)$, $(x:n) \nless (y:m)$, by proposition 1.10'

$\underset{\sim}{T}[(x:n) \xleftarrow{f_1} \underset{\sim}{U}]/(y:m) = \underset{\sim}{T}/(y:m) = \Phi$ , and by proposition 1.9'

$$\underset{\sim}{T}[(x:n) \xleftarrow{f_1} \underset{\sim}{U}][(y:m) \xleftarrow{f(r)} \Psi] = \underset{\sim}{T}[(y:m) \xleftarrow{f(r)} \Psi][(x:n) \xleftarrow{f_1} \underset{\sim}{U}]$$
$$= \underset{\sim}{Q}\left[(x:n) \xleftarrow{f_1} \underset{\sim}{U}\right],$$

then $\underset{\sim}{T}[(x:n) \xleftarrow{f_1} \underset{\sim}{U}] \underset{(y:m),(r)}{\overset{f(r)}{\Longrightarrow}} \underset{\sim}{Q}[(x:n) \xleftarrow{f_1} \underset{\sim}{U}]$. ¤

Corollary: $\underset{\sim}{T}$, $\underset{\sim}{Q}$, $\underset{\sim}{U}$, $\underset{\sim}{V} \in \mathcal{J}(V_N \cup V_T)$, $(x_1:n_1)$, $(x_2:n_2) \in \underset{\sim}{U}$, $(x_3:n_3) \in \underset{\sim}{T}$ and $(x_1:n_1) \nleq (x_2:n_2)$, $(x_2:n_2) \nleq (x_1:n_1)$,

$\underset{\sim}{T} \xrightarrow[(x_3:n_3)(r_3)]{f(r_3)} \underset{\sim}{Q}$, $\underset{\sim}{U} \xrightarrow[(x_2:n_2)(r_2)]{f(r_2)} \underset{\sim}{V}$, $f_1 \in (0,1]$, then $\underset{\sim}{U}[(x_1:n_1) \xleftarrow{f_1} \underset{\sim}{T}]$

$\xrightarrow[r_3, r_2]{f(r_1), f(r_2)} \underset{\sim}{V}\ [(x_1:n_1) \xleftarrow{f_1} \underset{\sim}{Q}]$ and $\underset{\sim}{U}\ (x_1:n_1) \xleftarrow{f_1} \underset{\sim}{T}] \xrightarrow[r_2, r_3]{f(r_2), f(r_3)} \underset{\sim}{V}[(x_1:n_1)$

$\xleftarrow{f_1} \underset{\sim}{Q}]$.

Definition 2.4: A fuzzy tree language (FTL) generated by a fuzzy tree grammar $G = (V_N, V_T, P, \Gamma, J, f)$ is a fuzzy subset $L(G)$ of set $\mathcal{J}(V_T)$, where $\mu_{L(G)}(\underset{\sim}{Q}) = \mu_G(\underset{\sim}{Q})$ for all $\underset{\sim}{Q} \in \mathcal{J}(V_T)$. Also, $L(G)$ may be called a fuzzy tree language over $V_T$.

Let the fuzzy subsets $L$, $L_1$ of $(V_T)$, where

$$\mu_{L_1}(\underset{\sim}{T}) = \begin{cases} a\ (>0) & \underset{\sim}{T} = \underset{\sim}{Q} \\ 0 & \underset{\sim}{T} \neq \underset{\sim}{Q} \end{cases}, \quad T \in \mathcal{J}(V_T)$$

if $\mu_{L_1}(\underset{\sim}{Q}) = \mu_L(\underset{\sim}{Q})$, then we say $(\underset{\sim}{Q}, \mu_{L_1}(\underset{\sim}{Q})) \in L$ or immediately $\underset{\sim}{Q} \in L$.

Otherwise $(\underset{\sim}{Q}, \mu_{L_1}(\underset{\sim}{Q})) \bar{\in} L$ or $\underset{\sim}{Q} \bar{\in} L$. If the fuzzy subsets $L, M$ of $\mathcal{J}(V_T)$, satisfying $\forall \underset{\sim}{Q} \in L$ we have $\underset{\sim}{Q} \in M$, then we write $L \Subset M$. Clearly, $L \Subset M$, $M \Subset L$ iff $L=M$; if $L \Subset M$, we get $L \subset M$.

The fuzzy tree grammars $G_1$ and $G_2$ are equivalent, if $L(G_1) = L(G_2)$.

The fuzzy tree grammars are isomorphism, if there exists a map $g$ from $G_1 = (V_{N_1}, V_{T_1}, P_1, \Gamma_1, J_1, f_1)$ into $G_2 = (V_{N_2}, V_{T_2}, P_2, \Gamma_2, J_2, f_2)$, such that $g(\Gamma_1) = \Gamma_2$, $g(J_1) = J_2$, $g(f_1) = f_2$ and $g$ is bijective.

If we also have $g(V_{N_1}) = V_{N_2}$, $g(V_{T_1}) = V_{T_2}$, then $G_1$ and $G_2$ are str-isomorphism, we don't distinguish them.

If $G_1$ and $G_2$ are isomorphism (or str-isomorphism), then $L(G_1)$ and $L(G_2)$ are also isomorphism (or str-isomorphism).

Theorem 2.4: All of fuzzy tree languages over set $\Sigma$ is called a family of fuzzy tree languages over $\Sigma, \pi$, a family of fuzzy tree languages, is closed under the union operation.

Proof: $\forall L_1, L_2 \in \pi$, $\exists G_1 = (V_{N_1}, \Sigma, P_1, \Gamma_1, J_1, f_1)$, $G_2 = (V_{N_2},$

$\Sigma$, $P_2, \Gamma_2, J_2, f_2)$. $G_1$, $G_2$ are FTG. Let $L(G_1) = L_1$, $L(G_2) = L_2$ and $V_{N_1} \cap V_{N_2} = J_1 \cap J_2 = \phi$ . Being supposed for $\forall [(r_i)\Phi_i \to \Psi_i \; f_i(r_i)] \varepsilon P$, we have $\Phi_i \bar{\varepsilon} \mathcal{J}(\Sigma)$. Otherwise, adding # into $V_{Ni}$, where $\# \bar{\varepsilon} V_{n_i} \cup \Sigma$ . If $R(\Phi_i) = a \varepsilon \Sigma$, then replaced a by # in $\Gamma_i$ and $P_i$, and put $[\# \to a] \varepsilon P_i$, thus this grammar and given grammar $G_i$ are equivalent (i=1,2).

Now, take $G = (V_{N_1} \cup V_{N_2}, \Sigma, P_1 \cup P_2, \Gamma_1 \cup \Gamma_2, J_1 \cup J_2, f)$ where

$$f(r) = \begin{cases} f_1(r) & r \varepsilon J_1 \\ f_2(r) & r \varepsilon J_2 \end{cases},$$

then G is an FTG too.

Let $L = L(G)$, then $L \varepsilon \pi$.

$Q \varepsilon \mathcal{J}(\Sigma)$, if $Q \varepsilon L$, then $\mu_L(Q) = \mu_G(Q) > 0$, i.e., there exists an f-deduction chain $T \Rightarrow \# Q$, where $\mu_G(Q) > 0$, $T \varepsilon \Gamma_1 \cup \Gamma_2$, the set of productions is $A = \{(r_{ij})\Phi_{ij} \to \Psi_{ij} \; f_i(r_{ij})\} \subset P_1 \cup P_2$, i=1 or 2, j=1,...,k.

If $T \varepsilon \Gamma_1$, by the construction of G, $A \subset P_1$, thus, this f-deduction chain is f-deduction chain under $G_1$. Thus $(Q, \mu_L(Q)) \varepsilon L_1$. Clearly, $\mu_{L_2}(Q) \le \mu_L(Q)$. Then $Q \varepsilon L_1 \cup L_2$.

If $T \varepsilon \Gamma_2$, it may be proven as same as foregoing reason. Thus $L \Subset L_1 \cup L_2$.

Since it may be converse, then we have $L_1 \cup L_2 \Subset L$ too. ¤

Definition 2.5: A concatenated tree of a fuzzy tree $T_1$ joined by a fuzzy tree $T_2$ at node (x:n) with concatenated grade $f_1$, or $T_1$ concatenated by $T_2$ at (x:n), is a fuzzy tree $C(T_1;T_2)$

$$C(T_1;T_2) = T_1[(x:n) \xrightarrow{f_1} T_2] = \{W(s_1,y) \mid W(s_1,y) \subset T_1\} \cup$$

$$\{W(s_1,x) \cup W(x,s_2) \cup W(s_2,z) \mid W(s_1,x) \subset T_1,\ W(s_2,z) \subset T_2,$$

$$W(x,s_2) = \langle (x:n),\ ((x:n)(s_2:1),\ f_1),\ (s_2:1) \rangle\}$$

where $T_1 \cap T_2 = \Lambda$, $(x:n) \varepsilon T_1$, $R(T_1) = (s_1:1)$, $R(T_2) = (s_2:1)$, $f_1 \varepsilon (0,1]$, above node (x:n) is called concarenated point. We assign $T_1[(x:n) \xrightarrow{0} T_2] = T_1$.

The concatenation concept may be extended to several trees.

A concatenated tree of a fuzzy tree $T$ joined by fuzzy trees

$\underset{\sim}{T}_1, \underset{\sim}{T}_2, \ldots, \underset{\sim}{T}_k$ at nodes $(x_1{:}n_1), (x_2{:}n_2), \ldots, (x_k{:}n_k)$ with concatenated grade $f_1, f_2, \ldots, f_k$, or $\underset{\sim}{T}$ concatenated by $\underset{\sim}{T}_1, \underset{\sim}{T}_2, \ldots, \underset{\sim}{T}_k$ at $(x_1{:}n_1), (x_2{:}n_2), \ldots, (x_k{:}n_k)$, is a fuzzy tree $C(\underset{\sim}{T}{:}\underset{\sim}{T}_1, \ldots, \underset{\sim}{T}_k) = \underset{\sim}{T}[(x_1{:}n_1) \xrightarrow{f_1} \underset{\sim}{T}_1][(x_2{:}n_2) \xrightarrow{f_2} \underset{\sim}{T}_2] \cdots [(x_k{:}n_k) \xrightarrow{f_k} \underset{\sim}{T}_k]$, where $\underset{\sim}{T} \cap \underset{\sim}{T}_1 = \underset{\sim}{T} \cap \underset{\sim}{T}_2 = \cdots = \underset{\sim}{T}_{k-1} \cap \underset{\sim}{T}_k = \Lambda$, $(x_i{:}n_i) \neq (x_j{:}n_j)$, $1 \leq i \neq j \leq k$, $f_1, \ldots, f_k \in (0,1]$. But, if it doesn't state to the contrary, we usually assume $f_i > 0$ $(i=1,\ldots,k)$.

$f = (f_1, \ldots, f_k)$ is called concatenated grade vector.

Proposition 2.5: (1) Let $\underset{\sim}{T}_1, \underset{\sim}{T}_2$ be fuzzy trees, $(y_1{:}m_1)$, $(y_2{:}m_2)$, $(x{:}n) \in \underset{\sim}{T}_1$, $(z{:}t) \in \underset{\sim}{T}_2$, $f_1 \in (0,1]$, if $(y_1{:}m_1) < (x{:}n)$, $(y_2{:}m_2) \nless (x{:}n)$, then $(y_1{:}m_1) < (z{:}t)$ and $(y_2{:}m_2) \nless (z{:}t)$ in $T_1[(x{:}n) \xrightarrow{f_1} T_2]$;

(2) Let $\underset{\sim}{T}$, $\underset{\sim}{T}_1$, $\underset{\sim}{T}_2$ be fuzzy trees, $(x_1{:}n_1), (x_2{:}n_2) \in \underset{\sim}{T}$, $f_1$, $f_2 \in (0,1]$, then $\underset{\sim}{T}[(x_1{:}n_1) \xrightarrow{f_1} \underset{\sim}{T}_1][(x_2{:}n_2) \xrightarrow{f_2} \underset{\sim}{T}_2] = \underset{\sim}{T}[(x_2{:}n_2) \xrightarrow{f_2} \underset{\sim}{T}_2][(x_1{:}n_1) \xrightarrow{f_1} \underset{\sim}{T}_1]$;

(3) Let $\underset{\sim}{T}$, $\underset{\sim}{T}_1$, $\underset{\sim}{T}_2$ be fuzzy trees, $(x{:}n) \in \underset{\sim}{T}$, $(y{:}m) \in \underset{\sim}{T}_1$, $f_1$, $f_2 \in (0,1]$, then $\underset{\sim}{T}[(x{:}n) \xrightarrow{f_1} \underset{\sim}{T}_1[(y{:}m) \xleftarrow{f_2} \underset{\sim}{T}_2]] = \underset{\sim}{T}[(x{:}n) \xrightarrow{f_1} \underset{\sim}{T}_1][(y{:}m) \xleftarrow{f_2} \underset{\sim}{T}_2]$;

(4) Let $\underset{\sim}{T}$, $\underset{\sim}{T}_1$, $\underset{\sim}{T}_2$ be fuzzy trees, $(x_1{:}n_1), (x_2{:}n_2) \in \underset{\sim}{T}$ $(x_1{:}n_1) \nless (x_2{:}n_2)$, $f_1, f_2 \in (0,1]$, then $\underset{\sim}{T}[(x_1{:}n_1) \xleftarrow{f_1} \underset{\sim}{T}_1][(x_2{:}n_2) \xrightarrow{f_2} \underset{\sim}{T}_2] = \underset{\sim}{T}[(x_2{:}n_2) \xrightarrow{f_2} \underset{\sim}{T}_2][(x_1{:}n_1) \xleftarrow{f_1} \underset{\sim}{T}_1]$.

These conclusions are obvious.

Definition 2.6: $L_1$ is concatenated by $L_2$, which means there exists a fuzzy subset $C_{<f>}(L_1;L_2)$ of $(\Sigma)$, where $L_1$, $L_2$ are fuzzy trees over $\Sigma$, $C_{<f>}(L_1;L_2) = \{(C(\underset{\sim}{T}_1;\underset{\sim}{T}_2,\ldots,\underset{\sim}{T}_\ell), \mu_{c_{<f>}(L_1;L_2)}(C(\underset{\sim}{T}_1;\underset{\sim}{T}_2,\ldots,\underset{\sim}{T}_\ell))) \mid \underset{\sim}{T}_1 \in L_1, \underset{\sim}{T}_2,\ldots,\underset{\sim}{T}_\ell \in L_2, 1 \leq \ell-1 \leq$ number of nodes of $\underset{\sim}{T}_1$, $f = (f_2, f_3, \ldots, f_\ell)$ is concatenated grade vector of $C(\underset{\sim}{T}_1; \underset{\sim}{T}_2, \ldots, \underset{\sim}{T}_\ell)$, these concatenated grade vector are different in the dimension and the value fo component if we take the different fuzzy trees $\underset{\sim}{T}_i$ $(i=1,2,\ldots,1)$ and the different nodes of $\underset{\sim}{T}_1$ as the concatenated points, but we suppose that the set of all $f_i$, when taking the different concatenated grade vector, is a finite set $<f>$ in (0,1 , and $L_1 \Subset C_{<f>}(L_1;L_2)$, $\mu_{c_{<f>}(L_1;L_2)}(\underset{\sim}{Q}) = \mu_{c_{<f>}(L_1;L_2)}(C(T_1;T_2,\ldots,T_\ell))$

$$= \bigvee_{\substack{Q=C(T_1;T_2,\ldots,T_\ell)\\ \forall T_1 \in L_1, T_2,\ldots,T_\ell \in L_2}} \{\mu_{L_1}(T_1) \wedge [\bigwedge_{2\le i\le \ell} \mu_{L_2}(T_i) \wedge f_i]\}.$$

$C_{<f>}$ is called concatencated operation between $L_1$ and $L_2$, too.

Theorem 2.6: $\pi$, the family of fuzzy tree languages over $\Sigma$, is closed under the concatenated operation.

Proof: $\forall L_1, L_2 \in \pi, \exists G_1 = (V_{N_1}, \Sigma, P_1, \Gamma_1, J_1, f_1)$, $G_2 = (V_{N_2}, \Sigma, P_2, \Gamma_2, J_2, f_2)$, $G_1, G_2$ are FTG. Let $L(G_1) = L_1$, $L(G_2) = L_2$, and $V_{N_1} \cap V_{N_2} = J_1 \cap J_2 = \phi$. Also, same as in Theorem 2.4 we suppose

$$\forall [(r_i)\Phi_i \to \Psi_i \; f_i(r_i)] \in P_i \;, \text{ where } \Phi_i \,\bar{\varepsilon}\, \mathcal{J}(\Sigma),\; i=1,2.$$

Let $G = (V_{N_1} \cup V_{N_2}, \Sigma, P_1 \cup P_2 \cup P', \Gamma_1 \cup \Gamma', J_1 \cup J_2 \cup J', f)$, where $\Gamma' = \{T^{(1)}[(x:n) \xrightarrow{f_i} T^{(2)}] \mid \forall T^{(1)} \in \Gamma_1, T^{(2)} \in \Gamma_2, \forall (x:n) \in T^{(1)}, \forall f_i \in <f>\}$, and $[(r^{(1)}, T^{(2)}, (x:n), f_i) T^{(1)} \to T_1^{(1)}[(x:n) \xrightarrow{f_i} T^{(2)}] f(r')$, where $(r') = (r^{(1)}, T^{(2)}, (x:n), f_i)$, $f_i \varepsilon <f>$, $(x:n) \in T_1^{(1)}] \in P'$ iff $[(r^{(1)}) T^{(1)} \to T_1^{(1)} f_i(r^{(1)})] \in P_1$ and $T^{(2)} \in \Gamma_2$,

$J' = \{r' \mid (r') = (r^{(1)}, T^{(2)}, (x:n), f_i), r^{(1)} \in J_1, T^{(2)} \in \Gamma_2$, $(x:n) \in T^{(2)}, f_i \in <f>\}$

$$f(r) = \begin{cases} f_1(r) & r \in J_1 \\ f_2(r) & r \in J_2 \\ f_1(r^{(1)}) \wedge \mu(T^{(2)}) \wedge f_i, & (r) = (r^{(1)}, T^{(2)}, (x:n), f_i) r \varepsilon J' \end{cases}$$

then G is an FTG. Let L=L(G), then $L \in \pi$.

$\forall Q \in L$, $\mu_L(Q) = \mu_G(Q) > 0$, $\exists$ f-deduction chain $\beta$ :

$$T \overset{G}{\Longrightarrow} T_1 \overset{G}{\Longrightarrow} \ldots \overset{G}{\Longrightarrow} T_k \overset{G}{\Longrightarrow} Q, \ni \mu_\beta(Q) = \mu_L(Q).$$

Let the set of productions of chain $\beta$ be A. If $A \cap P' = \phi$ and $T \in \Gamma$, then $\beta$ is an f-deduction chain under $G_1$, thus $(Q, \mu_G(Q)) \in L_1 \subseteq C_{<f>}(L_1;L_2)$.

If it, $A \cap P' = \{(r_0^{(1)}, T_0^{(2)}, (x_0:n_0), f_0) T_0^{(1)} \to T_2^{(1)}[(x_0:n_0) \xrightarrow{f_0} T_0^{(2)}] f(f_0')\}$, has one and only one production, then $A = A_1 \cup A_2$ {production $r_0'$}, where $A_1 \subset P_1$, $A_2 \subset P_2$. Thus f-deduction chain $\beta$ may be resolved into f-deduction $\beta_1$ under $G_1$ with the start tree $T$, the end tree $Q_1 \varepsilon L_1$ and the set of productions $A_1$, and f-deduction $\beta_2$ under $G_2$ with the start tree $T_0^{(2)}$, the end tree $Q_2 \varepsilon L_2$ and the set of productions $A_2$.

Thus $Q = Q_1[(x_0:n_0) \xrightarrow{f_0} Q_2]$, $\mu_L(Q) = \mu_{L_1}(Q_1) \wedge f_0 \wedge \mu_{L_2}(Q_2)$ $= \mu_{C_{<f>}(L_1;L_2)}(Q)$, i.e., $Q \varepsilon C_{<f>}(L_1;L_2)$.

By induction, we may prove that the conclusion is true, when the productions of set $A \cap P'$ are more than one.

Its proof is same as above, for $A \cap P' = \phi$, $T \varepsilon \Gamma'$. Thus, $L \Subset C_{<f>}(L_1;L_2)$. It is obvious for $C_{<f>}(L_1;L_2) \Subset L$.

Now, we introduce the concept of tree recursion as follow:

Any fuzzy tree $T$ can be written as $R(T)[(T_1,\mu_1),\ldots,(T_n,\mu_n)]$, where n is a number of out-edges of $R(T)$, $R(T_i)(i=1,\ldots,n)$ is the direct descendants of $R(T)$, its signs are ordered from the left to the right, and $T_i = T/R(T_i)$, $\mu_i = \mu(R(T)R(T_i))$, $i=1,\ldots,n$.

The expression, $T=R(T)[(T_1,\mu_1),\ldots,(T_n,\mu_n)]$, is called tree recursion of $T$.

Definition 2.7: A drawn leaves operation about a fuzzy tree is a map $\mathbb{H}$ that is defined as follow:

$\mathbb{H}$ is a map from a fuzzy subset, $D = \{(T,\mu_D(T)) \mid T \varepsilon \mathcal{J}(B)$, $0 \le \mu_D(T) \le 1\}$, of set $\mathcal{J}(B)$ into a fuzzy subset E, $E = \{(\alpha, \mu_E(\alpha)) \mid \alpha \varepsilon B^*$, $0 \le \mu_E(\alpha) \le 1$, of set $B^*$, where the elements of set $B^*$ are constituted by all strings, with a finite length, over B. Satisfying

(1) $\mathbb{H}(\Lambda) = \varepsilon$ (none string)

$$\mathbb{H}(R(T)\ (T_1,\mu_1),\ldots,(T_n,\mu_n)\ ) = \begin{cases} R(T) & \text{if } T_1=T_2=\ldots=T_n=\Lambda \\ \mathbb{H}(T_1)\mathbb{H}(T_2)\ldots\mathbb{H}(T_n) & \text{otherwise} \end{cases}$$

and if $T_i = \Lambda$, $T_j \ne \Lambda (j=1,2,\ldots,i-1,i+1,\ldots,n)$, then $\mathbb{H}(T_1)\mathbb{H}(T_2)\ldots\mathbb{H}(T_{i-1})\mathbb{H}(T_i)\mathbb{H}(T_{i+1})\ldots\mathbb{H}(T_n) = \mathbb{H}(T_1)\ldots\mathbb{H}(T_{i-1})\mathbb{H}(T_{i+1})\ldots\mathbb{H}(T_n)$;

(2) $$\mu_E(\mathbb{H}(T)) = \bigvee_{\substack{\mathbb{H}(Q)=\mathbb{H}(T) \\ Q \varepsilon D}} \mu_D(Q)$$

E. S. Santos (1974) studied a sort of context-free fuzzy language, following definition 2.8, definition 2.9 and definition 2.10 are adopted there, but here we simplify some of them for convenience to our discussion.

Definition 2.8 (Santos, 1974): A context-free fuzzy grammar (CMG) is a system $G=(V_N, V_T, P,h,J,f)$ satisfying

(1) $V_N$ is a finite set of nonterminal symbols;

(2) $V_T$ is a finite set of terminal symbols, and $V_N \cap V_T = \phi$ ;

(3) P is a finite set of productions of the form $(r)A \rightarrow \alpha f(r)$, where $A \in V_N$, $\alpha \in (V_T \cup V_N)^*$;

(4) $J=\{r\}$ is a finite labeled set of productions in P;

(5) f is a map from J into (0,1] and is called a membership of production r;

(6) h is a map from $V_N$ into (0,1] and is called a membership of element of set $V_N$.

Definition 2.9 (Santos, 1974): $G=(V_N,V_T,P,h,J,f)$ is a CMG,$\alpha$, $\beta \in (V_N \cup V_T)^*$, a deduction $\alpha \overset{p}{\Longrightarrow}\# \beta$ from $\alpha$ to $\beta$ where the set of productions $[(r_i)A_i \rightarrow \alpha_i\ f(r_i) = p_i] \in P$ $(i=1,\ldots,n)$, $p=p_1 \wedge p_2 \wedge \ldots \wedge p_n$ iff $\gamma_i \in (V_T \cup V_N)^*$, $(i=1,\ldots,n-1)$,$\gamma_0 = \alpha$, $\gamma_n = \beta$, $\gamma_i$ is given by that $\alpha_i$ replaces A which is the place of number $k_i$ in $\gamma_{i-1}$, and called direct deduction of $\gamma_{i-1}$. If this direct deduction is existed, we denote it by $\gamma_{i-1} \overset{p_i}{\Longrightarrow} \gamma_i$ , otherwise by $\alpha \overset{0}{\Longrightarrow}\# \beta$ .

If $\alpha \overset{p}{\Longrightarrow}\# \beta$, we write $g(\beta|\alpha) = p$.

Definition 2.10 (Santos, 1974): $G=(V_N,V_T,P,h,J,f)$ is a CMG, a fuzzy subset L(G) of $V_{T*}$ is called a context-free fuzzy language (CFFL), if $L(G) = \{(\alpha,\mu_{L(G)}(\alpha)) | \alpha \in V_{T*}\ ,\ \mu_{L(G)}(\alpha) = \bigvee_{\substack{A \in V_N \\ A \Rightarrow \# \alpha}} [h(A) \wedge g(\alpha|A)]\}$.

Following theorem 2.7 and theorem 2.9 we can establish the connection between the family of fuzzy tree languages and the context-free fuzzy language as found by Santos.

Theorem 2.7: If a fuzzy subset $L_1$ over $\Sigma^*$ is a CFFL, then there exists a fuzzy subset L of $\mathcal{J}(\Sigma \cup \{\#\})$ such that L is a FTL and $\mathbb{H}(L) = L_1$.

Proof: Since L is a CFFL, thus we have $G_1 = (V_N,\Sigma, P_1,h,J_1,f_1)$ which is a CMG, and $L(G_1) = L_1$.

Let $\# \bar{\in} \Sigma \cup V_N$, # is labeled, but for the simplicity, we abbreviate its second label in the following discussion.

Set up $G=(V_N,\Sigma \cup \{\#\}, P,\Gamma, J_1,f)$, where $\Gamma = \{\#[(A,h(A))] | \forall A \in V_N$ , and $[(r)A \to \#[(u_1, f_1(r)), (u_2, f_1(r)), \ldots, (u_k, f_1(r))]\ f(r) = 1] \in P$ iff $[(r)A \to \alpha\ f_1(r), \alpha = u_1, u_2 \ldots u_k, u_i (i=1,\ldots,k) \in V_N \cup \Sigma] \in P_1$, then $G$ is a FTG. Let $L=L(G)$. $\forall Q \in L$, we may suppose that

$$\mu_{\mathbb{H}(L)}(\mathbb{H}(Q)) = \mu_L(Q) = \mu_G(Q),$$

then $\exists Q$ f-deduction chain $\beta$ under G:

$$A'_1 \overset{1}{\underset{(r_1)}{\Longrightarrow}} T'_1 \overset{1}{\underset{(r_2)}{\Longrightarrow}} \cdots \overset{1}{\underset{(r_{k-1})}{\Longrightarrow}} T'_{k-1} \overset{1}{\underset{(r_k)}{\Longrightarrow}} Q,$$

where $A' = \#[(A_1, h(A_1))]$, $A_1 \in V_N$, and the set of productions is $\{(r_j)A_j \to \#[(u_{j,1}, f_1(r_j)), \ldots, (u_j, t_j, f_1(r_j))]\ j=1,\ldots,k\}$, $\mu_\beta(Q) = \mu_G(Q) = \bigwedge_{1 \le j \le k} [h(A_1) \wedge f_1(r_j)]$.

By the constructions of $G_1$ and G, we can see from above that the set of productions in P may be corresponded the set productions in $P_1$ one by one as follow:

$$[(r_j)A_j \to \#[(u_{j,1}, f_1(r_j)), \ldots, (u_j, t_j, f_1(r_j))]]$$

$$\Longleftrightarrow [(r_j)A_j \to \alpha_j\ f_1(r_j),\ \alpha_j = u_{j,1}, \ldots, u_j, t_j]\ (j=1,\ldots,k),$$

thus, we have $A_1 \overset{f_1(r_1)}{\underset{(r_1)}{\Longrightarrow}} \alpha'_1 \overset{f_1(r_2)}{\underset{(r_2)}{\Longrightarrow}} \alpha'_2 \overset{f_1(r_3)}{\underset{(r_3)}{\Longrightarrow}} \cdots \overset{f_1(r_{k-1})}{\underset{(r_{k-1})}{\Longrightarrow}} \alpha'_{k-1} \overset{f_1(r_k)}{\underset{(r_k)}{\Longrightarrow}} \alpha'_k$

and $\mu_{L_1}(\alpha'_k) = h(A_1) \wedge g(\alpha'_k \mid A) = \bigwedge_{1 \le j \le k} [h(A_1) \wedge f_1(r_j)]$.

Now, by the constructions of $\Gamma$ and productions in P, thus $\mathbb{H}(Q) = \alpha'_k$, i.e., $\mathbb{H}(L) \Subset L_1$. Since this process is converse, we have $\mathbb{H}(L) = L_1$. ¤

Brainerd (1969) had proven next theorem to common tree grammar (see the theorem 3.16 in (2)).

Theorem 2.8: For each tree grammar, we can construct an equivalent expansive tree grammar (i.e., one which has a single nonterminal axiom and whose productions are all of the form $x \to a[x_1, \ldots, x_n]$, where $x_1, \ldots, x_n \in V_N \cup \{f\}$, $a \in V_T$, $x \in V_N$, and providing $a[x_1, \ldots, x_{i-1}, f, x_{i+1}, \ldots, x_n] = a[x_1, \ldots, x_{i-1}, x_{i+1}, \ldots, x_n]$).

For the fuzzy tree grammar, repeating the method raised by Brainerd, we have

Theorem 2.8': For each fuzzy tree grammar, we can construct an equivalent expansive fuzzy tree grammar (i.e., one which has a single nonterminal axiom and whose productions are all of the form

$(r)x \to a[(x_1,\mu_1), (x_2,\mu_2),\ldots,(x_n,\mu_n)]f(r)$, where $\mu_1,\ldots,\mu_n \in (0,1]$, others are the same as Theorem 2.8).

Theorem 2.9: A fuzzy subset L of $\mathcal{J}(\Sigma)$ is an FTL, then there exists a fuzzy subset $L_1$ of $\Sigma^*$, satisfying $L_1$ is a CFFL and $\mathbb{H}(L) = L_1$.

Proof: Since L is an FTL, we may establish an FTG: $G = (V, \Sigma, P, \Gamma, J, f)$.

By Theorem 2.8', we may suppose that G is an expansive fuzzy tree grammar.

Let $s \in \Gamma$, $0 < \mu(s) \leq 1$, $G_1 = (V,\Sigma,P_1,h,J_1,f_1)$, where $P_1 = \{(r)x \to x_1\ldots x_n \; f_1(r) \mid \forall [(r)x \to a[(x_1,\mu_1),\ldots,(x_n,\mu_n)],\ f(r)] \in P,$ $f_1(r) = \bigwedge_{1 \leq i \leq n} [\mu_i \wedge f(r)]\} \cup \{(r')x \to s \; f_1(r') = 1 \mid x \in V,\ x \neq s,\ r' = \langle\langle x \rangle\rangle\}$,
$h: V \to (0,1]$, satisfying

$$h(x) = \begin{cases} \mu(s) & \text{if } x = s \\ 1 & \text{if } x \in V \text{ and } x \neq s, \end{cases}$$

$J_1 = J \cup \{r' \mid r' = \langle\langle x \rangle\rangle,\ x \in V\}$, then $G_1$ is a CMG. Let $L(G_1) = L_1$. The following is only a repeatition about the proof of Theorem 2.7, so we have $\mathbb{H}(L) = L_1$. ¤

For the context-free fuzzy language under the sense of Lee and Zadeh, the Theorem 2.7 and Theorem 2.9 can be established as the same as above.

Definition 2.11: For $\lambda \in [0,1)$, a $\lambda$-cut of the fuzzy subset L of $\mathcal{J}(A)$ is a subset $C(L,\lambda)$ of $A\#$, satisfying $T \in C(L,\lambda)$ iff T is a basic tree of $\underset{\sim}{T}$, $\underset{\sim}{T} \in L$ and $\mu_L(\underset{\sim}{T}) > \lambda$.

Proposition 2.10: If $0 \leq \lambda_1 < \lambda_2 < 1$, then $C(L,\lambda_1) \supset C(L,\lambda_2)$.

The proof is obvious.

Definition 2.12 (Brainerd, 1969): A tree grammar is a system $G_t = (V_N, V_T, P, \Gamma)$ satisfying

(1) $V_N$ is a finite set of nonterminal symbols;

(2) $V_T$ is a finite set of terminal symbols, and $V_N \cap V_T = \phi$;

(3) P is a finite set of productions of the form $\Phi \to \Psi$, where $\Phi, \Psi \in (V_N \cup V_T)\#$;

(4) $\Gamma \subset (V_N \cup V_T)\#$ is a finite set of axioms.

Definition 2.13 (Brainerd, 1969): A direct deduction $T \Rightarrow Q$, from T to Q under $G_t$, means $\exists (\Phi \to \Psi) \in P$, $T/x = \Phi$, $T[x \to \Psi] = Q$.

A deduction $T \overset{\#}{\Rightarrow} Q$, from T to Q under $G_t$, means $\exists\ T_1,\ldots,T_k \in (V_N \cup V_T)\#$, $\ni T \Rightarrow T \Rightarrow \ldots \Rightarrow T_k \Rightarrow Q$.

Definition 2.14 (Brainerd, 1969): A tree language $L(G_t)$ generated by $G_t$ means $L(G_t) = \{T \in V_T^{\#} | \exists T_0 \in \Gamma, \ni T_0 \Rightarrow^{\#} T\}$. Also, it is called a tree language over $V_T$.

The following are obvious.

Proposition 2.11: Each tree grammar is a fuzzy tree grammar.

Theorem 2.12: If a fuzzy subset L of $\mathcal{T}(V_T)$ is an FTL, then the $\lambda$-cut $C(L,\lambda)$ of L is a tree language for any $\lambda \in [0,1)$.

Proof: Since L is an FTL, thus $\exists G = (V_N, V_T, P, \Gamma, J, f)$ is an FTG, $\ni L(G) = L$.

Let $G_1 = (V_N . V_T, P_1, \Gamma_1)$ where $(\Phi_1 \to \Psi_1) \in P_1$ iff $[(r)\ \Phi \to \Psi f(r)] \in P$ and $\mu(\Phi) \wedge \mu(\Psi) \wedge f(r) > \lambda$, $\Phi_1$ and $\Psi_1$ are basic trees of $\Phi$ and $\Psi$ separately.

Let us introduce a concept of basic productions, it is convenient for discussion. By production $(r)\ \Phi \to \Psi\ f(r)$, a production $\Phi_1 \to \Psi_1$ may be established, where $\Phi_1, \Psi_1$ are basic trees of $\Phi$, $\Psi$ separately, then production $\Phi_1 \to \Psi_1$ is called basic production of $(r)\Phi \to \Psi$.

$P_1$ may be said again as follow:

$P_1$ is established by the basic productions of productions $(r)\ \Phi \to \Psi f(r)$, of P, satisfying $\mu(\Phi) \wedge \mu(\Psi) \wedge f(r) > \lambda$.

Since $T \in \Gamma_1$ iff $\underset{\sim}{T} \in \Gamma$ and $\mu(\underset{\sim}{T}) > \lambda$, T is the basic tree of $\underset{\sim}{T}$, then $G_1$ is a tree grammar.

$\forall T \in C(L,\lambda)$, $\exists \underset{\sim}{T} \in L$, $\mu_L(\underset{\sim}{T}) > \lambda$, and T is the basic tree of $\underset{\sim}{T}$.

Since $\underset{\sim}{T} \in L$, we have $\underset{\sim}{T}$ f-deduction chain $\beta$ under G: $\underset{\sim}{T}_0 \underset{(r_1)}{\Longrightarrow} \underset{\sim}{T}_1 \underset{(r_2)}{\Longrightarrow} \cdots \underset{(r_{k-1})}{\Longrightarrow} \underset{\sim}{T}_{k-1} \underset{(r_k)}{\Longrightarrow} \underset{\sim}{T}$, where $\underset{\sim}{T}_0 \in \Gamma$, the set of productions is $\{(r_i)\Phi^{(i)} \to \Psi^{(i)} f(r_i),\ i=1,\dots,k\}$, and $\mu_\beta(\underset{\sim}{T}) =$

$$\mu_G(\underset{\sim}{T}) = \mu_L(\underset{\sim}{T}) = \bigwedge_{1 \le i \le k} [\mu(\Phi^{(i)}) \wedge \mu(\Psi^{(i)}) \wedge f(r_i) \wedge \mu(\underset{\sim}{T}_0)] > \lambda.$$

Thus, we get the set of basic productions, $\{\Phi_1^{(i)} \to \Psi_1^{(i)},\ i=1,\dots,k\} \in P_1$, of the set of productions $\{(r_i)\Phi^{(i)} \to \Psi^{(i)} f(r_i),\ i=1,\dots,k\}$, and the basic tree, $T_0 \in \Gamma_1$, of $\underset{\sim}{T}_0$.

By these conclusions, a $T_0 - T$ deduction under $G_1$ is generated, i.e., $T \in L(G_1)$, $C(L,\lambda) \subset L(G_1)$.

Because the nodes of a fuzzy tree are same as the nodes of its basic tree, the inverse of above process is obvious, i.e., $C(L,\lambda) = L(G_1)$. ⌑

Corollary: A set A of trees over $\Sigma$ is a tree language iff for any $\lambda \in [0,1)$, there exists a fuzzy tree language L over $\Sigma$ such that $A = C(L,\lambda)$.

## IV. FUZZY FOREST GRAMMAR

Definition 3.1: A finite subset, $\alpha = \{T_1,\dots,T_k | T_i \in A^{\#}, 1 \le i \le k\}$, of the set $A^{\#}$ of trees over A is called a forest over A. It is written by $\alpha = [T_1,\dots,T_k]$, where $T_i (1 \le i \le k)$ is tree and is called component tree of a forest $\alpha$, it is written by $T_i \in \alpha$ $(i=1,\dots,k)$. We denote the number k of component tree of forest $\alpha$ by $B(\alpha) = k$, and all of the forests over A is written by $\mathcal{P}A^{\#}$.

Let $\alpha = [T_1,\dots,T_k]$, $\beta = [Q_1,\dots,Q_\ell]$. If $\forall Q_j \in \beta$ $(1 \le j \le l)$, there exists $T_i \in \alpha$, $x_j \in T_i$, $1 \le i \le k$, such that $Q_j = T_i // x_j$, then $\beta$ is called a subforest of $\alpha$. If $Q_j = T_i / x_j$, then $\beta$ is called a full subforest of $\alpha$.

Definition 3.2: A forest grammar is a system $F_t = (V_N, V_T, P, \Gamma)$ satisfying

(1) $V_N$ is a finite set of nonterminal symbols;

(2) $V_T$ is a finite set of terminal symbols, and $V_N \cap V_T = \phi$;

(3) P is a finite set of productions of the form $\Phi \to \Psi$, where $\Phi, \Psi \in \mathcal{P}(V_N \cup V_T)^{\#}$;

(4) $\Gamma \subset \mathcal{P}(V_N \cup V_T)^{\#}$, is a finite set of axioms.

Definition 3.3: For $\alpha,\beta \in \mathcal{P}(V_N \cup V_T)^{\#}$, a direct deduction $\alpha \Rightarrow \beta$ from $\alpha$ to $\beta$ under the forest grammar $F_t$ means there exists $[\Phi \to \Psi] \in P$, the forest $\beta$ can be gotten substituting full subforest $\Phi$ of $\alpha$ by $\Psi$. A deduction $\alpha \Longrightarrow^{\#} \beta$ from $\alpha$ to $\beta$ under $F_t$ means there exist $\alpha_1,\dots,\alpha_m \in \mathcal{P}(V_N \cup V_T)^{\#}$, such that $\alpha \Longrightarrow \alpha_1 \Longrightarrow \dots \Longrightarrow \alpha_m \Longrightarrow \beta$.

Definition 3.4: A forest language $L(F_t)$ generated by the forest grammar $F_t$ is a subset of $\mathcal{P}(V_T)^{\#}$, where $L(F_t) = \{\alpha \in (V_T)^{\#} | \beta \in \Gamma, \beta \Longrightarrow^{\#} \alpha\}$. Also, $L(F_t)$ is called a forest language over $V_T$.

Definition 3.5: A finite subset $\underset{\sim}{\alpha} = \{\underset{\sim}{T}_1,\dots,\underset{\sim}{T}_k | \underset{\sim}{T}_i \in \mathcal{J}(A), i=1,2,\dots,k\}$ of the set of $\mathcal{J}(A)$ is called a fuzzy forest over A, and may be denoted by $\underset{\sim}{\alpha} = [\underset{\sim}{T}_1,\dots,\underset{\sim}{T}_k]$. Each $\underset{\sim}{T}_i$ $(1 \le i \le k)$ is called an f-component tree and written by $\underset{\sim}{T}_i \in \underset{\sim}{\alpha}$.

The number k of f-component trees of $\underset{\sim}{\alpha}$ by $B(\underset{\sim}{\alpha}) = k$, and all of the forests over A is written by $\mathcal{P}\mathcal{J}(A)$.

Let $\underset{\sim}{\alpha} = [\underset{\sim}{T}_1,\ldots,\underset{\sim}{T}_k]$, $\underset{\sim}{\beta} = [Q_1,\ldots,Q_\ell]$. If $\forall \underset{\sim}{Q}_j \in \underset{\sim}{\beta}(1 \le j \le l)$, there exists $\underset{\sim}{T}_i \in \alpha$, $x_j \in \underset{\sim}{T}_i$, $1 \le i \le k$, such that $\underset{\sim}{Q}_j = \underset{\sim}{T}_i//x_j$, then $\underset{\sim}{\beta}$ is called a fuzzy subforest of $\underset{\sim}{\alpha}$. If $\underset{\sim}{Q}_j = \underset{\sim}{T}_i/x_j$, then $\underset{\sim}{\beta}$ is called a fuzzy full subforest and written by $\underset{\sim}{\beta} = \underset{\sim}{\alpha}/(x_1,\ldots,x_\ell)$.

Definition 3.6: A fuzzy forest grammar (FFG) is a system $\underset{\sim}{F}_t = (V_N, V_T, P, \Gamma, J, f)$ satisfying

(1) $V_N$ is a finite set of nonterminal symbols;

(2) $V_T$ is a finite set of terminal symbols, and $V_N \cap V_T = \phi$;

(3) P is a finite set of productions of the form $(r)\Phi \to \Psi f(r)$, where $\Phi, \Psi \in \mathcal{PJ}(V_N \cup V_T)$, $B(\Phi) = B(\Psi) = k$, $r = (r_1,\ldots,r_k)$;

(4) $\Gamma \subset \mathcal{PJ}(V_N \cup V_T)$, is a finite set of axioms;

(5) $J = \{r | r=(r_1,\ldots,r_k), k=1,\ldots,l\}$, is called the labeled set of productions r of P;

(6) f is a map from J into $(0,1]^\ell = (0,1] \times \ldots \times (0,1]$, and called the vector of membership of production r, where $f(r) = f((r_1,\ldots,r_k)) = (f(r_1),\ldots,f(r_k))$, $f(r_i) \in (0,1]$, $i=1,\ldots,k$.

Definition 3.7: Let $\underset{\sim}{\alpha} = [\underset{\sim}{s}_1,\ldots,\underset{\sim}{s}_t]$, $\underset{\sim}{\gamma} = [\underset{\sim}{Q}_1,\ldots,\underset{\sim}{Q}_i]$ be fuzzy forests, $x_1,\ldots,x_{j_1} \in \underset{\sim}{s}_1,\ldots,x_{j_k+1},\ldots,x_{j_{k+1}} (=x_i) \in \underset{\sim}{s}_{k+1}$, $1 \le j_1 < j_2 < \ldots < j_k < j_{k+1} = i$, $1 \le k+1 \le t$, then fuzzy forest $\alpha[(x_1,\ldots,x_i) \xleftarrow{(f_1,\ldots,f_i)} \underset{\sim}{\gamma}]$ is called a replacement of $\underset{\sim}{\gamma}$ in $\underset{\sim}{\alpha}$ at $(x_1,\ldots,x_i)$ with the vector of membership, $(f_1,\ldots,f_i)$, where $\underset{\sim}{\alpha}[(x_1,\ldots,x_i) \xleftarrow{(f_i,\ldots,f_i)} \underset{\sim}{\gamma}] = [\underset{\sim}{T}_1,\underset{\sim}{T}_2,\ldots,\underset{\sim}{T}_{k+1},\underset{\sim}{s}_{k+2},\ldots,\underset{\sim}{s}_t]$, $\underset{\sim}{T}_{\ell+1} = \underset{\sim}{s}_{\ell+1}[(x_{j_\ell+1}) \xleftarrow{f_{j_\ell+1}} Q_{j_\ell+1}][(x_{j_\ell+2}) \xleftarrow{f_{j_\ell+2}} Q_{j_\ell+2}]\ldots[(x_{j_\ell+1}) \xleftarrow{f_{j_\ell+1}} Q_{j_\ell+1}]$, $i=0,1,\ldots,k$. Clearly, $\underset{\sim}{\alpha}[(x_1,\ldots,x_i) \xleftarrow{(f_1,\ldots,f_i)} \underset{\sim}{\gamma}]/(R(\underset{\sim}{Q}_1),\ldots R(\underset{\sim}{Q}_i)) = \underset{\sim}{\gamma}$.

Definition 3.8: Let $\underset{\sim}{\alpha}, \underset{\sim}{\beta} \in \mathcal{PJ}(V_N \cup V_T)$, a direct f-deduction $\underset{\sim}{\alpha} \Longrightarrow \underset{\sim}{\beta}$ from $\underset{\sim}{\alpha}$ to $\underset{\sim}{\beta}$ under a fuzzy forest grammar $\underset{\sim}{F}_t$ means there exists $[(r)\Phi \to \Psi\, f(r)] \in P$ and nodes $x_1,\ldots,x_i$ of f-component trees of $\underset{\sim}{\alpha}$, such that $\underset{\sim}{\alpha}/(x_1,\ldots,x_i) = \Phi$, $\underset{\sim}{\alpha}[(x_1,\ldots,x_i) \xleftarrow{f(r)} \Psi] = \underset{\sim}{\beta}$.

A f-deduction $\underset{\sim}{\alpha} \Longrightarrow^{\#} \underset{\sim}{\beta}$ from $\underset{\sim}{\alpha}$ to $\underset{\sim}{\beta}$ means there exists $\underset{\sim}{\alpha}_1,\ldots,\underset{\sim}{\alpha}_n \in \mathcal{PJ}(V_N \cup V_T)$, such that $\underset{\sim}{\alpha} \Longrightarrow \underset{\sim}{\alpha}_1 \Longrightarrow \ldots \Longrightarrow \underset{\sim}{\alpha}_n \Longrightarrow \underset{\sim}{\beta}$. $\underset{\sim}{\alpha} \Longrightarrow \underset{\sim}{\alpha}_1 \Longrightarrow \ldots \Longrightarrow \underset{\sim}{\alpha}_n \Longrightarrow \underset{\sim}{\beta}$ is called a $\underset{\sim}{\alpha}$-$\underset{\sim}{\beta}$ f-deduction chain. In a $\underset{\sim}{\alpha}$-$\underset{\sim}{\beta}$ f-deduction chain, the set of all productions in each direct f-deduction, $\{(r_i)\Phi_i \to \Psi\, f(r_i),\ i=1,\ldots,n\}$, is called the set of productions of this chain, and $\underset{\sim}{\alpha}, \underset{\sim}{\beta}$ are called start forest, end forest separately, and $\underset{\sim}{\alpha}_1,\ldots,\underset{\sim}{\alpha}_n$ are called middle forests.

If $\underset{\sim}{\alpha} \varepsilon \Gamma$, $\underset{\sim}{\alpha}-\underset{\sim}{\beta}$ f-deduction chain is called $\underset{\sim}{\beta}$ f-deduction chain of $\underset{\sim}{F}_t$, $\underset{\sim}{\beta}$ is called end forest of $\underset{\sim}{F}_t$.

Clearly, each $\underset{\sim}{\alpha}-\underset{\sim}{\beta}$ f-deduction chain under a fuzzy forest grammar may be resolved into the k f-deduction chains, $\underset{\sim}{T}_1 - \underset{\sim}{Q}_1$, $\underset{\sim}{T}_2 - \underset{\sim}{Q}_2, \ldots, \underset{\sim}{T}_k - \underset{\sim}{Q}_k$, under the fuzzy tree grammar, where $\underset{\sim}{\alpha} = [\underset{\sim}{T}_1, \ldots, \underset{\sim}{T}_k]$, $\underset{\sim}{\beta} = [\underset{\sim}{Q}_1, \ldots, \underset{\sim}{Q}_k]$.

Definition 3.9: The grade of the generation of an end forest $\underset{\sim}{\beta}$ of a fuzzy forest grammar $\underset{\sim}{F}_t$ is

$$\mu_{\underset{\sim}{F}_t}(\underset{\sim}{\beta}) = \bigvee_{\substack{\underset{\sim}{\alpha} \Longrightarrow \#\underset{\sim}{\beta} \\ \underset{\sim}{\alpha} \varepsilon \Gamma}} \bigwedge_{\substack{1 \le j \le m_{\underset{\sim}{\alpha}} \\ 1 \le i \le k_j}} f(r^{(j,i)}) \wedge \mu(\underset{\sim}{s}_{j,i}) \wedge \mu(\underset{\sim}{Q}_{j,i}) \wedge \mu(\underset{\sim}{\alpha})$$

where if $\underset{\sim}{\alpha} = [\underset{\sim}{T}_1, \ldots, \underset{\sim}{T}_n]$, thus we define $\mu(\underset{\sim}{\alpha}) = \bigwedge_{1 \le w \le n}(\underset{\sim}{T}_w)$, the set of productions of $\underset{\sim}{\alpha}-\underset{\sim}{\beta}$ f-deduction chain is

$\{(r^{(j)})\Phi_j \to \Psi_j \ f(r^{(j)}), \ j=1,\ldots,m_{\underset{\sim}{\alpha}}\}$, where $\Phi_j = [\underset{\sim}{s}_{j,1}, \ldots, \underset{\sim}{s}_{j,k_j}]$, $\Psi_j = [\underset{\sim}{Q}_{j,1}, \ldots, \underset{\sim}{Q}_{j,k_j}]$, $k_j = B(\Phi_j) = B(\Psi_j)$, $f(r^{(j)}) = (f(r^{(j,1)}), \ldots, f(r^{(j,k_j)}))$.

If for $\forall \underset{\sim}{\alpha} \varepsilon \Gamma$, there is not any $\underset{\sim}{\alpha}-\underset{\sim}{\beta}$ f-deduction chain, then we define $\mu_{\underset{\sim}{F}_t}(\underset{\sim}{\beta}) = 0$.

As similar to the definition 2.3, the grade of the generation of a f-deduction chain $\pi$ is given by

$$\mu_\pi(\underset{\sim}{\beta}) = \bigwedge_{\substack{1 \le j \le m_{\underset{\sim}{\alpha}} \\ 1 \le i \le k_j}} [f(r^{(j,i)}) \wedge \mu(\underset{\sim}{s}_{j,i}) \wedge \mu(\underset{\sim}{Q}_{j,i}) \wedge \mu(\underset{\sim}{\alpha})],$$

where the sense of each term is same as that of definition 2.3.

Definition 3.10: A fuzzy forest language (FFL) generated by fuzzy forest grammar $\underset{\sim}{F}_t = (V_N, V_T, P, \Gamma, J, f)$ is a fuzzy subset $L(\underset{\sim}{F}_t)$ of $\mathcal{PJ}(V_T)$, the value of membership of $L(\underset{\sim}{F}_t)$ at $\underset{\sim}{\alpha} \varepsilon \mathcal{PJ}(V_T)$ equals to the grade of the generation of $\underset{\sim}{\alpha}$ under $\underset{\sim}{F}_t$, i.e., $\mu_{L(\underset{\sim}{F}_t)}(\underset{\sim}{\alpha}) = \mu_{\underset{\sim}{F}_t}(\underset{\sim}{\alpha})$. Also, $L(\underset{\sim}{F}_t)$ is called a fuzzy forest language over $V_T$, $\forall \underset{\sim}{\alpha} \varepsilon L(\underset{\sim}{F}_t)$, $\exists \underset{\sim}{\alpha}$ f-deduction chain $\pi$, i.e., $\underset{\sim}{\alpha}_0 - \underset{\sim}{\alpha}$, where $\underset{\sim}{\alpha}_0 \varepsilon \Gamma$, $\underset{\sim}{\alpha}_0 = [\underset{\sim}{s}_1, \ldots, \underset{\sim}{s}_k]$, $\underset{\sim}{\alpha} = [\underset{\sim}{T}_1, \ldots, \underset{\sim}{T}_k]$, satisfying $\mu_{L(\underset{\sim}{F}_t)}(\underset{\sim}{\alpha}) = \mu_\pi(\underset{\sim}{\alpha})$.

Now, let f-deduction chain $\pi$ be resolved into the k f-deduction chain, $\underset{\sim}{s}_1 - \underset{\sim}{T}_1$, $\underset{\sim}{s}_2 - \underset{\sim}{T}_2, \ldots, \underset{\sim}{s}_k - \underset{\sim}{T}_k$, under the fuzzy tree grammar, here, corresponding grade of the generation of the chains are $f_1, \ldots, f_k$, $f_i = \mu_{L(\underset{\sim}{F}_t)}(\underset{\sim}{T}_i) = \mu_{\underset{\sim}{F}_t}(\underset{\sim}{T}_i) = \mu_\pi(\underset{\sim}{T}_i)$, $i=1,\ldots,k$. Clearly,

$\mu_{L(\tilde{F}_t)}(\alpha) = \bigwedge_{1 \le i \le k} \mu_{L(\tilde{F}_t)}(\tilde{T}_i)$. Thus, for any $\tilde{\alpha} \in L(\tilde{F}_t)$, corresponding to a vector of membership $(\mu_{L(\tilde{F}_t)}(\tilde{T}_1), \ldots, {}_{L(\tilde{F}_t)}(\tilde{T}_k))$, and $\mu_{L(\tilde{F}_t)}(\tilde{\alpha}) = \bigwedge_{1 \le i \le k} {}_{L(\tilde{F}_t)}(T_i)$, where $\tilde{\alpha} = [\tilde{T}_1, \ldots, \tilde{T}_k]$.

Fuzzy forest grammars $\tilde{F}_t$ and $\tilde{F}'_t$ are equivalent iff $L(\tilde{F}_t) = L(\tilde{F}'_t)$.

The following theorem reveals a relation between fuzzy forest language and a sort of fuzzy tree language (fuzzy binary tree language), we use a tool of BIN function raised by Takahashi (1975).

A binary tree is a tree in which each node has two direct descendants exactly, if there are descendants.

Let $\# \bar{\in} A$, $\mathcal{J}_b(A \cup \{\#\}) = \{\tilde{T} \in \mathcal{J}(A \cup \{\#\}) \mid \forall\, a \in \tilde{T}$, if $a \in A$, then $r(a) = 2$; if $a = \#$, then $r(a) = 0\}$ is called a set of all binary trees over $A \cup \{\#\}$.

Theorem 3.1: A fuzzy subset M of $\mathcal{PJ}(A)$, $M = \{(\tilde{\alpha}, \mu_M(\tilde{\alpha})) \mid \tilde{\alpha} \in \mathcal{PJ}(A),\ \tilde{\alpha} = [\tilde{T}_1, \ldots, \tilde{T}_k], \mu_M(\tilde{\alpha}) = \bigwedge_{1 \le i \le k} \mu_M(\tilde{T}_i), \mu_M(\tilde{T}_i) \in (0,1]$ , is a fuzzy forest language iff there exists a fuzzy subset N of $\mathcal{J}_b(A \cup \{\#\})$, $N = \{(\tilde{\beta}, \mu_N(\tilde{\beta})) \mid \tilde{\beta} \in \mathcal{J}_b(A \cup \{\#\}), \mu_N(\tilde{\beta}) \in (0,1]\}$ is a fuzzy tree language, and there exists a bijective map B in from M to N, such that $\mu_M(\tilde{\alpha}) = \mu_N(\tilde{\beta})$, if $\beta = \mathrm{Bin}(\tilde{\alpha})$, $\tilde{\alpha} \in M$, $\tilde{\beta} \in N$.

Proof: For fuzzy subset M of $\mathcal{PJ}(A)$, we define a map Bin from M into $\mathcal{J}_b(A \cup \{\#\})$ as follow:

$\forall \tilde{\alpha} \in M$, $\tilde{\alpha} = [\tilde{T}_1, \ldots, \tilde{T}_k]$, the vector of membership of $\tilde{\alpha}$ is $(\mu_M(\tilde{T}_1), \ldots, \mu_M(\tilde{T}_k))$, $\mu_M(\tilde{\alpha}) = \bigwedge_{1 \le i \le k} \mu_m(\tilde{T}_i)$.

Let $\tilde{T}_i = a_i[(\tilde{T}_{i,1}, \mu_{i,1}), \ldots, (\tilde{T}_i, \ell_i, \mu_i, \ell_i)]$, take $\tilde{\alpha}_i = [\tilde{T}_{i,1}, \ldots, \tilde{T}_{i,}\ell_i]$, the vector of membership of $\tilde{\alpha}_i$ is $(\mu_{i,1}, \ldots, \mu_{i,}\ell_i)$, $i = 1, \ldots, k$, then $\mathrm{Bin}(\tilde{\alpha}) = a_1[(\mathrm{Bin}(\alpha_1), 1), (\tilde{T}'_1, \mu_M(\tilde{T}_1))]$, where $T'_1 = a_2[(\mathrm{Bin}(\tilde{\alpha}_2), 1), (T'_2, \mu_M(T_2))], \ldots, T'_{k-2} = a_{k-1}[(\mathrm{Bin}(\tilde{\alpha}_{k-1}), 1), (\tilde{T}'_{k-1}, \mu_M(T_{k-1}))$, $\tilde{T}'_{k-1} = a_k[(\mathrm{Bin}(\tilde{\alpha}_k), 1), (\#, \mu_M(\tilde{T}_k))]$, and $\mathrm{Bin}(\Lambda) = \#$.

Now, let $\tilde{\beta} = \mathrm{Bin}(\tilde{\alpha})$, $N = \{(\tilde{\beta}, \mu_N(\tilde{\beta})) \mid \forall \tilde{\alpha} \in M$, $\tilde{\beta} = \mathrm{Bin}(\tilde{\alpha})$, and for $\tilde{\alpha} = [\tilde{T}_1, \ldots, \tilde{T}_k]$, take $\mu_N(\tilde{\beta}) = \bigwedge_{1 \le i \le k} \mu_M(T_i)\}$, then N is a fuzzy subset of $\mathcal{J}_b(A \cup \{\#\})$, the map Bin is bijective. If M is an FFL, then there exists $\tilde{F}_t = (V_N, A, P, \Gamma, J, f)$ which is an FFG, such that $L(\tilde{F}_t) = M$.

$\forall \tilde{\alpha} \in \Gamma$, $\tilde{\alpha} = [\tilde{T}_1, \ldots, \tilde{T}_k]$, $\mu(\tilde{\alpha}) = \bigwedge_{1 \le i \le k} \mu(\tilde{T}_i)$, and $(\mu(\tilde{T}_1), \ldots,$

$\mu(\underset{\sim}{T}_k))$ is the vector of membership of $\underset{\sim}{\alpha}$.

Now, let $\Gamma' = \text{Bin}(\Gamma)$, then this map is also a bijective.

$\forall[(r)\Phi \to \Psi\ f(r)] \in P$, where $\Phi = [\Phi_1,\ldots,\Phi_k]$, $\Psi = [\Psi_1,\ldots,\Psi_k]$, $\Phi_i, \Psi_i \in \mathcal{J}(V_N \cup A)$, $(i=1,\ldots,k)$, $f(r) \triangleq (f(r_1),\ldots,f(r_k))$, we establish $\Phi' = \text{Bin}(\Phi)$, $\Psi' = \text{Bin}(\Psi)$, where the vector of membership of $\Phi$ is $(1,\ldots,1)$, of $\Psi$ is $(f(r_1),\ldots,f(r_k))$, and $P' = \{(r)\Phi' \to \Psi'\ f'(r) \mid \forall[(r)\Phi \to \Psi\ f(r)] \in P$ where $f'(r) = \bigwedge_{1 \le i \le k} f(r_i)$.

We define $P' = \text{Bin}(P)$ too, then this map is bijective from P to P'.

Finally, we get a fuzzy binary tree grammar $G = (V_N, A \cup \{\#\}, P', \Gamma', J, f')$.

By the construction of G, we have $L(G) = N$.

Above discussing process may be converse, thus the conclusion is true. ¤

Definition 3.11: A $\lambda$-cut of the fuzzy subset M of $\mathcal{P}\mathcal{J}(A)$ is a subset $C(M,\lambda)$ of $\mathcal{P}A^{\#}$, satisfying

$$\alpha \in C(M,\lambda) \text{ iff } \underset{\sim}{\alpha} \in M, \lambda \in [0,1), \mu_M(\underset{\sim}{\alpha}) > \lambda,\ \underset{\sim}{\alpha} = [\underset{\sim}{T}_1,\ldots,\underset{\sim}{T}_k],\ \underset{\sim}{T}_i \in \mathcal{J}(A),$$

$$i=1,\ldots,k, \text{ and } T_i \text{ is a basic tree of } \underset{\sim}{T}_i,\ \alpha = [\underset{\sim}{T}_1,\ldots,\underset{\sim}{T}_k].$$

The proofs of the following theorems are same as the corresponding theorems in section 3.

Theorem 3.2: A set B of forests over $\Sigma$ is a forest language iff for any $\lambda \in [0,1)$, there exists a fuzzy forest language M over $\Sigma$, such that $B=C(M,\lambda)$.

Theorem 3.3: All of the fuzzy forest languages over $\Sigma$ may be called a family of the fuzzy forest languages, the family $\Delta$ of the fuzzy forest languages is closed under the operation of union.

Definition 3.12: A fuzzy forest $\alpha = [\underset{\sim}{T}_1,\ldots,\underset{\sim}{T}_k]$ is concatenated by fuzzy forests $\underset{\sim}{\beta}_i = [\underset{\sim}{Q}_{i,1},\ldots,\underset{\sim}{Q}_{i,k}]$ with a concatenated grade matrix $\{f_{i,j}\}(f_{i,j} \in (0,1]; i=1,\ldots,m; j=1,\ldots,k)$, where $m \le$ the number of nodes of each fuzzy tree within $\underset{\sim}{T}_1,\ldots,\underset{\sim}{T}_k$, means to establish a fuzzy forest as follow:

$$C(\underset{\sim}{\alpha};\underset{\sim}{\beta}_1,\ldots,\underset{\sim}{\beta}_m) = [\underset{\sim}{S}_1,\ldots,\underset{\sim}{S}_k], \text{ where } \underset{\sim}{S}_\ell = C(\underset{\sim}{T}_\ell;\ \underset{\sim}{Q}_{1,\ell},\ldots,$$

$$\underset{\sim}{Q}_{m,\ell}) = \underset{\sim}{T}_\ell[(X_{\ell,1})\xrightarrow{f_{1,\ell}}\underset{\sim}{Q}_{1,\ell}][(X_{\ell,2})\xrightarrow{f_{2,}}\underset{\sim}{Q}_{2,\ell}]\ldots[(X_{\ell,m})\xrightarrow{f_{m,}}\underset{\sim}{Q}_{m,\ell}],$$

$$1=1,\ldots,k.$$

A fuzzy forest language $L_1$ over $\Sigma$ is concatenated by a fuzzy forest language $L_2$, means to establish a fuzzy subset $C_{<f>}(L_1;L_2)$ as follow:

$$C_{<f>}(L_1;L_2) = \{(C(\underset{\sim}{\alpha};\underset{\sim}{\beta}_1,\ldots,\underset{\sim}{\beta}_m),\ \mu_{C_{<f>}}(L_1;L_2)(C(\underset{\sim}{\alpha};\underset{\sim}{\beta}_1,\ldots,\underset{\sim}{\beta}_m)))$$

$\underset{\sim}{\alpha} \in L_1, \underset{\sim}{\beta}_1,\ldots,\underset{\sim}{\beta}_m \in L_2$, $1 \leq m \leq$ the number of nodes of each f-component tree of $\underset{\sim}{\alpha}$, $B(\underset{\sim}{\alpha}) = k$, $\{f_{i,j}\}(i=1,\ldots,m;\ j=1,\ldots,k)$ is a concatenated grade matrix of $C(\underset{\sim}{\alpha};\underset{\sim}{\beta}_1,\ldots,\underset{\sim}{\beta}_m)$, where if $\underset{\sim}{\alpha}$ may be different in $L_1$, then the number of rows, columns and elements of the related matrix may be different too, but it is suppose that the set of elements of all $\{f_{i,j}\}$ is a finite set $<f>$ valued in $(0,1]$, and

$$\mu_{C_{<f>}(L_1,L_2)}(C(\underset{\sim}{\alpha};\underset{\sim}{\beta}_1,\ldots,\underset{\sim}{\beta}_m)) = \mu_{C_{<f>}(L_1,L_2)}(r) = \bigvee_{\substack{\underset{\sim}{r}=C(\underset{\sim}{\alpha};\underset{\sim}{\beta}_1,\ldots,\underset{\sim}{\beta}_m) \\ \underset{\sim}{\alpha} \in L_1;\underset{\sim}{\beta}_1,\ldots,\underset{\sim}{\beta}_m \in L_2}} \bigwedge_{\substack{1\leq i\leq m \\ 1\leq j\leq k}} [\mu_{L_2}(\underset{\sim}{\beta}_i) \wedge f_{i,j} \wedge \mu_{L_1}(\underset{\sim}{\alpha})]\}.$$

Also, $C_{<f>}$ is called a concatenated operation between $L_1$ and $L_2$.

Theorem 3.4: The family $\Delta$ of fuzzy forest languages over $\Sigma$ is closed under the concatenated operation.

Finally, we set up a natural connection between fuzzy forest language and n-fold fuzzy language raised by Mizumoto, Toyoda and Tanaka (1973).

Definition 3.13 (Mizumoto, etc., 1973): $N(\geq 1)$-fold fuzzy grammar is a system

$$\text{n-FG} = (V_N, V_T, P, S, J, \{f_0, f_1, \ldots, f_n\})$$

where $V_N$ is a finite set of nonterminal symbols, $V_T$ is a finite set of terminal symbols, $V_N \cap V_T = \phi$, $S \in V_N$ is a starting symbol, P is a finite set of productions of the form $(r)u \to V$, where $r \in J$, $u \to V$ is an ordinary production, $u \in V_N^* - \{f\}$, $V \in (V_N \cup V_T)^*$, $J = \{r\}$ is a set of labels of productions, $f_i(i=0,1,\ldots,n)$ is a function of membership defined as follow:

(1) If i=0, then the value $f_0(r)$ in $[0,1]$ is the grade of the application of an initial production r in $J_s$.

$$f_0 : J_s \to [0,1]$$

where $J_s$ is the set of all labels whose productions are initial productions;

(2) If $1 \leq i \leq n$, then the value $f_i(r_1,r_2,\ldots,r_i;r_{i+1})$ in $[0,1]$ is the grade of the application of the production $r_{i+1}$ after the i productions $r_1,r_2,\ldots,r_i$ were applied sequentially to the middle string in a f-deduction.

Also, the $f_i$ (i=0,1,...,n) may be called i-fold fuzzy function.

If all of the productions are the productions of type i (i=0, 1,2,3), then n-fold fuzzy grammar is called type i.

Definition 3.14 (Mizumoto, etc., 1973): A deduction chain, $S \xrightarrow[r_1]{} \alpha_1 \xrightarrow[r_2]{} \alpha_2 \xrightarrow[r_3]{} \cdots \xrightarrow[r_m]{} \alpha_m$, of length m by the productions $r_1, r_2, \ldots, r_m$ from S to $\mu_m$ .

If $m \leq n$, let $f_0(r_1) = \mu_1$, $f_1(r_1;r_2) = \mu_2, \ldots, f_{m-1}(r_1, \ldots, r_{m-1};r_m) = \mu_m$, then $S \xrightarrow[r_1]{\mu_1} \alpha_1 \xrightarrow[r_2]{\mu_2} \alpha_2 \xrightarrow[r_3]{\mu_3} \cdots \xrightarrow[r_m]{\mu_m} \alpha_m$ .

If $m > n$, let $m=n+j$, $f_n(r_j, r_{j+1}, \ldots, r_{n+j-1}; r_{n+j}) = \mu_{n+j}$, $j \geq 1$, then $S \xrightarrow[r_1]{\mu_1} \alpha_1 \xrightarrow[r_2]{\mu_2} \cdots \xrightarrow[r_n]{\mu_n} \alpha_n \xrightarrow[r_{n+1}]{\mu_{n+1}} \alpha_{n+1} \xrightarrow[r_{n+2}]{\mu_{n+2}} \cdots \xrightarrow[r_{n+j}]{\mu_{n+j}} \alpha_{n+j} = \alpha_m$.

These chains are called n-fold fuzzy deduction chain, or n-fold chain of $\alpha_m$ briefly. Now, we define

$$\mu_{n\text{-}FG}(x) = \bigvee_{\forall n\text{-fold chain } t} \bigwedge_{1 \leq i \leq k_t} \mu_i \text{, where } x \in V_T^*\text{, } t\text{: } S \xrightarrow[r_1]{\mu_1} \alpha_1 \xrightarrow[r_2]{\mu_2} \cdots \xrightarrow[r_{k_t}]{\mu_{k_t}} d_{k_t} = x.$$

Definition 3.15 (Mizumoto, etc., 1973): A fuzzy language generated by N-FG of type i is a fuzzy subset L(N-FG) of $V_T^*$ with above defined $\mu_{N-FG}(x)$ as membership function, i.e., L(N-FG) = $\{(x, \mu_{N-FG}(x)) \mid x \in V_T^*\}$. The above defined fuzzy language may be called an N-fold fuzzy language of type i.

Definition 3.16: A drawn leaves operation about fuzzy forest is a map $\mathbb{H}^*$ which is defined as follows:

$\mathbb{H}^*$ is a map from a fuzzy forest language L over A into a fuzzy subset of A*, satisfying

$$\mathbb{H}^*(\underset{\sim}{\alpha}) = \mathbb{H}(\underset{\sim}{T}_1)\mathbb{H}(\underset{\sim}{T}_2)\ldots\mathbb{H}(\underset{\sim}{T}_k)$$

$$\mu_{\mathbb{H}^*(L)}(\mathbb{H}^*(\underset{\sim}{\alpha})) = \bigvee_{\substack{\mathbb{H}^*(\underset{\sim}{\beta}) = \mathbb{H}^*(\underset{\sim}{\alpha}) \\ \forall \underset{\sim}{\beta} \in L}} \mu_L(\underset{\sim}{\beta})$$

where $\underset{\sim}{\alpha} \in \mathcal{PJ}(A)$, $\underset{\sim}{\alpha} = [\underset{\sim}{T}_1, \ldots, \underset{\sim}{T}_k]$, $\underset{\sim}{T}_i \in \mathcal{J}(A)$, i=1,...,k. $\mathbb{H}$ is drawn leaves operation about fuzzy tree defined by definition 2.7.

Theorem 3.5: For any n-fold fuzzy language $L_1$ of type 2 over $\Sigma$, there exists a fuzzy forest language L over $\Sigma \cup \{\#\}$, such that

$\mathbb{H}^*(L) = L_1$.

Proof: Let $L_1$ be an n-fold fuzzy language of type 2 over $\Sigma$, then there exists an n-fold fuzzy grammar of type 2

$$n\text{-FG} = (V_N, \Sigma, P, S, J, \{f_0, f_1, \ldots, f_n\}),$$

such that $L(n\text{-FG}) = L_1$.

We take $\# \bar{\varepsilon}\ V_N \cup \Sigma$.

Now, establish a system $\underset{\sim}{F}^* = (V_N, \Sigma \cup \{\#\}, P_1, S, J_1, 1)$, where the $P_1$ can be founded as follow:

(1) For $\forall[(r)u \to \alpha] \varepsilon P$, where $u \varepsilon V_N$, $\alpha = V_1, V_2 \ldots V_\ell$, $V_i \varepsilon V_N \cup \Sigma$ $(1 \le i \le 1)$, to build marked tree $\#^{(r)}[(V_1, < >), \ldots, (V_\ell, < >)]$ of $\alpha$.

Clearly, this correspondence is one by one between $\alpha$ and its marked tree, and the value in $< >$ can be obtained from the finite set $A = \{f_i(r') \mid i=0,1,\ldots,n,\ r'=(r_1,\ldots,r_{i-1};r_i),\ r_1,\ldots,r_i \varepsilon J\}$;

(2) For $\forall r_1 \varepsilon J_s$, if $[(r_1)S \to \alpha f_0(r_1)] \varepsilon P$, where $\alpha = V_1 V_2 \ldots V_\ell$, $V_i \varepsilon V_N \cup \Sigma$, $i=1,\ldots,1$, then we make $[(r_1)S \to \#^{(r_1)}[(V_1, f_0(r_1)), \ldots, (V_\ell, f_0(r_1))]] \varepsilon P_1$.

If $2 \le m \le n+1$, and $[(r_m)V \to \alpha f_{m-1}(r_1, r_2, \ldots, r_{m-1}; r_m)] \varepsilon P$, where $V \varepsilon V_N$, $\alpha = V_1 V_2 \ldots V_\ell$, $V_i \varepsilon V_N \cup \Sigma^m$, $i=1,\ldots,1$, then we make $[(r_1, r_2, \ldots, r_m)[\underset{\sim}{T}_1, \ldots, \underset{\sim}{T}_{i-1}, \underset{\sim}{T}_i, \underset{\sim}{T}_{i+1}, \ldots, \underset{\sim}{T}_{m-1}] \to [\underset{\sim}{T}_1, \ldots, \underset{\sim}{T}_{i-1}, \underset{\sim}{T}_i[V \leftarrow \#^{(r_m)}[(V_1, \mu_m), \ldots, (V_\ell, \mu_m)]], \underset{\sim}{T}_{i+1}, \ldots, \underset{\sim}{T}_{m-1}]] \varepsilon P_1$, where

$$\mu_m = f_{m-1}(r_1, r_2, \ldots, r_{m-1}; r_m), \text{ and}$$

$$\underset{\sim}{T}_j = \#^{(r_j)}[(S_{j,1}, < >_j), (S_{j,2}, < >_j), \ldots, (S_{j,k_j}, < >_j),$$

$j=1,\ldots,m-1$. $S_{j,t} \varepsilon V_N \cup \Sigma \cup \{\#\}$, $t=1,\ldots,K_j$. $S_{i,t_0} = V$,

$1 \le t_0 \le k$.

(3) For $\forall r_1 \varepsilon J - J_s$, if $[(r_{n+1})V \to \alpha f_n(r_1, \ldots, r_n; r_{n+1})] \varepsilon P$, where $\alpha = V_1 V_2 \ldots V$, $V_i \varepsilon V_N \cup \Sigma$, $i=1,\ldots,1$, $V \varepsilon V_N$, then we make

$$[(r_1, r_2, \ldots, r_n; r_{n+1})[\underset{\sim}{T}_1, \ldots, \underset{\sim}{T}_{i-1}, \underset{\sim}{T}_i, \underset{\sim}{T}_{i+1}, \ldots, \underset{\sim}{T}_n]$$

$$\to [\underset{\sim}{T}_1, \ldots, \underset{\sim}{T}_{i-1}, \underset{\sim}{T}_i[V \leftarrow \#^{(r_{n+1})}[(V_1, \mu_{n+1}), \ldots, (V_\ell, \mu_{n+1})]],$$

$\underset{\sim}{T}_{i+1}, \ldots, \underset{\sim}{T}_n]] \varepsilon P_1$, where $\mu_{n+1} = f_n(r_1, r_2, \ldots, r_n : r_{n+1})$ and

$$\underset{\sim}{T}_j = \#^{(r_j)}[(S_{j,1}, < >_j), (S_{j,2}, < >_j), \ldots, (S_j, k_j, < >_j), \ j=1,\ldots,n.$$

$S_{j,t} \varepsilon V_N \cup \Sigma \cup \{\#\}$, $S_{i,t_0} = V$, $1 \le t_0 \le k$.

Let $J_1 = J_s \cup J_s \times (J-J_s) \cup \ldots \cup J_s \times (J-J_s)^{n-1} \cup J \times (J-J_s)^n$.

Thus $\underset{\sim}{F}_t$ is an FFG, and we take $L=L(\underset{\sim}{F}_t)$.

As for verifying $\mathbb{H}^*(L) = L_1$, it is only repeated simply by related content of theorem 2.7. ¤

REFERENCES

Brainerd, W. S. (1968), The Minimalization of Tree Automata, Inform. Control, 13, 484-491.

Brainerd, W. S. (1969), Tree Generating Regular Systems, Inform. Control, 14, 217-231.

Doner, J. E. (1970), Tree Acceptors and Some of Their Applications, J. Comput. System Sci. 4, 406-451.

Fu, K. S., Huang, T. (1972), Stochastic Grammars and Languages, Int. J. Comput. and Inform. Sci. 1.

Hopcroft, J. E., Ullman, J. D. (1969), Formal Languages and Their Relation to Automata, Addison-Wesley Publishing Company, Ch. 2, 10-32, Ch. 3, 34-60, Ch. 9., 157-175.

Lee, E. T., Zadeh, L. A. (1969), Note on Fuzzy Languages, Inform. Sci. 1, 421-434.

Mizumoto, M., Toyoda, J., Tanaka, K. (1973), N-Fold Fuzzy Grammars, Inform. Sci., 5, 25-43.

Negoita, C. V., Ralescu, D. A. (1975), Applications of Fuzzy Sets to Systems Analysis, Birkhaser Verlag, Basel und Stuttgart, Ch. 5, 135-142.

Santos, E. S. (1974), Context-Free Fuzzy Languages, Inform. Control, 26, 1-11.

Takahashi, M. (1975), Generalizations of Regular Sets and Their Application to a Study of Context-Free Languages, Inform. Control, 27, 1-36.

Zadeh, L. A. (1965), Fuzzy Sets, Inform. Control, 8, 338-353.

# FUZZY PRODUCTION RULES: A LEARNING METHODOLOGY

L. Lesmo, L. Saitta and P. Torasso

Istituto di Scienze dell'Informazione
Università di Torino
Corso Massimo D'Azeglio 42
10125 - Torino (Italy)

## INTRODUCTION

In many research fields it is possible to obtain good scientific results only after large amounts of data have been collected and analyzed; the analysis allows the researcher to detect regularities, similarities and discriminant features which may be useful to characterize different classes of objects. On the other hand, the manual examination of a large set of data is slow and error prone, so that many techniques have been proposed and are actually used to perform that analysis automatically (e.g. discriminant analysis); unfortunately, most of those techniques are based on mathematical methodologies which impose strong constraints on the kinds of data that can be analyzed.

For example, many statistical methods require that the features are independent of each other, that they assume only numeric values belonging to a given interval, that the distribution of the values of these features may be fitted by an a-priori defined mathematical function (e.g. normal distribution) and so on.

Even if these conditions hold, this kind of approach suffers from another significant drawback, i.e. its limited possibility of exploiting the semantics underlying the domain under consideration. In fact, the decision theoretic approach produces descriptions that are too mathematical and difficult to comprehend for a human expert. This prevents the human user from intervening in the analysis of the rules and in their modification. More important is the fact that this approach makes difficult the sharing of the responsibility in the learning of the rules between the human expert and the system. In fact, in the classical decision theoretic approach the designer must only supply to the system a set of labelled samples, i.e. a set of vectors each of which contains the values of a predefined set of characteristic features and is assigned to a particular class on the basis of the expert's knowledge. Moreover, because of the assumptions made about the parameters characterizing the sample, the expert is forced to select the characteristic features in an accurate way, in order to avoid as much as possible the breaking of these assumptions; this prevents him, for example, from using features which are known to be strongly dependent on each other, even if these features and the relationships existing among them are actually used by the expert during the classification process.

A quite different approach consists in replacing the mathematical tools used in the classical decision theory by a deductive machinery mainly based on logics and on the associated inference rules. In the recent years a particular kind of formalism has become very popular, i.e. the production rules; the systems which use production rules are in fact oriented to a sort of heuristic decision making (more than to a formal logic approach) in that it allows the designer to write ad-hoc inference rules which may be used to model a particular piece of knowledge put at disposal by the expert; nevertheless the deductive mechanisms which allow the system to control the interactions among the different rules are independent of the contents of the rules used by the system for modelling the particular semantic domain[1,2].

A rather simple control structure assumes that the system has the overall goal of confirming or disconfirming hypotheses, on the basis of the feature values (or findings) characterizing the individual under examination. In case a particular hypothesis can not be directly verified on the basis of the available data, a set of sub-hypotheses is emitted and the process is recursively applied to them. The main difficulty of this mechanism is its inability to cope with competing hypotheses; in other words, at the very end of the deductive process a particular hypothesis is either true or false, thus preventing the user from having an evaluation of its evidence.

In order to overcome this difficulty, the production rules have been augmented by a mechanism which combines the evidence degrees associated with the input data (or with the verified sub-hypotheses) to obtain the degree of evidence which must be assigned to a given hypothesis. In most systems this mechanism is purely heuristic, in that the weights and coefficients used to compute the resulting evidence are determined manually on the basis of the available knowledge. On the other hand, some investigation has been carried on recently in order to make that mechanism sound from a mathematical point of view[3,4,5].

One of the most promising approaches seems to be the one based on fuzzy logics and fuzzy linguistic variables. An evident advantage of this approach consists in the possibility of modelling in a uniform way different kinds of features (numeric, binary, linguistic, etc.) and of using the same formalism (fuzzy linguistic variables) to represent the degree of evidence associated with a particular hypothesis.

But the fuzzy approach seems very appealing also to face another problem which is common to all production rule systems, i.e. the one of knowledge acquisition. In fact, the purely heuristic approaches require a great deal of effort to adjust the weights occurring in the rules, because the lack of an adequate theory makes difficult the design of efficient algorithms for the automatic learning. On the other hand, the theoretically based approaches make very difficult the interaction with the human expert which could allow the introduction into the system of knowledge which is already available. On the contrary, the already noticed homogeneity of the fuzzy approach allows to represent in a formally sound way the knowledge available to the system (so that efficient learning algorithms may be designed), but this representation, and in particular the use of fuzzy linguistic variables, shows a strict correspondence to the way people express their knowledge of the deduction rules they use (so that it is also possible to directly translate the information given by the expert into internal fuzzy rules).

It could be questioned why there is the need of adopting an approach allowing both kinds of knowledge acquisition. It could in fact happen that the same piece of information is suitable for being automatically learned and is well clear to the expert in the field; however, it seems possible to draw a line which splits (very roughly) the overall knowledge into two parts. The first part concerns the selection of the features which must appear in a particular production rule and the determination of the way they must be combined to emit an hypothesis together with its evidence degree. This part corresponds to the knowledge needed by a simple system which directly goes from available findings to the emission of the most highly rated hypothesis and is very suitable for automatic learning. This suitability is particularly appealing because an expert is normally not able to explain in a precise way how the evidence associated with the available input data must be used to obtain the evidence of the hypothesis.

The second part is normally more clear to the expert, in that it concerns global properties of the system under investigation (e.g. causality relations among hypotheses, interconnections among findings); these properties are not the basis of the deduction mechanism (which must go from findings to conclusions) but they strongly affect the behavior of the system by establishing what could be called horizontal relationships.

It has to be noticed that, whereas it is possible to develop efficient and accurate learning algorithms to handle the knowledge composing the first part introduced above, it would be very difficult to cope in an automatic way with the complexity introduced by the existence of causality relations or, more generally, of mutual dependencies among hypotheses, so that a cooperation between the system (its automatic procedures) and the expert (its knowledge) greatly reduces the time required to develop the set of rules without affecting the accuracy and, consequently, the reliability of the system itself.

The ultimate goal of a learning algorithm devoted to the acquisition of rules to be used for classification purposes is to infer, for each class, a description (or a set of possible descriptions) of the objects belonging to that class, which allows the system to discriminate between the different classes. This implies that the learning algorithm must be able to select those features which are characteristic of a given class, i.e. which distinguish it from the other ones.

The automatic learning of structural descriptions has received an increasing attention in recent years[6,7]; however, further research efforts are required to develop efficient algorithms, in particular when the description of the input objects may be affected by noise, errors or uncertainty.

This work describes a learning algorithm which accepts in input a set of individuals, each of which is described by a large number of features of different kinds and is assigned to a particular class, and infers a set of production rules. These production rules may be used by a system which emit hypotheses about the class to which an individual could belong and associates to each hypothesis a linguistic term representing its evidence degree. It should be noticed, however, that the input data are not directly submitted to the learning algorithm; in fact they are initially processed by a set of rules provided by the expert which converts them into a linguistic representation, by taking into account also the possible interdependencies existing among the different data. This first step, which shows how the cooperation among the expert and the system is easily achieved, has the purpose of making the feature representation more abstract and homogeneous, thus reducing the sensitivity of the system to variations in the distribution of the individuals of the sample.

The second section of the paper contains the details of the learning algorithm and the third one the definition of the linguistic variables. Section 4 reports an example of application.

## LEARNING OF FUZZY PRODUCTION RULES

As it has been noticed in the introduction, production rules are nowadays a widely used tool for representing knowledge in systems where human-like capabilities are required.

A production rule has the following general form:

$$\text{Condition} \xrightarrow{w} \text{Decision}$$

The condition appearing in the left-hand side of a rule consists in a set of constraints which must be satisfied by the input data in order to take the decision occurring in the right-hand side. The parameter w represents the confidence degree in taking the decision when the condition part is satisfied. It should be noticed, however, that in practical cases the satisfaction of the constraints constituting the condition part of the production rule cannot be represented by boolean values (i.e. True or False), because the agreement between the input data and the required conditions may be only partial; for this reason an evaluation of this matching is required and it seems reasonable to express it by means of a value ranging from 0 to 1.

The fuzzy set approach is able to handle in a homogeneous way both the evaluation of the matching conditions containing ill-defined and uncertain terms and the degree of certainty associated with the production rules. In both cases, in fact, the corresponding values may be modelled by fuzzy linguistic variables.

In order to describe the learning algorithm in a more precise way, a formalization of the concepts introduced above is required. A production rule may be defined as having the following structure:

$$C \xrightarrow{w} H \qquad (1)$$

where: C is the <u>condition part</u> of the rule

H is the <u>hypothesis</u> (i.e. the right-hand side which, in general, represents a decision which should be taken - or action which should be performed - when the condition is verified; in our case refers to an hypothesis whose emission depends on the evaluation of the condition part of the rule)

w is the <u>strength of the implication</u>, i.e. a value expressing how well the satisfaction of the condition supports the emission of the hypothesis.

The internal structure of the condition part C is:

$$C = ((V_1 = \lambda_{1k_1}) \ \underline{\text{and}} \ (V_2 = \lambda_{2k_2}) \ \underline{\text{and}} \ \dots \ \underline{\text{and}} \ (V_m = \lambda_{mk_m})) \qquad (2)$$

where $V_j (1 \leq j \leq m)$ is a linguistic variable which may assume values belonging to an alphabet $\Lambda_j$ of linguistic terms $\lambda_{jk_j}$ (possibly modified by edges[9]).

The quantitative evaluation of the evidence degree associated with each hypothesis is obtained by taking into account the existence of different production rules having the same right-hand side. Let us assume that the hypothesis H is supported by Q different production rules:

$$
\begin{array}{l}
C_1 \xrightarrow{w_1} H \\
C_2 \xrightarrow{w_2} H \\
\quad \vdots \\
C_Q \xrightarrow{w_Q} H
\end{array}
\qquad (3)
$$

The possibility of the hypothesis H (i.e. its evidence degree) is evaluated as[8]:

$$\Pi(H) = \underset{1 \leq q \leq Q}{\operatorname{Max}}\{\min[\Pi(C_q), w_q]\} \qquad (4)$$

where $\Pi(C_q)$ represents the possibility value associated with the condition part of the q-th production rule (i.e. an evaluation of its agreement with the input data) and is computed as:

$$\Pi(C_q) = \min_{1 \leq j \leq m_q} \mu_{j\,k_j} \qquad (5)$$

The parameters $\mu_{jk_j}$ are the possibility values representing the degree of compatibility between the predicate $V_j = \lambda_{jk_j}$ and the actual experimental data characterizing the individual under examination.

In order to introduce the learning algorithm some definitions are useful. Let us define $\mathcal{V}_H$ as the set of all the linguistic variables associated with the parameters which are "possibly interesting" for evaluating the hypothesis H. Notice that the use of $\mathcal{V}_H$ in the learning phase implies that pre-selection of "possibly interesting" features has been performed by the human expert. The obvious purpose of this pre-selection is to limit the number of variables which will be taken into account by the learning algorithm; however, the word "possibly" indicates that a feature should be selected even if its correlation with the hypothesis is not sure in that the actual statement of the relationships between data and hypotheses will be made by the learning algorithm (this means that in cases where the number of available features is low enough, the pre-selection is unnecessary).

Another task which is left to the user is to define, for each $V_j \varepsilon \mathcal{V}_H$, a set of linguistic terms $\Lambda_j$; this allows the user to communicate to the system the meaning of the different terms which will be used to refer to a given feature, thus establishing a common dictionary in the interactions between the user and the system.

The main goal of the learning algorithm is to determine what combinations of features (with the associated linguistic term) actually support a given hypothesis; in other words, it must choose, among all the possible condition parts, that subset which will be used to build the rules associated with the hypothesis under consideration. Finally, it must determine the strength of the rules which have been created.

The set of all possible production rules (after the above mentioned pre-selection step) may be partitioned depending on the hypothesis a particular rule refers to. In fact we may define $\mathcal{H} = \{H_1, H_2, \ldots, H_T\}$ to be the set of all possible hypotheses; the subset $\mathcal{S}_t$ associated with the hypothesis $H_t$ contains rules of the following form:

$$C_s^{(t)} \xrightarrow{w_s^{(t)}} H_t \tag{6}$$

where

$$C_s^{(t)} \equiv (V_{p_1}^{(t)} = \lambda_{p_1 k_1}^{(t)}) \ \underline{\text{and}} \ \ldots \ \underline{\text{and}} \ (V_{p_r}^{(t)} = \lambda_{p_r k_r}^{(t)}) \tag{7}$$

In (7) $V_{p_j}^{(t)}$ belongs to the set $\mathcal{V}_{H_t}$ whose cardinality is $R_t$ and each $V_{p_j}^{(t)}$ assumes linguistic values belonging to the set $\Lambda_{p_j}^{(t)}$, whose cardinality is $m_{p_j}^{(t)}$. Moreover r ranges from 1 to $R_t$ and the generic $k_j$ ranges from 1 to $m_{p_j}^{(t)}$.

The number of different rules of the form (6), which have potentially to be taken into account, obviously depends on the cardinality of the $\mathcal{V}_{H_t}$ and of the different $\Lambda_j^{(t)} (1 \leq j \leq R_t)$. In fact, the number of a-priori possible rules referring to the hypothesis $H_t$ is:

$$M^{(t)} = 2^{K^{(t)}} - 1 \tag{8}$$

where

$$K^{(t)} = \sum_{i=1}^{R_t} m_i^{(t)} \tag{9}$$

and $K^{(t)}$ represents the total number of different terms which can occur in the condition part of the rules. Notice that $M^{(t)}$ represents the cardinality of $\mathcal{G}_t$.

It is intuitive that not all these rules are equally relevant to the hypothesization of $H_t$; in fact, the relevance of the rule having as condition part $C_s^{(t)}$ is expressed by its strength $w_s^{(t)}$; a rule of no relevance will have $w_s^{(t)} = 0$ and this is equivalent to say that the particular set of terms (linguistic variable - linguistic term) appearing in the condition part of that rule is not correlated with the given hypothesis, so that there is no reason for including that rule into the final set. In fact the learning algorithm has the final goal of excluding all the rules whose weight is equal to zero (or below a given threshold, as it will be explained in the following).

The learning algorithm is described in detail in[10,11] and an outline of it is presented below.

Let $Z = \{z_1, z_2, \ldots , z_N\}$ be the training set used to infer the strength of the rules; for each element $z_i$ of Z one and only one hypothesis $H_t \varepsilon \mathcal{X}$ is supposed to be true and it is communicated by the expert to the system. For a particular element $z_i \varepsilon Z$ it is possible to determine the possibility value $\mu^{(t)}_{p_j k_j}(z_i)$ representing the compatibility degree between the term $(V^{(t)}_{p_j} = \lambda^{(t)}_{p_j k_j})$ and the experimental data associated with $z_i$. It follows that, according to (5), the evidence degree of $C_s^{(t)}$ (see (7)) is:

$$\Pi(C_s^{(t)} | z_i) = \underset{1 \le j \le r}{\text{Min}} \; \mu^{(t)}_{p_j k_j}(z_i) \tag{10}$$

Finally, the evidence degree which will be assigned to the hypothesis on the basis of the data concerning the element $z_i$ is, according to (4):

$$\Pi(H_t | z_i) = \underset{1 \le s \le M^t}{\text{Max}} \{\text{Min}[\Pi(C_s^{(t)} | z_i), \; w_s^{(t)}]\} \tag{11}$$

The strengths $W_s^{(t)}$ ($1 \leq s \leq M^t$; $1 \leq t \leq T$) are determined in such a way that the average number of times (average over different samples) the correct hypothesis obtains a possibility value not greater than that of a wrong one (error probability) is minimized. This is accomplished[10] by explicitly expressing the error probability in terms of the strengths $w_s(t)$ and then by minimizing it by means of standard methods.

In order to avoid a combinatorial explosion in the number of rules, a pre-selection of these rules is performed (not to be confused with the pre-selection of the linguistic variables mentioned above). It turns out, in fact, that many of the conditions $C_s^{(t)}$ (see (6) and (7)) are never (or very seldom) satisfied by the data associated with an individual (in the training set) for which the hypothesis $H_t$ is true; for this reason the strength of the production rules containing these conditions in the left-hand side can be set equal to zero.

Once the strengths $w_s^{(t)}$ have been inferred, a further selection of the rules is (possibly) performed by discarding all those rules whose strength is under a given threshold; this may be accomplished (in general) without any significant increase in the error rate[11].

It is important to notice that, even if after all the selection steps the number of selected rules remains quite large, not all of them must be evaluated to classify an individual. In fact, it can be proved[10] that (both in case the individual is new and in case it belongs to the training set) the conditions of exactly $K^{(t)}$ rules must be evaluated, where the choice of the rules to be considered is performed on the basis of the input data characterizing that individual.

Even if the learning algorithm outlined above operates in a numeric way, i.e. both the strengths and the evidence degrees are represented by real numbers ranging in the interval [0,1], the final results of the algorithm can be easily expressed in linguistic form; this is accomplished by associating fuzzy restrictions, corresponding to liguistic terms, with the strengths and the evidence degrees (see figures 1 and 2). In particular, the strength of a rule is replaced by a corresponding linguistic term, which represents its "relevance" in supporting the associated hypothesis, whereas the membership degrees concerning the different terms which appear in the condition parts are expressed by means of linguistic values representing the "evidence" with which the available data match the term.

Obviously, also the Min and Max operations must be replaced by equivalent operators which take as arguments linguistic values. The semantics of these operators is represented by means of tables (an example is reported in figure 3) so that the evaluation of a hypothesis may be obtained by simple and fast table look-ups.

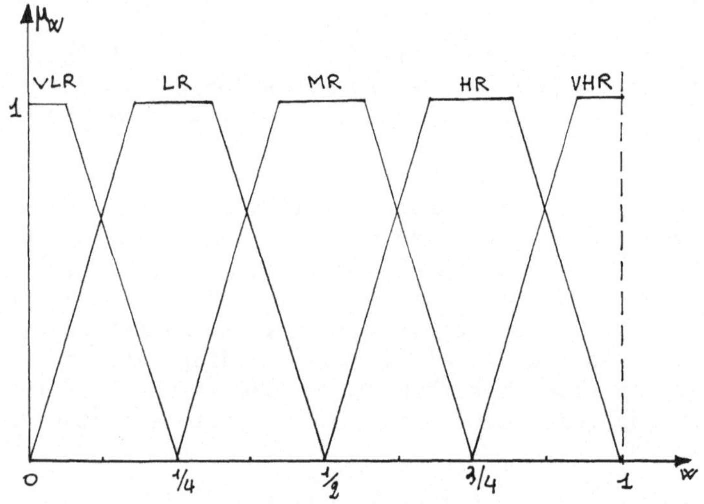


Fig. 1. Definition of fuzzy restrictions corresponding to the linguistic terms: Very Low Evidence (VLE); Low Evidence (LE); Medium Evidence (ME); High Evidence (HE); Very High Evidence (VHE).

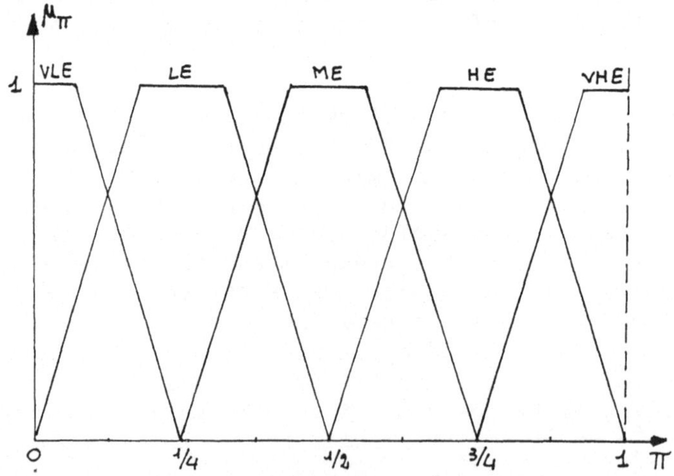


Fig. 2. Definition of fuzzy restrictions corresponding to the linguistic terms: Very Low Relevance (VLR); Low Relevance (LR); Medium Relevance (MR); High Relevance (HR); Very High Relevance (VHR).

| $\mu_\pi$ \ $\mu_w$ | VLR | LR | MR | HR | VHR |
|---|---|---|---|---|---|
| VLE | VLE | VLE | VLE | VLE | VLE |
| LE | VLE | LE | LE | LE | LE |
| ME | VLE | LE | ME | ME | ME |
| HE | VLE | LE | ME | HE | HE |
| VHE | VLE | LE | ME | HE | VHE |

Fig. 3. Table look-up for determining the linguistic value of the evidence of a hypothesis H, given that one of the condition part of the rule:

$$C \xrightarrow{w} H$$

and the linguistic value of the strength w.

## DEFINITION OF THE LINGUISTIC VARIABLES

As stated in the previous section, each linguistic variable $V_j$ may assume any linguistic value $\lambda_{jk_j}$ belonging to the set $\Lambda_j$. This set may be defined as a set of arbitrary fuzzy restrictions on the range of values assumed by a numerical base variable $\xi_j$ corresponding to $V_j$.

Two different criteria can be followed in the definition of $\Lambda_j$, according to the desire of making the set of fuzzy restrictions dependent or not on the actual distribution of the $\xi_j$ values in the training set.

In order to make the set $\Lambda_j$ independent of the training set, the restrictions corresponding to the terms $\lambda_{jk_j}$ must be defined a-priori, possibly taking into account the expert's knowledge. In this case the expert will take care of defining the terms, trying to approximate as well as possible their meaning in the everyday practice. As an example, let us consider the definition of the fuzzy restrictions concerning the base variable $\xi_i$ = "Serum Albumine Concentration". In this case, the knowledge of an expert physician may suggest the introduction of the following linguistic terms.

$$\Lambda_i = \{\text{Normal, Slightly Altered, Altered, Very Altered}\} \qquad (12)$$

In figure 4 is reported a possible definition of these linguistic terms, where the fuzzy restrictions have been placed according to the practice of expert physicians.

On the contrary, it may seem reasonable to take into account the probability distribution of $\xi_j$ values in the training set, thus making the definition of $\Lambda_j$ more sensitive to the training set itself. In fact it could seem that this second alternative allows the system to better exploit the information carried by the training set.

Let us consider again as base variable the "Serum Albumine Concentration" introduced above; in figure 5 are reported its probability densities for subjects belonging to four predefined classes (continuous lines); each class contains the subjects for whom the hypothesis of a given severity degree of liver impairment is true. Figure 5 presents also the fuzzy restrictions (dashed lines) which are defined on the basis of the four probability distributions and which are used to describe four linguistic terms which will constitute the set $\Lambda_j$.

Formally, the fuzzy restrictions are defined as:

$$\mu_k(\xi) = \begin{cases} \xi \geq \bar{\xi}_k: & \begin{cases} 1 & \text{if } \bar{\xi}_k = \underset{1\leq j\leq 4}{\text{Max}}\ \bar{\xi}_j \\ e^{-\frac{1}{2}\left(\frac{\xi-\bar{\xi}_k}{\sigma_k}\right)^2} & \text{if } \bar{\xi}_k \neq \underset{1\leq j\leq 4}{\text{Max}}\ \bar{\xi}_j \end{cases} \\ \xi < \bar{\xi}_k: & \begin{cases} 1 & \text{if } \bar{\xi}_k = \underset{1\leq j\leq 4}{\text{Min}}\ \bar{\xi}_j \\ e^{-\frac{1}{2}\left(\frac{\xi-\bar{\xi}_k}{\sigma_k}\right)^2} & \text{if } \bar{\xi}_k \neq \underset{1\leq j\leq 4}{\text{Min}}\ \bar{\xi}_j \end{cases} \end{cases}$$

where $\bar{\xi}_k$ and $\sigma_k$ represent the mean value and the standard deviation computed for the individuals belonging to the k-th class ($1\leq k\leq 4$).

Some experimental studies have been undertaken recently in order to compare the two criteria outlined above. Even if the second criterium takes into account the properties of the training set, the first one has produced results showing a significantly lower misclassification error; it should be noticed however, that the use of linguistic terms defined independently of the training set makes the learning algorithm produce a slightly larger set of rules.

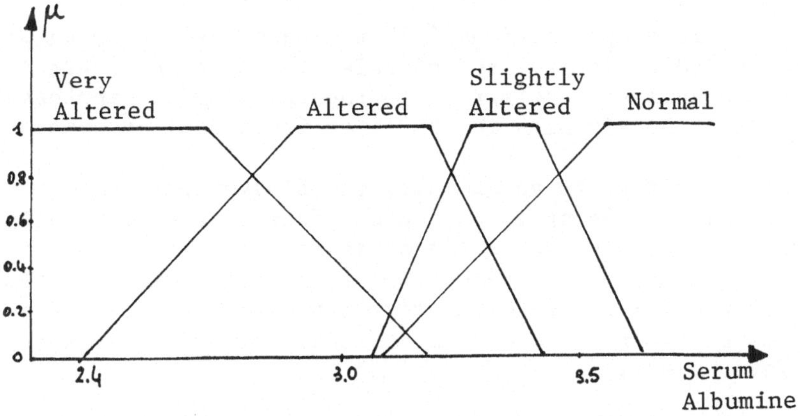


Fig. 4. Definition of the linguistic terms qualifying the linguistic variable corresponding to the base variable "Serum Albumine". This definition takes into account the medical experience.

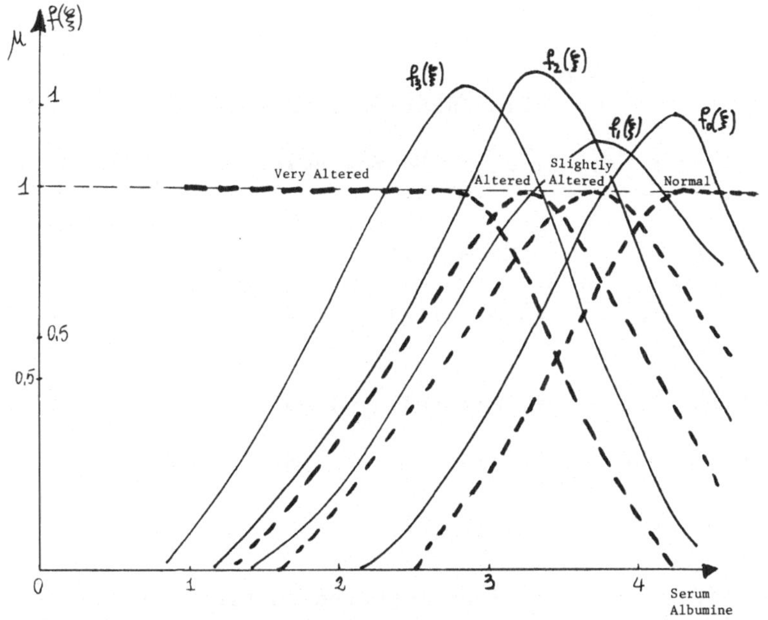


Fig. 5. Probability densities of "Serum Albumine" in a training set containing subjects belonging to four pre-defined classes. The dashed lines represent the fuzzy restrictions defined according to (13).

EXAMPLE OF APPLICATION

Among various applications[11,12] the described learning procedure has been applied to infer a set of rules (used in the context of liver disease analysis) referring to the emission of hypotheses about the presence of impairments in the liver functional assessment.

Let us consider, as an example, the Biosynthetic impairment, which is normally present in patients suffering from cirrhosis. This impairment is hypothesized on the basis of three laboratory tests (ALB,CHE,PT)*, each one corresponding to a linguistic variable, assuming values belonging to the term set (12). For each impairment four levels of severity have been defined, namely: Absent, Low, Severe, Very Severe.

Two rules referring to the Biosynthetic impairment are reported in the following:

$C_1$ ≡ (ALB=Very Altered)<u>and</u>(CHE=Very Altered)<u>and</u>(PT=Very Altered)

$$\xrightarrow{\text{Very Highly Relevant}} \text{BIOSYNTHETIC IMPAIRMENT VERY SEVERE} \tag{14}$$

$C_2$ ≡ (CHE=Very Altered)

$$\xrightarrow{\text{Highly Relevant}} \text{BIOSYNTHETIC IMPAIRMENT VERY SEVERE} \tag{15}$$

Let us consider a patient with the following values for the laboratory tests:

ALB = 2.6 CHE = 2125 PT = 52.3

From Figure 4 (and analogous definitions for the other two laboratory tests) we obtain:

$\mu_1$ = Poss (ALB = Very altered) = 1.00

$\mu_2$ = Poss (CHE = Very altered) = 0.95

$\mu_3$ = Poss (PT = Very altered) = 0.93

By using the fuzzy restrictions reported in Figure 1 to translate $\mu_1$ and $\mu_2$ and $\mu_3$ into the corresponding linguistic terms, we obtain:

*The three laboratory tests are: ALB = Serum Albumine; CHE = Serum Cholin Esterase; PT = Prototombin Time.

(ALB = Very Altered) = Very High Evidence (VHE)

(CHE = Very Altered) = Very High Evidence (VHE)

(PT = Very Altered) = Very High Evidence (VHE)

Then, taking into account the whole condition $C_1$:

$C_1$ is VHE (16)

In the same way it is obtained:

$C_2$ is VHE (17)

By accessing the table of Figure 3 with the pair <VHE,VHR> for rule (14) and with the pair <VHE,HR> for rule (15) we get respectively:

$H_{Df}$ = BIOSYNTHETIC IMPAIRMENT VERY SEVERE is VHE

and:

$H_{Df}$ = BIOSYNTHETIC IMPAIRMENT VERY SEVERE is HE

By taking the Max over the two values, we finally obtain:

H is VHE (Very High Evidence)

To show that also rule (15) may be useful for supporting the associated hypothesis, consider the following set of values:

ALB = 3.08 CHE = 2120 PT = 55.1

The corresponding membership values are:

$\mu_1 = 0.24$ $\mu_2 = 0.96$ $\mu_3 = 0.55$

The associated linguistic terms are:

(ALB = Very Altered) = Low Evidence (LE)

(CHE = Very Altered) = Very High Evidence (VHE)

(PT = Very Altered) = Medium Evidence (ME)

The condition parts of the two rules report:

$C_1$ is LE

$C_2$ is VHE

By taking into account the strengths of implication, we obtain:

BIOSYNTHETIC IMPAIRMENT VERY SEVERE is LE (from rule (14))

BIOSYNTHETIC IMPAIRMENT VERY SEVERE is HE (from rule (15))

So that the rule (15) allows the system to conclude that the hypothesis H is HE (High Evidence).

CONCLUSIONS

The approach of fuzzy production rules for decision making represents an effort towards uniformity and effectiveness. The term uniformity may be regarded as referring to a twofold aspect of the adopted formalism. Firstly, there is an internal uniformity, which concerns the different parameters (to use a general term) occurring in the elementary structures composing the system, i.e. the production rules; in fact, all these parameters (evidence associated to the input data - of whatever kind they are -, confidence degrees in taking the decision, relevance of the different patterns for emitting a give hypothesis) are represented by the same formalism, allowing thus a uniform treatment. Second, there is an external uniformity, in that the linguistic terms used by the system to take the decisions are defined as having a meaning which corresponds rather strictly to the meaning in the everyday practice.

As regards the effectiveness of the methodology, it should firstly be noticed that the above mentioned "external" uniformity makes much easier the explanation of the operations performed by the system[14]; that is, it is immediate to write procedures which explain to the user why a particular decision has been taken (in fact, it is sufficient to list the rules which have been satisfied together with the various linguistic terms describing the evidence of the condition part and the relevance of the rule).

Beyond this first benefit, fuzzy production rules offer also a mathematical formalism which allows the development of efficient learning algorithms; a particular learning algorithm has been outlined in the paper and its main advantages have been described.

A last remark, concerning the behavior of this algorithm, refers to the results obtained in some applications in real world problems. The algorithm has been applied to learn classification rules in different medical domains[11,12,13]; in these cases,characterized by small training sets and by large numbers of features of different kinds, the misclassification errors were lower than the ones obtained by using more classical decision theoretic methodologies.

ACKNOWLEDGEMENTS

The authors are indebted to Prof. G. Molino and Dr. C. Cravetto, of the Second Medical Clinic of the University of Turin, for providing the experimental data used in testing the algorithm.

REFERENCES

1. W. Van Melle: "A Domain Independent Production Rule System for Consultation Program," Proc. 6th IJCAI (Tokyo, 1979), pp. 923-925.
2. S. M. Weiss, C. A. Kulikowski: "EXPERT: A System for Developing Consultation Models," Proc. 6th IJCAI (Tokyo, 1979), pp. 942-947.
3. R. O. Duda, P. E. Hart, N. J. Nilsson: "Subjective Bayesian Methods for Rule Based Inference System," SRI Tech. Report 124 (Menlo Park, 1976).
4. M. Ishizuka, K. S. Fu, J. T. Yao: "Inexact Inference for Rule Based Damage Assessment of Existing Structures," Proc. 7th IJCAI (Vancouver, 1981), pp. 837-842.
5. J. A. Barnett: "Computational Methods for a Mathematical Theory of Evidence," Proc. 7th IJCAI (Vancouver, 1981), pp. 868-875.
6. T. G. Dietterich, R. S. Michalski: "Inductive Learning of Structural Descriptions: Evaluation Criteria and Comparative Review of Selected Methods," Artificial Intelligence, vol. 16 (1981), pp. 257-294.
7. R. S. Michalski: "Pattern Recognition as Rule Guided Inductive Inference," IEEE Trans. on PAMI, vol. PAMI-2, (1980), pp. 349-361.
8. L. A. Zadeh: "Fuzzy Sets as a Basis for a Theory of Possibility," Fuzzy Sets and Systems, Vol. 1, (1978), pp. 3-28.
9. L. A. Zadeh: "A Fuzzy Set Theoretic Interpretation of Linguistic Edges," J. of Cybernetics, Vol. 2 (1972), pp. 4-34.
10. R. De Mori, L. Saitta: "Automatic Learning of Fuzzy Naming Relations over Finite Languages," Information Sciences, Vol. 21 (1980), pp. 93-139.
11. L. Saitta, P. Torasso: "Fuzzy Characterization of Coronary Disease," Fuzzy Sets and Systems, Vol. 5 (1981), pp. 245-258.
12. L. Lesmo, L. Saitta, P. Torasso: "Computer Aided Evaluation of Liver Functional Assessment," Proc. 4th Annual Symp. on Computer Applications in Medical Care (Washington, 1980), pp. 181-189.
13. L. Lesmo, L. Saitta, P. Torasso: "Learning of Fuzzy Production Rules for Medical Diagnosis," in 'Approximate Reasoning in Decision Analysis,' M. Gupta and E. Sanchez, Ed.s, North-Holland Publ. Co. (1982), pp. 249-260.

14. L. Lesmo, M. Marzuoli, G. Molino, P. Torasso: "An Expert System for the Evaluation of Liver Functional Assessment," Proc. First Annual Conference of American Association for Medical Systems and Informatics (Washington, D.C., 1982), pp. 23-27.

# DECISION SUPPORT WITH FUZZY PRODUCTION SYSTEMS

Thomas Whalen and Brian Schott

Decision Sciences Laboratory
Georgia State University
Atlanta, Georgia

## I. PRODUCTION SYSTEMS

### Introduction

A production system is primarily a formalism for representing knowledge about any particular area of problem solving (Farley, 1980). A program written as a production system is a collection of "production rules," which take the form "if left-hand-side proposition is true, then right-hand-side proposition is true." Such a program may operate in one or both of two modes. In the "left-hand-side driven" or "modus ponens" mode, a rule compares the database with the left-hand-side proposition; if the comparison succeeds, action is taken in the database and/or the environment to make the right-hand-side proposition true. In the "right-hand-side driven," "goal driven," or "modus tollens" mode, the program attempts to find or create circumstances which satisfy the left-hand-side of some rule whose right-hand-side moves the system closer to its goal.

The propositions on the left-hand-side and the right-hand-side of each production rule may be simple propositions about the value of one particular element of the database. More complex propositions can be built up using the Boolean operators AND, OR, and NOT. Propositions may be descriptive ones such as "last quarter's average price is above $5," they may be goal propositions such as "your goal is to maximize market share," or they may be action propositions such as "you should charge $6 next quarter."

One of the most important features of a production system is that individual IF-THEN production rules interact with each other by means of the data base rather than by means of traditional programming control structures. Thus the only way that one rule is connected to another is if the right-hand-side variables of one rule match the left-hand-side variables of the other. (Note that in an interactive production system the database includes input from the user in response to production-rule output.)

## Advantages

One of the most important advantages of production rule formalisms for representing knowledge in computer-executable form is that production systems are highly modular. Knowledge is elicited in chunks from an expert in the area to be covered by the production system. Each chunk becomes a production rule which represents the expert's answer to a "what if" question; the question itself may be posed by the expert, by the systems designer, or by the system itself in dialogue with the expert.

This way of organizing knowledge in discrete chunks which interact with each other only by means of a data base is a very natural way of modeling human cognitive processes, because it comes closer than any other feasible programming technique to mirroring the actual process by which a human expert performs his decision-making task (Newell and Simon, 1972).

A production system program comes into being through a rule creation process which focuses on expert knowledge in the form in which the expert is most able to provide it, rather than upon clever programming and coding techniques. Many of the most successful production system applications use a translation utility which allows natural language input to be converted automatically or nearly automatically into executable code.

Because a production system has such a high degree of modularity, it is also easy to extend by adding new production rules and altering only those existing rules which have been superceded. This has allowed systems such as MYCIN and DENDRAL to continue to evolve in a smooth, natural progression since their initial implementation.

Another advantage is that the simple, standardized structure of the rules themselves, combined with their one-to-one correspondence with sentences in a natural language, gives a production system an important capability to explain its own actions and "reasoning." Every step in a chain of reasoning can be documented by the machine in terms of the IF-THEN rules in conjunction with the facts from the data base. This gives a production system the potential for a high degree of clarity and naturalness at the level of the individual steps. It also provides the ability to produce an execution trace or "protocol" designed to make sense to anyone knowledgable in the subject matter of the problems the system is designed to solve, without any necessity for special training in computer science.

Disadvantages

A computer program embodying a production system does not have a beginning, a middle, or an end, but only a collection of rules communicating with each other through the data base. This makes the overall execution sequence of a production system extremely difficult to predict in advance, despite the simplicity of the micro-level behavior of a production system and the readability of the post-execution trace. A production system is neither truly sequential nor truly parallel from a programming point of view, since the execution of one production rule changes the data base and the data base determines which production rule will be executed next.

Closely related to this difficulty is the problem of conflict resolution, which occurs when the contents of the data base match the left-hand-side of more than one rule at some particular point in processing and the effect on the data base is different depending upon which rule is executed first. In fact, a very common case of this arises when two rules are eligible to be executed at time $t$ but either rule will change the data base in such a way that the other rule will no longer be eligible to be executed at time $t + 1$. Most successful programs which use the production system structure resolve this problem by adding a procedurally coded executive routine, or else by using nonintuitive "housekeeping" rules to maintain various switches and flags in the data base. Since these mechanisms have no real world significance, they are not meaningful to the subject-matter expert who provides the basic rules.

Because classical production systems are rooted in Aristotelian logic, production rules must be executed in an "all or nothing" manner; rules are executed if the pattern of their left-hand-side proposition matches a pattern found in the database and are not executed otherwise. In effect, all variables are treated as if they were measurable only on nominal scales: in order for a classical production system to make use of a variable measured on a numerical scale, the scale must be divided into regions and production rules written whose left-hand-sides ask whether the variable is or is not located within a given region. In other words, there is no concept of "closeness".

In order to mitigate the problems caused by "all or nothing" execution, some successful applications of production systems such as MYCIN introduce various sorts of uncertainty or approximation measures on top of the basic production system structure. However, these accommodations, like the routines for conflict resolution, have a tendency to cause the loss of some of the advantages of pure production systems.

The fuzzy production systems methodology which is described in this paper takes a more radical approach to these disadvantages by examining the potential of building a new kind of production system upon a framework of non-Aristotelian logic.

## II. FUZZY LOGIC

Fuzzy logic is built upon the concepts of many valued logic and fuzzy mathematics. In many valued logic, statements are not restricted to the "TRUE" and "FALSE" of classical logic, but may have other values, such as "UNDEFINED," "UNKNOWN," etc. (Lucasiewicz, 1970), (Haak, 1974). Many valued logic has drawn considerable interest from the computer science community (Rine, 1977). For example, the CASE statement of ALGOL and PASCAL can usefully be studied as a generalization of the two-valued IF-THEN-ELSE statement.

The elements of fuzzy logic which play the largest role in fuzzy production systems are fuzzy variables, linguistic variables, fuzzy propositions, and fuzzy implications (Zadeh, 1973), (Zadeh, 1975). (Two other concepts, truth value restrictions and inverse truth value restrictions, are important in some fuzzy logics, but are not used in the present study, and thus will not be discussed at length.)

A <u>fuzzy variable</u> is a variable whose value in any one particular instance is a fuzzy subset of a universe of discourse. The "base variable" of a fuzzy variable is a crisp (nonfuzzy) variable which takes on individual values from the same universe of discourse. Thus, the domain of the base variable is the universe of discourse itself, whereas the domain of the fuzzy variable is the collection of possible fuzzy subsets of this universe of discourse, known as the fuzzy powerset of that universe of discourse.

A very useful way of representing a fuzzy variable is as a vector variable. The number of dimensions in the vector variable is equal to the number of elements in the universe of discourse of the fuzzy variable; the value of each dimension gives the degree to which the corresponding element of the universe of discourse belongs to the fuzzy value of the fuzzy variable.

A <u>linguistic variable</u> is a special kind of fuzzy variable; it is distinguished by being associated with an appropriate linguistic syntax and semantics. The syntax generates a collection of words and phrases to describe all possible fuzzy subsets of the variable's universe of discourse; the semantics provides a unique name from the syntax for any fuzzy value of the variable, and a unique fuzzy value from the fuzzy powerset of the variable's universe of discourse for any syntactically-correct phrase. (Several synonymous phrases will generate the same fuzzy set and several similar fuzzy sets will generate the same phrase, so the two functions are not strictly mutually inverse.) As a result, the value of a linguistic variable may be given either directly by a fuzzy subset of that variable's universe of discourse, or indirectly by a linguistic description of that fuzzy subset. For example, a linguistic variable such as PRICE might take the value "very high" or the value "more or less medium," as well as a vector representation.

Fuzzy propositions are statements which assert that the value of a linguistic variable is one particular fuzzy subset in the domain of that variable (fuzzy powerset of that variable's universe of discourse). An example of a fuzzy proposition is "next month's sales will be better than average". There are two possible interpretations of the fuzziness of this proposition: either it expresses our uncertainty about what the exact value of next month's sales will be, or it expresses the inherent vagueness of the concept "better than average." This distinction is in the province of the fuzzy mathematical theory of truth value restrictions and inverse truth value restrictions; however, in the present study the two interpretations are treated alike within the system, and consideration of which of the two is meant is left to the expert and the user.

The propositions of fuzzy logic, like those of other logical systems, allow compound propositions to be constructed from simple propositions. Fuzzy logic provides five operators for this purpose; one corresponding to the Boolean NOT, two versions of AND ($\cap$ and $\cdot$ ), and two versions of OR ( $\cup$ and + ). The two "ANDs" and the two "ORs" give the same results as the respective Boolean operators when the propositions on which they operate are crisp. However, the operators can give results different from one another in the case of fuzzy propositions in which the values of the fuzzy variables are fuzzy sets. The universe of discourse of a compound fuzzy proposition is the Cartesian product of the universes of discourse of the simple fuzzy propositions from which it is constructed.

A fuzzy implication is a fuzzy relation between two fuzzy propositions; it takes the form: "IF P1 THEN P2," where P1 and P2 are both fuzzy propositions. This relation defines two functions, generalized modus ponens and generalized modus tollens, which are fuzzy extensions of two of the fundamental concepts of Aristotelian logic. Generalized modus ponens is a function from the domain of the fuzzy variable of proposition P1 into the domain of the fuzzy variable of proposition P2. In generalized modus ponens the hypothetical value of the fuzzy variable referred to by P1 is compared with the actual current value of that fuzzy variable. If the actual and hypothetical values of the fuzzy variable in P1 resemble one another closely, the output of the function is similar to that specified by P2. On the other hand, if the actual value of the fuzzy variable in P1 does not resemble the hypothetical value, then the output of the generalized modus ponens function resembles the fuzzy subset called "UNKNOWN", whose vector representation consists entirely of ones. In other words, "anything is possible" if the condition of the implication is very poorly satisfied; the implication provides no information as to what values of the fuzzy variable referred to in P2 are possible and what values are not.

Our current experimental fuzzy production system uses only generalized modus ponens; thus it is the fuzzy equivalent of a left-hand-side driven classical production system. Many successful production system applications, however, rely heavily on the classical logical principle of modus tollens, in which information about the variable on the right-hand-side of a production rule is used to draw conclusions or to generate goals regarding variables on the left-hand-side of the production rule. Preliminary investigation of the application of generalized modus tollens to a right-hand-side driven fuzzy production system is under way.

Over half a dozen definitions of fuzzy implication have been put forth. They differ in the details of the process whereby P1, P2, and the additional "given" proposition interact to produce the fuzzy proposition which is deduced (Baldwin and Pilsworth, 1980), (Bandler and Kohout, 1980), (Wilmott, 1980). Our research at present relies on the definition of implication which is based on the work of Lukasiewicz (1970) in many valued logic research.

## III. IMPLEMENTATION OF A FUZZY PRODUCTION SYSTEM

### A. Fuzzy Logic System

The computerized fuzzy logic system and the natural-language interface used in the present research both make extensive use of a collection of APL functions published by Wenstop (1980). Aside from the availability of these published functions, APL offers other advantages to implementation of fuzzy logic systems. APL functions and variables can both be represented by English words which reflect their meanings, and the default conditions of APL syntax often allow a number of variables and functions to be combined into a "sentence" whose English and APL interpretations are compatible. Thus the following sentences

1. SHARE IS LOW TO MEDIUM IMPLIES
   NEWPRICE SHOULDBE BELOW OTHERSNEWPRICE

2. SHARE IS LOWER LOWER MEDIUM
3. OTHERSNEWPRICE IS NOT (UPPER MEDIUM) TO HIGH
4. NEWPRICE SHOULDBE LOW TO MEDIUM

correspond to well-formed APL statements within the system we have implemented on a small computer (IBM 5100). The fact that APL is interpreted rather than compiled makes it easy to add or modify individual production rules and see the results almost immediately.

There are three kinds of functions in APL: constant functions, monadic functions, and dyadic functions. These three kinds of functions correspond to many of the structures in the English language. For example, constant functions (which require no arguments to produce a result) very often play the role of adjectives; examples in the present system (after Wenstop) are HIGH, MEDIUM, and LOW. Monadic functions, which take a single argument, may represent adverbs such as VERY and FUZZILY or prepositions such as ABOVE and BELOW. The principal dyadic or two-argument functions used in fuzzy logic are the verb IMPLIES and the conjunctions AND and OR. Another dyadic function, SYLLOGISM, is used internally to link production rules to elements of the database but is transparent to both the user and the subject-matter expert. The English-language interface of an APL-based system of fuzzy logic is greatly facilitated by the fact that monadic functions always precede their arguments and dyatic functions are placed between their arguments.

Wenstop's quasi-natural language forms the basis from which the linguistic system used in this research was developed. The linguistic system used in this subsection will be discussed in five parts: variables and constants, vocabulary (functions), syntax, semantics, and linguistic approximation.

Variables and Constants. In this system the value of a fuzzy variable is represented by a vector of eleven dimensions. Each dimension corresponds to one "base value," a point on a discretized mathematical or psychological continuum. The 11-dimensional format is desirable because of its association with deciles. Thus, often the base values range from 0% to 100% in increments of 10%.

The universes of discourse in the example used in this paper are given in Table 1, in which the numbers represent the discrete base-variable values defining each of the eleven points in the universe of discourse: The range and position of the base values are set by the context of the problem at hand. In the present case the base values represent an arithmetic progression whenever possible; however, in the case of PCT a power function progression was selected, since when n firms are competing the "medium" market share is 1/n rather than 1/2. The base values shown are percent market shares when n = 4. In general, base values are computed by

Table 1. Base Values for Fuzzy Variables

| Metric for Universe of Discourse | Base Values | | | | | | | | | | |
|---|---|---|---|---|---|---|---|---|---|---|---|
| simple deciles: | 0 | 10 | 20 | 30 | 40 | 50 | 60 | 70 | 80 | 90 | 100 |
| PCT (market share): | 0 | 1 | 4 | 9 | 16 | 25 | 36 | 49 | 64 | 81 | 100 |
| YEAR (time): | 0 | 1 | 2 | 3 | 4 | 5 | 6 | 7 | 8 | 9 | 10 |
| DOLLAR (price): | 0 | 2 | 4 | 6 | 8 | 10 | 12 | 14 | 16 | 18 | 20 |

raising each of the eleven decile fractions 0, .1, .2, .3, ..., .9, 1.0 to the power $\log_2(n)$ and expressing the result as a percentage. This function has the requisite property of mapping 0 to 0, .5 to 1/n, and 1 to 1, and it allocates the other values in a reasonably smooth manner.

In the Wenstop system five primary fuzzy sets (primary terms) are the basis of the other linguistic values generated and used in the system. Table 2 gives the vector representation which defines each of the five fuzzy sets: the numbers give the compatibility between the named set and the corresponding base values on the scale appropriate to the context in which the name is used.

Table 2. Five Primary Fuzzy Sets

| Linguistic Name | Vector Representation | | | | | | | | | | |
|---|---|---|---|---|---|---|---|---|---|---|---|
| LOW: | 1 | 1 | 0.7 | 0.3 | 0.1 | 0 | 0 | 0 | 0 | 0 | 0 |
| MEDIUM: | 0 | 0 | 0 | 0 | 0.4 | 1 | 0.4 | 0 | 0 | 0 | 0 |
| HIGH: | 0 | 0 | 0 | 0 | 0 | 0 | 0.1 | 0.3 | 0.7 | 1 | 1 |
| UNKNOWN: | 1 | 1 | 1 | 1 | 1 | 1 | 1 | 1 | 1 | 1 | 1 |
| UNDEFINED: | 0 | 0 | 0 | 0 | 0 | 0 | 0 | 0 | 0 | 0 | 0 |

Table 3. Example of the function "CRISP"

| | Scale values of base variable DOLLARS: | | | | | | | | | | |
|---|---|---|---|---|---|---|---|---|---|---|---|
| DOLLAR | $0 | $2 | $4 | $6 | $8 | $10 | $12 | $14 | $16 | $18 | $20 |
| | Values and Corresponding Vector Representations: | | | | | | | | | | |
| $8 | 0 | 0 | 0 | 0 | 1 | 0 | 0 | 0 | 0 | 0 | 0 |
| $9 | 0 | 0 | 0 | 0 | 1 | 1 | 0 | 0 | 0 | 0 | 0 |
| $9.25 | 0 | 0 | 0 | 0 | 1/3 | 1 | 0 | 0 | 0 | 0 | 0 |
| MEDIUM[a] | 0 | 0 | 0 | 0 | .4 | 1 | .4 | 0 | 0 | 0 | 0 |

[a]MEDIUM is from a psychological continuum and does not use CRISP.

Sometimes a crisp value is available for a particular fuzzy variable. If the crisp value happens to equal one of the discretized base values of the fuzzy variable, the representation is a vector with a 1 in the corresponding dimension and zeroes elsewhere. However, for intermediate values a mathematical continuum is needed to supplement Wenstop's psychological continuum. For this reason, we have developed an APL function named CRISP which uses linear interpolation to compute a fuzzy approximation to a crisp datum which lies between two adjacent discretized base values. For example, using the DOLLAR base value Table 3 explains the operator CRISP. The APL function CRISP does a linear interpolation and then normalizes the interpolation factors so that at least one of the two compatibility values equals 1.0.

Vocabulary. The vocabulary of the linguistic system is adapted from (Wenstop, 1980).

| Lexical Category | Generic Symbol | Category Members |
|---|---|---|
| variable name | X | any variable-name (e.g. PRICE) |
| primary terms | T | LOW, MEDIUM, HIGH, UNKNOWN, UNDEFINED |
| hedges (modifiers) | H | ABOVE, EQUALTO, BELOW, AROUND, RATHER, MOREORLESS, UPPER, LOWER, VERY, NOT, NEITHER, POSSIBLY, TRULY, INDEED, FUZZILY |

| | | |
|---|---|---|
| connectives | C | AND, OR, TO, EXCEPT |
| convenience verb | CV | SHOULDBE, IS (this is different from Wenstop's IS) |
| logical implication | I | IMPLIES |

Syntax. Rules of syntax are used to combine vocabulary into fuzzy propositions and implications. Let the symbol V stand for any linguistic value. The symbol X stands for the symbolic name of a linguistic variable. Then an assignment is "X is assigned V". Also any linguistic value V can be substituted for any of the following compound symbols (derived from the "generic symbols" in the section on "Vocabulary" above):

(X) (T) (H V) (V C V) (V CV V)

Next we define a Basic Linguistic Phrase (BLP) as having one of the following two syntactical forms.

$X_1$ CV H $X_2$

$X_1$ CV H V

Corresponding examples of these two BLPs are

PRICE IS BELOW OTHERSPRICE

NEWPRICE SHOULDBE EQUALTO MEDIUM.

Also define AOR to be either the connective AND or OR. Then the syntax for representing expert rules as fuzzy production rules is as follows.

($BLP_1$ AOR $BLP_2$ AOR ... AOR $BLP_k$) IMPLIES $BLP_0$

Semantics. Semantics refers to the meaning of the words in the system vocabulary and the meaning of the system syntax. To maintain brevity only a few examples of the semantics of this system are given here. More complete and explicit definitions of the system's semantics are implicit in the APL functions which define it.

| | |
|---|---|
| NOT V | the complement of V |
| $V_1$ AND $V_2$ | fuzzy intersection of $V_1$ and $V_2$ |
| $V_1$ OR $V_2$: | fuzzy union of $V_1$ and $V_2$ |

| | |
|---|---|
| $V_1$ TO $V_2$: | $V_1$ OR $V_2$ or anything in between |
| $BLP_1$ IMPLIES $BLP_0$: | fuzzy implication relation |
| $V_1$ SYLLOGISM $BLP_1$ IMPLIES $BLP_0$: | generalized modus ponens function |

The fuzzy implication relation, IMPLIES, is expressed as a fuzzy subset of the Cartesian product of the universes of discourse for the variable of $BLP_1$ and the variable of $BLP_0$. Membership in the relation varies directly with the membership of the variable of $BLP_0$ and inversely with the membership of the variable of $BLP_1$. The fuzzy relation is an 11 by 11 matrix whose x,y element,

$$m_{BLP_1 \text{ IMPLIES } BLP_0}(x,y),$$

equals

$$\text{minumum } (1,\ 1-m_{BLP_1}(x) + m_{BLP_0}(y)).$$

(The symbol $m_F(\cdot)$ is the fuzzy characteristic function of the fuzzy set F.)

SYLLOGISM is the name given to the action of applying a production rule (the fuzzy relation $BLP_1$ IMPLIES $BLP_0$) to a database value ($V_1$). The result is a suggested value of the variable contained in $BLP_0$.

Linguistic Approximation. An intermediate result of the execution of each rule is an 11-dimension vector. Since the ultimate user is a human being, it is desirable to convert the 11-dimension vector into a natural language equivalent. Wenstop's APL functions include a routine named LABEL which converts the vector representation of a fuzzy value into a linguistic description of the value's location and degree of imprecision. The process of conversion is called linguistic approximation. For example, an 11-dimension vector for PRICE which is 11 ones would be converted by linguistic approximation into UNKNOWN. For details on the operation of the LABEL function, see (Wenstop, 1980).

B. Expert Time and User Time in the System

Expert Time. The "expert" in a fuzzy production system is a person who possesses knowledge about the subject-matter of the decisions the system is to support; the goal of the system is to elicit this knowledge from the expert at a convenient time, represent it in a collection of fuzzy production rules, and use these rules together with the user's knowledge of the specific decision situation to come up with suggestions in "user time." Because of the separation in time, the system makes sense whether the "expert" and the "user" are two different people or the same person at different times; in the latter case the system serves as an aide-memoire rather than as an independent expert.

Before sitting down at the computer, the expert prepares a list of answers to "What if..." questions. Once this tentative list of rules has been prepared, the expert enters them into the computer during a session in which the computer provides guidance with prompts. The next step is to assemble the entered rules into a form which the APL system can operate on. At this point we discuss the assembly procedure used to transform the linguistic implication entered by the expert into a form acceptable by the APL system. Although rules with compound antecedents and variables in the consequent are possible in the system, we will discuss the following simple rule:

PRICE IS BELOW OTHERSPRICE IMPLIES NEWPRICE SHOULDBE MEDIUM

In the explanation which follows we distinguish between a database value and its corresponding generic value by underlining the specific data base value. For example, the firm's actual price is PRICE; the hypothetical price in the expert's implication rule is PRICE. In order to execute the example fuzzy production function, the variables PRICE and OTHERSPRICE must have actual values in the data base. The result of the execution of the example production rule yields a particular fuzzy set which will be suggested to the user as the value of NEWPRICE. The above linguistic implication is transformed into the following structure for APL.

---

NEWPRICE SHOULDBE

(PRICE)

syllogism

(PRICE IS BELOW OTHERSPRICE) IMPLIES (NEWPRICE SHOULDBE MEDIUM)

---

More generally the syntax of a fuzzy production function as it is being executed is as follows

---

$\underline{X}_0$ SHOULDBE

($\underline{X}_1$ IS $\underline{H\ V}_1$)

syllogism

($X_1$ IS H $V_1$) IMPLIES ($X_0$ SHOULDBE $V_0$)

---

The present system allows for fuzzy production rules whose antecedent (left-hand-side) is a compound fuzzy proposition if all the conjunctions are of the same type (all ANDs or all ORs); the consequent, or right-hand-side, must be a simple fuzzy proposition. The authors have proved theorems which allow deductions with compound antecedents to be decomposed into a set of simpler deductions (Whalen, 1981). If ANDs connect the propositions in the antecedent, then the operation of fuzzy union can be used to connect the results of the disaggregated deductions; if ORs connect the propositions then fuzzy intersection can be used to combine the deductions. The explanation for this counterintuitive result lies in the logical equivalence of "If A then B" with "B or not A".

User Time. "User time" refers to the actual use of a fuzzy production system prepared in "expert time" to support decisions in a changing and ill-defined environment. A fuzzy production system operates with a high degree of conceptual parallelism, avoiding the paradoxes and programming difficulties presented by the conflict resolution problem in classical production systems. This is accomplished in the present implementation by a device called "round scheduling." A "round" begins when the user initializes or updates the database to represent his current state of knowledge at whatever degree of precision or vagueness is appropriate. Each production rule then is executed, and its output is displayed to the user. (This output will be more or less vague depending on the degree to which the contents of the data base match the fuzzy pattern on the left-hand side of the rule. When large numbers of rules are involved, it is convenient to screen out rules whose output is so vague as to be useless in the current situation; this corresponds roughly to the case of rules in a classical production system which are not executed because their left-hand-sides do not match the database.) The user evaluates the collection of fuzzy advice given him by the outputs of all the rules, and he may take some action in the environment as a result. The next round begins when the user reports the results of his evaluations and actions to the system by updating the database.

## IV. THE PRICE INTERACTION GAME EXAMPLE

The price interaction game is a computerized simulation developed by Frazer (1975). In the game the student/manager has only one explicit decision variable, PRICE. Implicitly, the student/manager also controls his firm's objective; the objective is usually to balance the conflicting profit and market share goals of a company. In the game's competitive marketplace several firms all try to sell a single product with a known per-unit fixed cost and per-unit variable cost. The game is played for a fixed number of annual periods. The firm's expert has suggested the following data base for the decision support system.

Database

| variable name | initial value | base value |
|---|---|---|
| PRICE | $10 | DOLLAR |
| AVGPRICE | $10 | DOLLAR |
| OTHERSNEWPRICE | LOWER MEDIUM | DOLLAR |
| *NEWPRICE | UNKNOWN | DOLLAR |
| TIME | 0 | YEAR |
| SHARE | 25% | PCT |
| SHAREGOAL | (LOWER MEDIUM) to UPPER MEDIUM | PCT |
| OTHERSSHAREGOAL | NOT ABOVE UPPER MEDIUM | PCT |
| *NEWSHAREGOAL | UNKNOWN | PCT |

In the data base the two variable names which are distinguished by an asterisk (*) are meant to appear only on the right-hand-side of a fuzzy production. Variable names in the database which include the word NEW in the compound variable name all refer to variables in the upcoming decision period. Variable names which do not include the name NEW assume the values from the decision period just completed. The initial values show the assumed values of the variables before the game is played. It can be seen from the initial value column that some of the variables are rather crisp and others are rather fuzzy. The variables are grouped according to common base value scales.

Six fuzzy productions are listed below. The first four suggest values for the variable NEWPRICE (price to be charged next quarter), and the last two make suggestions regarding NEWSHAREGOAL (importance of the goal of building market share relative to the goal of short-term profit).

1. SHARE IS LOW TO MEDIUM
   IMPLIES NEWPRICE SHOULDBE BELOW OTHERSNEWPRICE.

2. TIME IS LOW TO LOWER LOWER MEDIUM
   IMPLIES NEWPRICE SHOULD BE LOWER LOWER MEDIUM.

3. SHARE IS LOW TO MEDIUM AND TIME IS BELOW LOWER MEDIUM
   IMPLIES NEWPRICE SHOULDBE MOREORLESS LOW.

4. TIME IS FUZZILY HIGH AND SHAREGOAL IS FUZZILY LOW
   IMPLIES NEWPRICE SHOULDBE VERY HIGH.

5. SHARE IS BELOW MEDIUM AND TIME IS BELOW MEDIUM IMPLIES NEWSHAREGOAL SHOULD BE VERY HIGH.

6. SHARE IS BELOW MEDIUM OR TIME IS BELOW MEDIUM IMPLIES NEWSHAREGOAL SHOULD BE ABOVE OTHERSSHAREGOAL.

The result of executing the six fuzzy productions when the database is in the initial state is shown below.

(1) NEWPRICE SHOULDBE BELOW LOWER MEDIUM

| | | | | | | | | | | |
|---|---|---|---|---|---|---|---|---|---|---|
| 1 | 1 | 1 | .6 | 0 | 0 | 0 | 0 | 0 | 0 | 0 |

(2) NEWPRICE SHOULDBE MOREORLESS LOW

| | | | | | | | | | | |
|---|---|---|---|---|---|---|---|---|---|---|
| 1 | 1 | .6 | 0 | 0 | 0 | 0 | 0 | 0 | 0 | 0 |

(3) NEWPRICE SHOULDBE MOREORLESS LOW

| | | | | | | | | | | |
|---|---|---|---|---|---|---|---|---|---|---|
| 1 | .9 | .6 | .2 | 0 | 0 | 0 | 0 | 0 | 0 | 0 |

(4) NEWPRICE SHOULDBE UNKNOWN

| | | | | | | | | | | |
|---|---|---|---|---|---|---|---|---|---|---|
| 1 | 1 | 1 | 1 | 1 | 1 | 1 | 1 | 1 | 1 | 1 |

(5) NEWSHAREGOAL SHOULDBE UNKNOWN

| | | | | | | | | | | |
|---|---|---|---|---|---|---|---|---|---|---|
| 1 | 1 | 1 | 1 | 1 | 1 | 1 | 1 | 1 | 1 | 1 |

(6) NEWSHAREGOAL SHOULDBE ABOVE UPPER MEDIUM

| | | | | | | | | | | |
|---|---|---|---|---|---|---|---|---|---|---|
| 0 | 0 | 0 | 0 | 0 | 0 | 0 | .6 | 1 | 1 | 1 |

Rules 1 and 2 give similar but not identical results. Rule 3 gives a result for NEWPRICE which is close to the result of rule 2, but it also takes into account rule 1. Rules number 5 and 6 both suggest values for NEWSHAREGOAL. Notice that since in rule 5 the antecedents are connected by AND, the result of rule 5 is UNKNOWN. However, in rule 6 the antecedent parts are connected by OR; the result of rule 6 is more precise, ABOVE UPPER MEDIUM.

Presumably the user would select the least fuzzy suggestions from among the outputs of the various rules which adhere to his other criteria for acceptance. In the absence of other evidence the user might select the result of production rule 3 as the best guide to price, and the result of rule 6 as the best guide to the firm's balance of goals.

## V. COMPARISON WITH THE MODEL OF INEXACT REASONING IN MYCIN

One of the most successful current practical applications of inexact reasoning through computerized production systems is MYCIN (Shortliffe and Buchanan, 1975). Thus, it is instructive to compare the fuzzy production system approach with the model of inexact reasoning which is embodied in MYCIN.

While there is a very sophisticated representation of quasi-statistical uncertainty in MYCIN, the word "close" remains undefined in MYCIN as in classical production systems. The contribution of MYCIN is to wed a very powerful production system with a definition of "maybe" which is closely based on an understanding of human cognitive processes. The left-hand-side of a MYCIN production rule is an exact and certain proposition, almost identical to the left-hand-side of a classical production rule. The right-hand-side of some MYCIN production rules, however, only assert that a proposition might be true, with a particular degree of uncertainty. Similarly, the database values in MYCIN may also have some degree of quasi-statistical uncertainty. However, the question of whether the rule is executed at all depends on whether the left-hand-side of the rule matches the data or not, in an all-or-nothing sense.

Uncertainty in MYCIN is based on separate measures of belief and disbelief in order to model human cognitive processes rather than theoretical statistics. Thus, MYCIN's deduction procedures do not exactly follow the rules of classical or Bayesian statistics. Interestingly, the results of MYCIN do approximate those of Bayesian statistics when the latter methodology is provided with appropriate simplifying assumptions.

In contrast to the quasi-statistical approach of MYCIN, fuzzy production systems are primarily oriented to approximate, vague and qualitatively diffuse concepts. However, the uncertainty of MYCIN can be stretched to accomodate the concept of approximation, and the nonstatistical uncertainty of fuzzy production systems is capable of a quasi-probabilistic interpretation in certain cases.

## VI. SUMMARY

A fuzzy production system is a production system built upon fuzzy logic. The system uses a data base of linguistic variables: i.e., fuzzy variables processed through a linguistic analysis and synthesis system to allow a more natural interaction with the user in terms of the linguistic names of fuzzy sets rather than their vector representations.

Another major component of a fuzzy production system is the set of fuzzy implications which form the production rules of the system. The intrinsic lattice structure which underlies fuzzy mathematics makes it easy to support ordinal or even interval scale data in a fuzzy production system, as well as nominal or qualitative types of values. In addition, the concept of "close" is readily defined; if the data comes close to matching the left-hand-side of a particular rule without actually matching it completely, then the outcome of the rule is somewhat less definite than it would have been if the match had been better.

The concept of fuzzy production systems provides a direct model of human-like deductions about variables ranging from crisp, exact concepts to very vague and diffuse concepts, all within a unified system of approximate reasoning. The conflict resolution problem of classical production systems is avoided because fuzzy logic is tolerant of situations which would be paradoxical in the Aristotelian logic of classical production systems.

Our research on fuzzy production systems is strongly oriented toward a judgemental interaction between the user and a computerized decision support system. The primary role of the computer is as an "intelligent scratchpad" which keeps track of more chunks of information than is possible for human shortterm memory, while imposing the fewest possible computer-oriented restrictions and distortions on the way the knowledge is represented. The secondary role is to draw tentative conclusions from that knowledge by means of a set of fuzzy implication rules provided by a human expert, and to present these conclusions cogently as suggestions to the user in natural language.

## VII. REFERENCES


Baldwin, J. F. and Pilsworth, B. W., 1980, Axiomatic Approach to Implication for Approximate Reasoning with Fuzzy Logic, Fuzzy Sets and Systems, 3, 193-219.

Bandler, Wyllis and Kohout, Ladislav, 1980, Fuzzy Power Sets and Fuzzy Implication Operators, Fuzzy Sets and Systems, 4, 13-30.

Farley, Arthur M., 1980, Issues in Knowledge-Based Problem Solving, IEEE Transactions on Systems, Man and Cybernetics, SMC-10, 446-459.

Frazer, R. A., 1975, "Business Decision Simulation: A Time-Sharing Approach," Reston Publishing Co., Reston, Virginia.

Haak, Susan, 1974, "Deviant Logic," Cambridge University Press, London.

Lukasiewicz, Jan, 1970, "Selected Works" (edited by L. Borkowski), North-Holland, Amsterdam.

Newell, A. and Simon, Herbert, 1972, "Human Problem Solving," Prentice-Hall, Englewood Cliffs, New Jersey.

Rine, David C. (editor), 1977, "Computer Science and Multiple-Valued Logic," North-Holland, Amsterdam.

Shortliffe, E. and Buchanan, B., 1975, A Model of Inexact Reasoning in Medicine, Mathematical Biosciences, 23, 351-379.

Whalen, Thomas, 1981, Compound Conditionals Under Two Compositional Rules of Inference, Unpublished Working Paper, Georgia State University Decision Sciences Laboratory.

Wenstop, Fred, 1980, Quantitative Analysis with Linguistic Variables, Fuzzy Sets and Systems, 4, 99-115.

Wilmott, Richard, 1980, Two Fuzzy Implication Operators in the Theory of Fuzzy Power Sets, Fuzzy Sets and Systems, 4, 31-36.

Zadeh, Lotfi A., 1973, Outline of a New Approach to the Analysis of Complex Systems and Design Processes, IEEE Trans. Systems Man and Cybernetics, SMC-3, 28-44.

Zadeh, Lotfi A., 1975, The Concept of a Linguistic Variable and its Application to Approximate Reasoning - III, Information Sciences, 9, 43-80.

# IMPRECISION IN COMPUTER VISION*

Ramesh Jain and Susan Haynes

Intelligent Systems Laboratory
Department of Computer Science
Wayne State University
Detroit, Michigan 48202

## INTRODUCTION

Visual perception has perplexed researchers in philosophy and psychology for centuries. Now it's also perplexing computer scientists. Despite persistent efforts by many noted scientists, it is still unclear how our brain "sees" the visual signals received by our eyes [10]. Earlier it was believed that a good understanding of optics, the retinal image, and the anatomy and physiology of eye and brain would unravel the puzzle of visual perception. However, the many advances in these fields have not solved the problem.

The last two decades have seen a growing interest in building computer systems that can "see" using a camera as an eye. Computer vision systems have potential application in diverse fields. Most research endeavors in computer vision have been to develop systems which can understand an image. Such a task, called image understanding or scene analysis, usually involves extraction of the description of the scene in terms of the objects present in the image and the spatial relationships among them. The goal of these systems is to obtain 3-dimensional information about the scene from its 2-dimensional projection, the image.

Many factors, such as illumination, surface properties of objects, geometrical shape, viewing angles, occlusion of an object

*This research was supported by the National Science Foundation under grant number MCS81000148.

by its own parts or by some other objects, convolve in the process of image formation. Moreover, the projection from a 3-dimensional world to a 2-dimensional image is many-to-one; this makes the process of recovery of the information in the image in terms of objects and relationships between objects more difficult. It is well recognized by researchers in computer vision [19,21,23] that imprecise information from many sources must be combined to understand the contents of an image.

Difficulties caused by complexity and imprecision of the data are not limited to image understanding problems. Researchers in various fields, particularly those working with humanistic systems, have faced these problems for a long time. As is well known, probability and statistics provide powerful tools for dealing with the uncertainty in events; they play major roles in the design of many complex engineering systems and in the analysis of many intricate processes. In many situations, the complexity is due to imprecision rather than due to uncertainty. Statistics and other conventional mathematical tools have resulted in rigorous analysis of a phenomenon which has no relation to the problem. In fact, the growing concern with precise mathematical analyses of complex processes has resulted in over-formalization in many fields. This observation led Zadeh to propose fuzzy set theory [27,28]. He states [9]:

"In general, complexity and precision bear an inverse relation to one another in the sense that, as the complexity of a problem increases, the possibility of analyzing it in precise terms diminishes. Thus 'fuzzy' thinking may not be deplorable, after all, if it makes possible the solution of problems which are much too complex for precise analysis."

Gaines [9] shows that approximate reasoning has been considered to be a very important tool by scientists and philosophers for a long time.

We believe that fuzzy set theory proposed by Zadeh, offers a useful tool to represent imprecise information and permits the combination of imprecise information obtained from various sources. In this paper we show that many image understanding systems use methods which can be very naturally represented using fuzzy sets. We review some application of fuzzy set theory in image understanding and related areas. A dynamic scene analysis system, called Vili is described briefly. This system, currently in its infancy, will exploit imprecise features and will employ approximate reasoning for the analysis of complex dynamic scenes.

## COMPUTER VISION SYSTEMS

Since most computer vision systems are monochromatic and monocular, in this paper we will assume that the input to a computer vision system is one or more 2-dimensional arrays of numbers representing intensity values in the picture. The nature of the output of the system depends on the application. Generally, a system is required to recognize some or all of the objects in the scene and to extract the spatial relationship among them. Since most systems have been concerned with one image for the analysis of a static scene, unless otherwise stated, our discussion in this section will be concerned with a static scene analysis system.

A scene analysis system usually comprises two distinct phases: segmentation and interpretation. In segmentation, the input array is partitioned into disjoint regions; each region satisfies a predicate based on some property of the intensity values. During the interpretation, domain knowledge is applied to extract some specific information. Most researchers [14,19,23] (Fischler [8] is an exception) believe that interpretation follows segmentation. Starting with the intensity array, several operations, such as thresholding, region growing, and region merging, may be sequentially applied to transform the intensity array so that in the final segmented array all the pixels belonging to a region contain the same value or label. Some approaches for segmentation do not use goal-dependent information [2,14,21]; others are strongly influenced by task-dependent information [6]. A major problem with most current segmentation approaches is that each step in the sequence results in a definite output. An error made at an early step in the sequence will most likely result in wrong results. By employing task-dependent knowledge such error may be corrected. The seemingly simple task of segmentation has proven very difficult [23]. Some researchers believe that obtaining a perfect segmentation is not realizable.

Marr [21] observed that, as for other large symbolic computer programs, four principles for the organization of computer vision systems are:

1. principle of explicit naming,
2. principle of modular design,
3. principle of least commitment, and
4. principle of graceful degradation.

The principles of least commitment and graceful degradation are the most relevant to the tasks of segmentation and of object recognition; non-deterministic methods may play important roles in implementation of these principles. We will discuss these methods here.

At many stages in a vision system, a datum may be interpreted in many different ways: possibly none of the interpretations being significantly better than others or possibly there is not yet sufficient evidence that a particular interpretation is the correct interpretation. In such situations commitment to a particular interpretation may mislead subsequent processes. Following the principle of least commitment, no interpretation should be selected prematurely. One should retain all significant competing interpretations until one of them unequivocally describes the datum. The motto of the system should be "Do not commit if you are uncertain."

If a system follows the principle of least commitment then in realistic images containing degraded data it may never make a decision because it may never be certain. It would be very irratating, however, if a system never gave any result because it was never certain. The principle of graceful degradation states that degrading the data should not prevent one from delivering at least some of the answer. Thus, if there is no possibility of a certain answer, a "rough" answer based on the available evidence should be computed by the system. It might be helpful if the system also indicates its confidence in the results.

We believe that fuzzy set theory and related methods of representation of imprecision and of handling poor definition offer adequate and potentially powerful formalisms to implement the principles of least commitment and graceful degradation in computer vision system. We first briefly describe concepts of fuzzy sets and property sets and then show their applications in this domain.

## FUZZINESS

In classical set theory, an object O may or may not be a number in a set X of object classes. The characteristic function of an element in a set in classical set theory is binary, either 0 or 1. The restriction of the characteristic function to 0 and 1, limits the applicability of this set theory. In many practical applications the membership of an object in a class is not binary; classes of tall men, of intelligent persons, of elongated objects are some classes whose membership cannot be represented satisfactorily by only 0 and 1.

By allowing the characteristic function of a set to take values in the interval [0,1] we can dramatically extend the applicability of the set theory. Observing this, Zadeh introduced the concept of fuzzy sets [27,28]. A fuzzy set A is a subset of the universal set X such that the characteristic function $f_A(x)$, $x \in X$ takes any value in the interval [0,1]. Thus A is represented by:

$$A = \int_{x \in X} (f_A(x) \,|\, x) \tag{1}$$

where $f_A$: $x \to [0,1]$ is called membership function of A and represents the A-ness of x. Let us consider a set of Bald people in a department, which may be

$$\text{BALD} = \{(1.0|\text{Bill}), (0.8|\text{Tom}), (0.5|\text{Jake}), (0.9)|\text{Bob})\} \tag{2}$$

Thus Bill is completely bald but Jake is partially bald.

The union, intersection and complement operations for fuzzy sets, A and B, are defined, respectively, as follows:

$$A \cup B \triangleq \int_{x \in X} (\text{Max}(f_A(x),\ f_B(x)) \,|\, x) \tag{3}$$

$$A \cap B \triangleq \int_{x \in X} (\text{Min}(f_A(x),\ f_B(x)) \,|\, x) \tag{4}$$

and

$$A' \triangleq \int_{x \in X} ((1 - f_A(x)) \,|\, x) \tag{5}$$

Fuzzy set theory has been studied extensively. Methods have been developed for application of fuzzy sets in diverse fields [3,4,16,20].

In our qualitative dicussion of the application of fuzzy sets in computer vision, in addition to the above definitions, we need to define "property" of a fuzzy set or a group.

Suppose that we are concerned with baldness of a group {Bill, Tom, Jake, Bob} as given by equation 2. Given the individual properties of members of a group, the property of the entire group can be computed using many different approaches; the choice depends on the application. The simplest way is to find the average. Thus k-ness of a group A, $c_k(A)$, is given by

$$c_k(A) = \frac{1}{n} \sum_{i=1}^{n} f_A(x_i); \qquad x_i \in X \tag{6}$$

Using the above equation the baldness of the group {Bill, Tom, Jake, Bob} is 0.8. A detailed discussion of properties of a group is in [1].

## IMPRECISE PROPERTIES

As discussed in earlier sections, computer vision systems infer 3-dimensional information starting with imprecise evidence and employing approximate reasoning. The strong appeal of relaxation processes [5,24] for extracting information is due to their power of combining imprecise local evidence to yield globally consistent results. The ready availability of probabilistic tools encouraged researchers to put relaxation approaches in a probabilistic framework. Haar [13] has proposed a fuzzy set theoretic approach to relaxation. The most attractive aspect of this approach is the flexible and intuitive representation for spatial relationships and competing and co-operating forces. In terms of absolute power, possibly both probabilistic and fuzzy frameworks are equivalent, but the power of a representation is better judged by the convenience in its application [26]. It appears that fuzzy calculus offers a better representation of several processes essential in the representation of an imprecise (not certain) evidence, for combining several pieces of information obtained from diverse sources, and in approximate reasoning.

The recognition task requires identification or measurement of certain features. Pattern recognition techniques are based on measured values of features [6]. Computer vision systems, particularly model based vision systems, work on the basis of the presence or absence of a specific property. The task is complicated by the fact that the presence of the property may not be obvious in a picture. In such a situation one may compute all possible supports for the presence and the absence of such a property and then combine them to assign a fuzzy value for the presence of the property.

Let us consider an example from [16]. It was shown that monotonicity of a region in an accumulative difference picture indicates the direction of motion of an object. An object is moving in the direction of decreasing numbers. What do we mean by monotonicity of a region? Consider Figure 1. Is region 1 monotonic? It appears that we can define monotonicity for a sequence. We can easily classify the sequence

1 2 3 4 5 6 7 8 9

as a monotonic sequence and the sequence

1 2 3 4 5 2 3 4 5

as a non-monotonic sequence. Clearly an object corresponding to the first sequence is moving to the left. What about the second sequence? Our intuitive notion that there may again be an object moving to the left is correct in most cases. The rejection of the second sequence from the class of monotonic sequences may lead to absurd results.

```
    113
    2 7
    4 751
     26                  12223 1              (1)
              38987656565555443211
             889876554646664212555554321
           398685343321  68978765544221
          889876554321   8 498775554321
          75984  334287 398898766554321
          88898765555221898889876554321
         3889876425574  199731
              77    6
              75  1512
 34 6          95
15946          664  345
259488        451   11 11
159488           27666761
269388
468588
356 139
451 1
25

    (2)
```

Fig. 1. Three accumulative difference picture regions. The largest connected region is region 1.

Now let us define a function

F = {ABS(asc-desc)/(asc+desc)} ** 2

where asc and desc represent the number of ascending and descending subsequences of length greater than 2 in the given sequence, respectively. The function F represents monotonicity of a sequence. Note that this function will classify both sequences given above as having perfect (1.0) monotonicity, but will assign 0.0 monotonicity to the sequence

543212345.

By accepting fuzzy monotonicity of a region, we observe each sequence and though the sequence may not be strictly monotonic, assign monotonicity using F. If asc > desc then the sequence is considered an ascending sequence, and if asc < desc then the sequence is considered a descending sequence. Now suppose that there are N rows (or sequences) in a region, such as shown in Figure 1, out of which A are ascending and D are descending. Then the monotonicity of the region may be obtained by combining monotonicities of individual rows. Thus the monotonicity of a region may be defined as

$F_r$ = {ABS(A-D)/(A+D)} ** 2.

A high value of monotonicity for a region indicates strong evidence for the motion of an object and a low value is weak evidence. See [16] for more properties of a region. Incidentally, the monotonicity of the region in Figure 1 using the above equation is 0.384.

Granlund [11,12] uses a novel method of information representation for the special purpose computer designed for image processing, the GOP image processor. A set of operations is applied to each neighborhood in the image (see Figure 2). Each operation measures a particular feature of the neighborhood. The measured feature depends on the level in the hierarchy. Features thus computed are mapped into a 2-dimensional space and arranged in a circular fashion, such that conflicting or incompatible features are located at opposing positions. Using suitable weights, the measured features are combined to represent the event in the neighborhood by a vector. The direction of the vector indicates the dominant event and the magnitude gives the grade of membership of the choice. The GOP processor uses this representation at all levels. This processor which will soon be commercially available, is one of the most powerful image processors designed so far. Note how by allowing fuzzy memberships of events, the representation and combination of information at several levels is simplified.

An important feature of the above and other similar methods [20,25], is that they define an ideal property or feature, k, and then measure the similarity of the observed feature with the ideal. The similarity is used as the measure of k-ness.

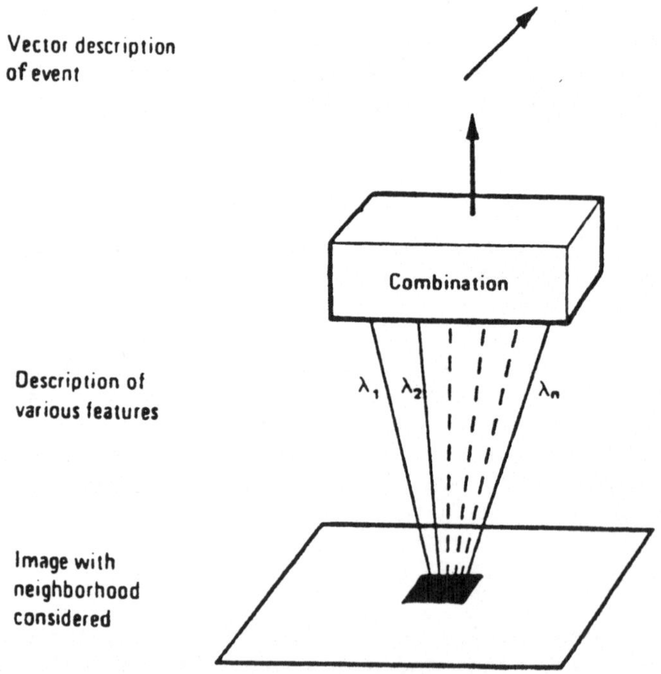


Fig. 2. Illustration of feature generation and information representation.

To see how the principle of least commitment can be implemented, let us consider the following scenario. We find a region in a picture which is a member of one of n classes, $c_1$, $c_2$, ..., $c_n$, of regions. We compare properties of this region with the desired properties of each class and represent the membership of the region as a fuzzy set, $M_1$, given by

$$M_1 = \{(f_{11}|c_1), (f_{12}|c_2), \ldots, (f_{1n}|c_n)\}$$

where $f_{1i}$ is the membership of the region in class $c_i$, obtained using a set $P_1$ of observed properties. Similarly, based on some other set $P_k$ of properties we may derive.

$$M_k = \{(f_{k1}|c_1), (f_{k2}|c_2), \ldots, (f_{kn}|c_n)\}$$

if we have m such sets of properties then these sets may be combined to yield

$$M = \{(f_1|c_1), (f_2|c_2), \ldots, (f_n|c_n)\}$$

where

$$f_j = f_{1j} \oplus f_{2j} \oplus \ldots \oplus f_{mj}$$

$\oplus$ is some operator.

The above method may be used to combine evidence from several sources or may be used in a dynamic scene by considering $P_k$ as a set of properties observed at $k^{th}$ time instant. The final decision about the classification is made based on the set M. The set M considers membership of a region in different classes. If the membership value $f_j$ is much higher than any other $f_i$, then the decision is obvious; otherwise more evidence is needed. In case a decision must be made, the set M has enough information to make a forced decision which, while not an unequivocal choice, does represent the best that can be done based on the available evidence.

This approach allows us to keep all options open so that even the weakest evidence based on $P_i$ has a chance to influence the final result. Another attractive feature of the approach is the possibility of a graceful decision if we are cornered.

In [16] the above approach was used for computing properties of SODP regions and to update the reference frame. This promising approach has not been investigated rigorously, but forms the basis of the system discussed in the next section.

VILI

The fact is that error recovery is a difficult, tedious process -- in general, the more convoluted the original processing, the more time-consuming (and difficult) the recovery from error (and backtracking). It is far better to avoid error in the first place. Now for scene analysis, this is especially true -- scene analysis is extremely complicated, involves many processes, any one of which or combination of which can cause error. By delaying decisions, one can hope to reduce the number and severity of errors. One of the reasons dynamic scenes are appealing (in addition of course to the sheer power and reliability of motion as a primitive) is that they allow yet another domain for application of the principle of least commitment. For example, it may well be that after a few more frames, things become much more clear. This opportunistic procrastination can be of several types: cumulative -- where successive frames each add increasing reliability to the decision, and intermittent -- where one may wish to simply skip over a few frames, to where a change is more drastic, or something has stabilized.

Since it is important to be able to delay over time (i.e. over frames) as well as over processes, we have felt it necessary to design a computer vision system which would incorporate and exploit this procrastination for dynamic scene analysis. We have named this system Vili. This system employs the principles proposed by Marr [21] extensively.

Martin and Aggarwal [22] in their review article on dynamic scene analysis, point out that much of the research in computer vision has followed the human perceptual divisions of peripheral, attentive and cognitive. We wish to make these distinctions explicit.

A thumb-nail sketch of the proposed system follows. The system is shown in Figure 3. Processing will occur on three levels in a roughly pipelined fashion. The lowest level, the peripheral level accepts input in the form of frames and possibly the camera's egomotion parameters (i.e. if the camera is moving, then its velocity). This level, using only rapid (quick and dirty) processing techniques, acts on the greylevels of the frame or frames, extracting regions of interest, for example, regions which show a large amount of change. Masks of these regions, with derived parameters are written to a buffer area which sits between the peripheral and the next level.

The next higher level, the attentive level, focuses a great deal of processing power on one or a very few of the interesting regions. Again, it acts on greylevels. A wide variety of processes will act on a given region extracting more accurate

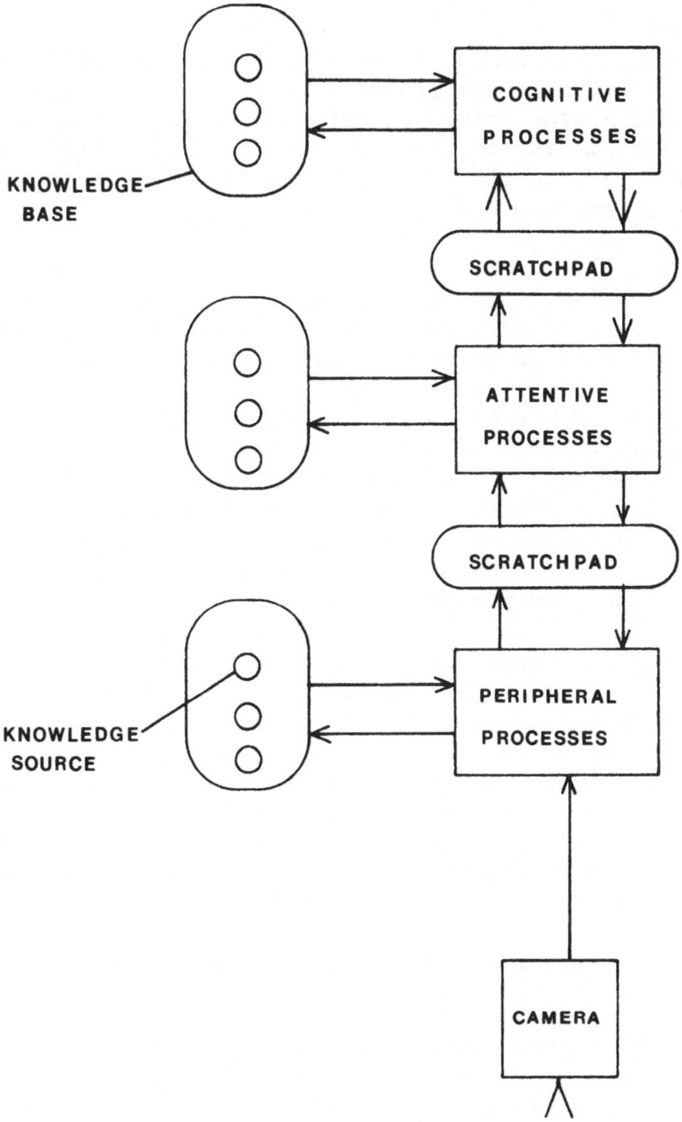


Fig. 3. Dynamic Scene Understanding System.

information about the region. A given process may use the results of another process to aid its computation, or even its own earlier results (for an earlier frame) may help (priming the pump). The processes make use of knowledge of imaging. These reasonably accurate region descriptions are written to a buffer sitting between the attentive level and the next higher level.

The next, and third (final) level is the cognitive level. It obtains, from the buffer, region descriptions. Using world specific knowledge then, it attempts to provide an interpretation of the scene. If it finds it needs more information about some aspect of the scene it can send this "request" to the AC buffer, possibly along with information leading to specific processing on what it's looking for. Through that operation it directs the operation of the attentive level.

At each level we envision the knowledge sources as separate from the processing. For the cognitive level this is an absolute requirement. At the peripheral level the knowledge sources will likely be so primitive as to not warrant the honorific "knowledge."

Now we will describe in greater detail how fuzziness is to be incorporated at each level and places where procrastination is crucial. Ultimately all information resides in the greylevels, the knowledge sources at each level and the camera egoparameters. Coming into the system we've got greyvalues. Going out we've got interpretation. The knowledge sources manipulate the greyvalues to provide these interpretations.

The peripheral level is supposed to obtain from the greyvalues of one or more frames, areas of interest, e.g. regions where changes are taking place. "Interesting" is usually a metric of something like difference (for motion). In [16,18] the difference value is thresholded. A 1 is placed in those pixel locations where the difference value exceeds the threshold, 0s are placed everywhere else. The size of the connected components of 1s are thresholded again to give significant sized regions. Then those regions which have 1s are "interesting". Here "interesting" is a combination of sufficient difference and sufficient size. By limiting the "interesting" metric to 1 or 0, we are throwing away a lot of useful information we've computed. Once we knew the magnitude of the difference at every point, once we knew the size of the region. These information will very likely need to be recomputed at some later time. Clearly the fuzzy concept of degree of membership in a class is useful here.

The advantages of not limiting metrics to 0 or 1 are demonstrated in [15]. Here the task is to detect edges which move. There are a number of ways to do this. One along the lines of the above moving region detection is:

1. compute an edginess value for every pixel, then threshold for "edge" vs. "no edge."
2. for those pixels which are marked "edge," compute the difference value between it and the corresponding value in the next frame. This is a motion detector. Threshold the difference value to indicate "sufficient motion of the edge point" vs. "insufficient motion."

This method clearly has many difficulties. In step 1 we lose low contrast edges, in step 2 we lose slowly moving edges. An alternative described in [15] is to compute the edginess value at every point and the difference at every point in parallel. Combine these two values in a reasonable way and the final value gives an idea about the time varying edginess of every point. Figure 4 shows two frames taken from the beginning and the end of a street scene.

(4a)

(4b)

Fig. 4. Two frames from a dynamic scene.

Early in the sequence only the white car is moving. The very low contrast car to the left is just beginning to enter the scene. The result of applying the operator described on the first two frames, is shown in Figure 5. Only the moving car gives good response. The spurious edges shown arise from registration error (that is, the frames are shifted with respect to each other). Thus the static edginess values are not thrown away before they are needed, but rather are made use of until the last possible minute before decision. This way edginess and difference can both be used to support a decision about time varying edginess.

All ready, when detecting points of interest (called feature points) people include information about how interesting a point is. This idea of letting interest take a range of values has proven very useful for subsequent matching operations (i.e. attentive process) between corresponding points of different frames. It is reasonable to expect that some regions will be more important than others. And at the very least they can be sorted by interest, with the most interesting getting a crack at first priority at the attentive level.

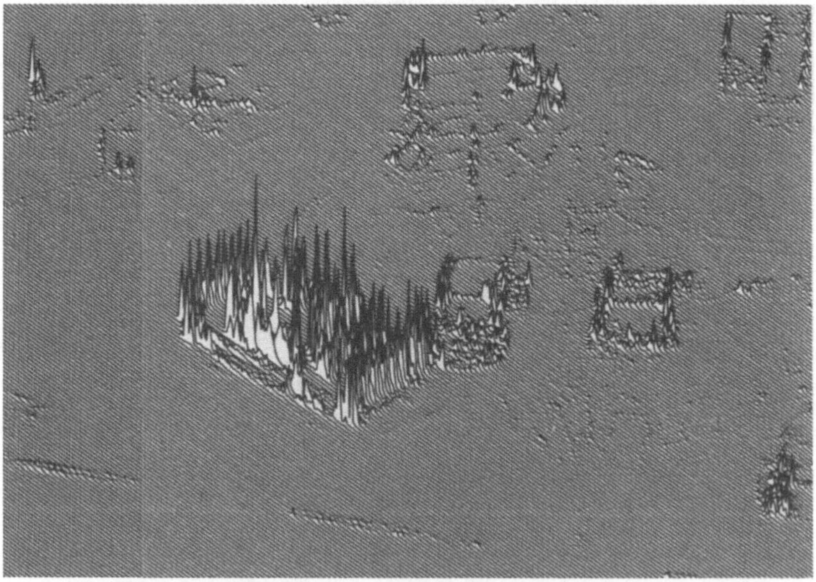

Fig. 5. A windowed perspective plot of time varying edge points of Figure 4a.

Many problems face us: what element has its interestingness measured -- a point, a region? If both, then how should one combine many points' interestingness to arrive at a region's interestingness -- perhaps some extension to fuzzy reasoning will be useful here. The most important point about the peripheral phase is that it is supposed to act as an early warning system [17]. It is required that it reduce the massive amounts of data to some more manageable amount. These manageable data are the regions of interest which are then passed to the attentive phase. We do not want the peripheral phase to make a decision on what areas the attentive phase should work on. We do want decision making delayed as long as possible.

At the attentive level the idea is to focus on a region of interest and do further processing on it. Many different computer vision techniques can be brought to bear on a region. These knowledge sources will consist of image specific knowledge, i.e., how a given real world surface images into greyvalues and greyvalue relationships. Some examples of methods are: texture analysis, shape description, perspective transformation, motion characteristics, shape from shading, stereopsis. Some techniques can make use of the results of others. We envision this level as consisting of several knowledge sources which can communicate results through a central buffer (blackboard) [7]. A given knowledge source may not have enough perfect information given to it to provide a perfect answer. But it may, even with imperfect information be able to give fairly reasonable results. (Of course, even with perfect information it may only arrive at reasonably certain results.) In turn, these fairly reasonable results may be used by a different knowledge source doing some other computation. This second knowledge source needs to know this information is unreliable. The natural way to set this up is to require each knowledge source to provide, along with its results, a measure of reliability or of certainty. This measure will depend on the quality of information provided to the knowledge source. This certainty factor fits very neatly into the fuzzy set paradigm [16]. Very little has been done on calculating certainty (i.e., fuzzy reasoning) given imprecise input from different sources in computer vision. Combining approximate information in a reasonable way is very important -- because these may be the only resources we have available to do the task.

At the attentive level one is dealing with two entities -- pixels and regions. Regions are built out of pixels, but many operations will act on regions, not on pixels. Deciding whether a pixel belongs to region A or B or both or neither becomes important. Combining the disparate answers to this question from different knowledge sources is difficult. It is expected that the fuzzy paradigm will be of use here.

Ultimate control comes from the cognitive level (delaying the decision of which region to analyze to the last moment). The more immediate control of the attentive knowledge sources however is contained within this level. Output is to the interfacing buffer. There is also an internal buffer for use at this level only (blackboard). The information a knowledge source can use comes from several places: the greyvalues, regions (with parameters) provided by the peripheral level, the blackboard, region (perhaps with parameters) from the cognitive level.

The cognitive level takes the results of the attentive level and provides an interpretation of the scene using world specific knowledge. We haven't specified the form the results from the attentive level will take except to say that it contains regions with attributes and likelihoods. It is important, though, that it be as self-contained as possible. If the cognitive level is given knowledge of the attentive level knowledge sources we are violating the principle of modularity. While tending to merge the attentive and cognitive phases does permit greater delay in decision making, control becomes much more difficult. The line between the two phases is fairly arbitrary -- but loosely: the attentive phase contains imaging specific knowledge (context-free), e.g., pixels to regions, regions to region compatibilities. The cognitive level contains world specific knowledge and forms the scene (i.e., region) to world interface. The decisions the cognitive level is making are not along the lines of: for region 1, is attribute A compatible with attribute B? Or even: if region 1 has attribute A, is this compatible with region 2 having attribute B? These are attentive level questions. The cognitive level, as a scene-world interface is concerned with questions like: is attribute A consistent with interpretation X for region 1? Or is interpretation X consistent with interpretation Y?

The world specific knowledge of the cognitive level is stored as models which are networks, procedures, or both. In a static world that is sufficient -- in a dynamic world it is crucial that there also be knowledge of which entities may interact, and in what manners they may interact. Note the importance of procrastination here -- the cognitive level may not be able to decide between two alternative interpretations until some of the entity interactions (motion parameters) are determined -- to do this requires observation over several frames.

To arrive finally at a scene interpretation, the cognitive level must search through the models removing impossibilities. This is not a search for the truth -- it's a search for the best answer(s) available, and the data we have to power the search is not exact. With imprecise attributes, many possible interpretations will result. Since we are arguing from imprecision, the interpretations will also be more or less uncertain. In the search through the model space, the removal of possibilities cannot, in general, be absolute. Some will be mostly eliminated, some merely less likely than others. If there is a chain of reasoning initiating from imprecise data, at every step conclusions are uncertain, and the possibilities will tend to multiply. All these many possible ways of making an error (most of the interpretations will be wrong) arise because we wished to avoid the high probability of definitely making an error when working with absolutes (rather than with "partly trues"). With all these maybes we are building in facilities for graceful degradation of performance and for easy error recovery. These many possible interpretations are not all equally likely -- the chain of interpretation proceeded from unequally reliable segmentations. It's a more efficient use of computer resources to focus on the most likely possibility. A consistent, good interpretation lies somewhere in the range of possible interpretations. If the most likely proves to be in error, the system doesn't stop, it continues with the next most likely alternative.

The principle of graceful degradation is in part, facilitated by the multiple possibilities. It can also be expressed as: the system is working on many possibilities. It is very unlikely that at the end of the processing (that is, when it provides an interpretation) that the interpretation is 100%, absolutely true. We hope for a situation where one possibility is very likely compared to a few (or several) others which have very low probabilities. This will not always happen. What may happen instead is that there will be three or four interpretations which do not have wildly differing likelihoods. When this happens, there is no point in just giving the single most likely interpretation. All the computation on the less likely interpretations will be wasted. We would prefer outputting all the more likely possibilities. This way whatever system is using the output (human, robot), has the information available. The information may still not be used, but at least a more graceful degradation of performance is possible.

Working through the model space with all these varying likelihoods is a stupefying task. It is very likely that the given results of the attentive phase are insufficient to resolve an ambiguity, though the attentive phase has the resources to provide the required datum. Perhaps the attentive phase simply chose to examine an area which was not of interest to the cognitive phase. When the system is entirely data-driven there is a lot of unnecessary work being done. Therefore we envision the cognitive level providing the top-down control of the attentive level. If the cognitive level finds that more information about a region will resolve some ambiguity it will send an instruction to the attentive level (via the interface). This instruction may contain the region, current information available on it, and what sort of information it needs. Note the cognitive level does not know about individual attentive level knowledge sources, so it cannot directly invoke a knowledge source. Rather, the attentive level controller (resource manager, translator, whatever it is) must interpret the command and the attentive level diverts its processing under its own control to the task set by the cognitive level.

When the system is first turned on, nearly all operation will be data-driven, the cognitive level may be tuned to the domain (e.g., it knows it's looking at a traffic scene), but giving very few directives. As frames come in, get processed, the attentive level acquires more and more reliable image attributes, we expect the cognitive level to take a more active role -- channeling more and more of the attentive level's activity as it knows, increasingly, what is important to analyze in the scene.

When there are many frames coming in to a system, there is no need to get hung up on any particular frame. If something is unanswerable or is not derivable at the moment, fine. Just wait a bit because things will change. The ability to procrastinate is a very powerful tool if handled with restraint and knowledge. Our belief is that this virtue, as well as the power of motion as a primitive makes dynamic scene analysis simpler than static scene analysis. And since hard and fast answers have shown themselves to be rare (if not impossible) entities in computer vision, some notation for representing approximate and loose answers is an absolute requirement.

## CONCLUSION

Computer vision is very difficult. In attempting to solve the problem researchers have drawn techniques and inspiration from many fields. Until recently, however, they have been handicapped by trying to obtain definite answers from degraded data and inherently approximate processes. With the recognition that answers cannot be definite comes the problem of representing the indefinite and complex (and for computer vision systems, stress the complex) data and results. The fuzzy sets representation and reasoning paradigm appear to offer a very convenient method of representation and of manipulation at many levels of processing in computer vision systems.

## ACKNOWLEDGEMENTS

We would like to thank the many people in the computer vision group at Wayne State University for their contributions in long discussions about pitfalls in vision in general and about Vili in particular.

The many very fruitful discussions on problems in real scene analysis with Professor Nagel are also gratefully acknowledged. The traffic scene was also proved by Professor Nagel.

REFERENCES

1. A. D. Allen, Measuring the empirical properties of a set, IEEE Trans. Systems, Man and Cybernetics, SMC-4, pp. 66-73, 1974.
2. H. G. Barrow and J. M. Tenenbaum, Recovering intrinsic scene characteristics for images, in "Computer Vision Systems," ed., A. Hanson and E. M. Riseman, Academic Press, 1978.
3. J. C. Bezdek and P. F. Castelaz, Prototype classification and feature selection with fuzzy sets, IEEE Trans. on Systems, Man and Cybernetics, SMC-7, #2, pp. 87-92, February 1977.
4. S. K. Chang, On the execution of fuzzy programs using finite-state machines, IEEE Trans. Computers, vol. C-21, #3, pp. 241-253, March 1972.
5. L. S. Davis and A. Rosenfeld, Hierarchial relaxation for wave-form parsing, in "Computer Vision Systems," ed. A. R. Hanson and E. M. Riseman, Academic Press, pp. 101-109, 1978.
6. R. O. Duda and P. Hart, "Pattern Classification and Scene Analysis," Wiley International, 1973.
7. L. D. Erman, F. Hayes-Roth, V. R. Lesser, and D. R. Reddy, "The Hearsay-II Speech-Understanding system: Integrating knowledge to resolve uncertainty," Computing Surveys, vol. 12, #2, pp. 213-253, June 1980.
8. M. A. Fischler, On the representation of natural scenes, in "Computer Vision Systems," ed. A. R. Hanson and E. M. Riseman, Academic Press, pp. 47-52, 1978.
9. B. R. Gaines, Foundations of fuzzy reasoning, Int. J. Man-Machine Studies, vol. 8, #6, pp. 623-668, 1976.
10. J. J. Gibson, "The Ecological Approach to Visual Perception," Houghton Miflin Company, Boston, 1979.
11. G. H. Granlund, In search of a general picture processing operator, Computer Graphics and Image Processing, vol. 8, #2, pp. 155-178, October 1978.
12. G. H. Granlund, D. Antonsson, J. Arvidsson, M. Hedlund, P. Henden, H. Knutsson, K. Lundgren, B. Nilsson, B. V. Post, and R. Wilson, The GOP Image Processor, in Proceedings of Computer Architecture for Pattern Analysis and Image Data-base Management, pp. 195-200, Hot Springs, Virginia, November 1981.
13. R. L. Haar, The representation and manipulation of position information using spatial relation, TR 923, Computer Science Center, University of Maryland, August 1980.
14. A. R. Hanson and E. M. Riseman, VISIONS: A computer system for interpreting scenes, in "Computer Vision Systems," ed. A. R. Hanson and E. M. Riseman, Academic Press, pp. 303-333, 1978.

15. S. M. Haynes and R. Jain, Detection of Moving Edges, Technical Report, TR 82-004, Department of Computer Science, Wayne State University, Detroit, Michigan, 1982.

16. R. Jain, Applications of fuzzy sets for the analysis of complex scenes, in Advances in Fuzzy Set Theory and Applications, ed. M.M. Gupta, North Holland, pp. 577-587, 1979.

17. R. Jain, Extraction of motion information from peripheral processes, Trans. on PAMI, vol. PAMI-3, #5, pp. 489-503, September 1981.

18. R. Jain and H. H. Nagel, On the analysis of accumulative difference pictures from image sequences of real world scenes, IEEE Trans. on Pattern Analysis and Machine Intelligence, vol. PAMI-1, #2, pp. 206-214, April 1979.

19. R. Jain and J. K. Aggarwal, Computer Analysis of Curved Objects, Proceedings of the IEEE, vol. 67, #5, pp. 805-812, March 1979.

20. E. T. Lee, The shape-oriented dissimilarity of polygons and its application to the classification of chromosome images, Pattern Recognition, vol. 6, pp. 47-60, 1974.

21. D. Marr, Early processing of visual information, A. I. Memo No. 340, MIT, December 1975.

22. W. N. Martin and J. K. Aggarwal, SURVEY-Dynamic Scene Analysis, Computer Graphics and Image Processing, vol. 7, #3, pp. 356-374, May 1978.

23. R. Reddy, Pragmatic aspects of machine vision, in "Computer Vision Systems," ed. A. R. Hanson and E. M. Riseman, Academic Press, 1978.

24. A. Rosenfeld, R. A. Hummel and S. W. Zucker, Scene Labeling by Relaxation Operations, IEEE Trans. Systems, Man, and Cybernetics, vol. SMC-6, pp. 420-433, 1976.

25. B. Widrow, The rubber mask technique I. Pattern Recognition, vol. 5, pp. 175-194, 1973.

26. P. H. Winston, "Artificial Intelligence," Addison-Wesley Publication, Reading, Massachusetts, 1977.

27. L. A. Zadeh, Fuzzy Sets, Information and Control, vol. 8, #3, pp. 338-353, June 1965.

28. L. A. Zadeh, The concept of a linguistic variable and its application to approximate reasoning, Part I, Information Sciences, vol. 8, pp. 199-249, Part II, Information Sciences, vol. 8, pp. 301-357, Part III, Information Sciences, vol. 9, pp. 43-80, 1975.

Authors current address: Electrical and Computer Engineering
University of Michigan
Ann Arbor, Michigan 48109

# FUZZY PROGRAMMING: WHY AND HOW? - SOME HINTS AND EXAMPLES*

Henri Prade

Langages et Systèmes Informatiques
Université Paul Sabatier
118 route de Narbonne
31062 TOULOUSE CEDEX FRANCE

## I. INTRODUCTION

The procedures used by the human mind to manipulate data are not often precisely specified. This situation is not generally due to an intrinsic imprecision of the natural language we used: in English or French, for instance, we are certainly able to express things in a precise way most of the time. It is not generally for the pleasure that people may be imprecise or vague in stating facts or methods, although some of them take advantage of it when they want to hide something! It is not even due to some intrinsic inability of the mind to put things into words. The deep reason of possible imprecision in specifying seems to be elsewhere. Sometimes precise data are simply not available because we have not the tools to perform precise measures or because it is impossible to do it: the processing time of an operation which has not been performed yet can be only evaluated on the basis of our a priori knowledge: this evaluation is more or less precise according to the situations; we have a possibility distribution concerning the value of the processing time. When the information is stored in our memory, we may be somewhat uncertain if we try to remember too precise facts. Other times, precision is not meaningful, especially in the human affairs; for example, the required profile of a candidate to an employment is

*Presented at the Invited Session 'Computer Applications of Fuzzy Set Theory' of the 4th IEEE Int. Computer Software & Applications Conference, Chicago, October 27-31, 1980.

linguistically specified, even if the elements of this profile can be more or less evaluated or graded in a numerical way. Although we may assess arbitrary values to what is only roughly specified, or we may refuse to consider what is imprecise, it seems more natural and useful to accept imprecise data or procedures and try to deal with them as such, because imprecision or fuzziness is often an unavoidable feature in humanistic systems. Moreover, Zadeh (1973) has pointed out that human mind often uses fuzzy labels in reasoning processes: "Thus, the ability to manipulate fuzzy sets and the consequent summarizing capability constitute one of the most important assets of the human mind as well as a fundamental characteristic that distinguishes human intelligence from the type of machine intelligence that is embodied in present-day computers". There is a balance between the imprecision of a procedure and its adaptability to a variety of situations.

Let us give two examples of fuzzy procedures to which it is referred in the following. First, a fuzzy decision procedure concerning selection of students: The considered examination criteria are for instance: Mathematics, English and French. For each of them, the global evaluation is built as a weighted mean of the grades got by the student during the past year. These grades are supposed to belong to the set of six levels A, B, C, D, E, F which may be linguistically interpreted respectively as 'very good', 'good', 'rather good', 'rather bad', 'bad', 'very bad' (This kind of grading has been actually used in French schools). What is required to be selected is for instance "to be at least good in mathematics and to be at least rather good in French or English". The second example is a fuzzy guidance procedure, such that: "Go about 300 meters towards North; you will see a narrow street; you will take it and almost immediately you will see a bank". Note that in this example the success of the procedure is guaranteed (if backtracking is allowed) by the existence of a landmark (i.e. the bank) which enables to control (in the sense of automatics) the position.

In the following, we first consider the various kinds of fuzzy instructions (we are not concerned here with the understanding and the translation (into fuzzy instructions) of imperatives expressed in natural language). A semantic pattern-matching procedure taking into account the uncertainty in meaning is then briefly sketched and its use is emphasized. Finally, preliminary discussions about fuzzy data types are presented.

## II. FUZZY INSTRUCTIONS

An instruction may be fuzzy because of the presence of fuzzy arguments (i.e. arguments whose values are fuzzy entities), or of fuzzy functions (i.e. functions whose values are fuzzy entities whatever their arguments) or of fuzzy predicates (i.e. predicates yielding truth-values other than true or false). An instruction may also be fuzzy because its arguments are fuzzily designated (even if the value of each possible argument is non-fuzzy).

Fuzzy arguments are, for instance, fuzzy sets on which are performed adapted versions of set-theoretic operations (Umano et al. 1978). Another important type of fuzzy data is 'fuzzy number'. For example, in the selection procedure stated above, we have to compute the weighted mean of grades which are fuzzy sets of the real interval [0,20], i.e. fuzzy numbers such the ones whose membership functions are pictured below.

Owing to a parametric representation of the membership functions, arithmetic operations extended to fuzzy numbers are easy to perform (Dubois and Prade 1980). Note that what we get is a parametric representation of a fuzzy set, not its semantic label; if we want a linguistic output, it is necessary to have a linguistic approximation function for labelling fuzzy sets of a given universe of discourse (Bonissone 1978, Eshragh and Mamdani 1979, Mandic and Mamdani 1980). The arithmetic operations performed on fuzzy numbers are not fuzzy functions because when restricted to ordinary numbers, they yield the usual results.

A fuzzy function such as "INCREASE SLIGHTLY" may be defined via usual arithmetic operations and fuzzy numbers (e.g. $x \to x + EPS$ where EPS is the label of a fuzzy number whose possible values are small) or directly by an analytical function mapping $x \in R$ on a fuzzy number (then, this function can be canonically extended to include fuzzy numbers in its domain).

Conditional instructions, which control the sequencing, may have premises (i.e. the "if part" of the instructions) which are fuzzy because of two reasons:

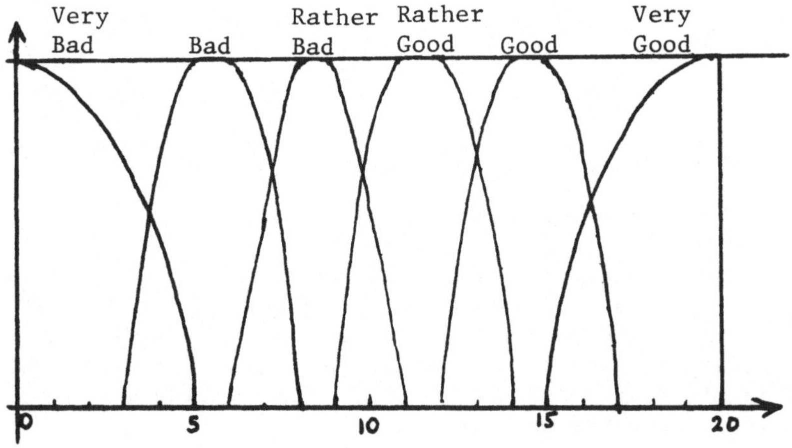

- the predicate is fuzzy (e.g. LARGE, APPROXIMATELY EQUAL)
- the values of the arguments are fuzzy, because they result from the execution of fuzzy instructions belonging to the kinds studied above.

For instance, in the selection procedure, we may add: "if the candidate is YOUNG, then, INCREASE SLIGHTLY his global grade in mathematics". But, YOUNG is a fuzzy predicate modeled by a fuzzy set on the universe of ages, therefore the consistency between the age of a considered candidate and Young may be intermediary between 0 and 1; in other words, it is neither completely true that the candidate is YOUNG nor completely false.

The height of the intersection of the fuzzy or non-fuzzy set modeling the value of the argument with the fuzzy or non-fuzzy set modeling the predicate will be interpreted as the truth-value of the premise (consistency degree of the values of the argument and of the predicate). This truth-value will be denoted by h $(AR \cap PRE)$. AR and PRE respectively label the fuzzy set of possible values of the argument and the fuzzy set of the elements satisfying the predicate.

A similar procedure can be used to determine the truth-value of the premise when the predicate is n-ary (Dubois and Prade 1980, Adamo 1980). After the evaluation of the truth-value of the premise two actions are possible: either execute the 'then part' of the instruction or go to the following instruction. A truth-value equal to h $(AR \cap PRE)$ is associated to the first action. The quantity h $(AR \cap \overline{PRE})$ is associated to the second one; $\overline{PRE}$ has a membership function which is deduced from this of PRE by complementation to 1.

Generally we have:

$$h\ (AR \cap \overline{PRE}) \neq 1 - h\ (AR \cap PRE)$$

The formula

$$h\ (AR \cap \overline{PRE}) = 1 - h\ (AR \cap PRE)$$

holds if the argument has a non-fuzzy value.

If we concurrently consider both so-valued actions, we can develop a graph of possibilities which represents a non-deterministic program, whose halting conditions are difficult to determine (Floyd 1967). Practically we are led to use the rule of the preponderant term of the alternative (Zadeh 1973): only the action corresponding to the greater truth-value is executed.

However, in case where the different branches of the control structure correspond to different modifications of the <u>same</u> set of variables and have the same sequencing after, it is possible to execute <u>one</u> intermediary instruction, which uses the conditional statements as a table giving a sampling of fuzzy points of a function, in order to "extrapolate" the result according to Zadeh's theory of approximate reasoning (1978). For instance, it would lead to modulate the slight increasing of the grade in mathematics according to the age of the candidate in the above example.

In the preceding, the manipulated fuzzy entities were non-fuzzily designated. When, entities are reachable by a fuzzy designation it may exist some ambiguity regarding the addressed entity, if its designation is too fuzzy in the context.

Examples: LIST-ALL BIG OBJECTS
ADD (WEIGHT BIG BOX) (WEIGHT LITTLE BOX)

In the first example, the system has to find out all the more or less big objects recorded in its associative memory. In the second example, the system has to find out the box whose qualification by BIG has a maximal meaningfulness, (Goguen 1976), the box whose qualification by LITTLE has a maximal meaningfulness and finally to add their weights. In both cases, a procedure of fuzzy pattern-matching (see next section) is needed.

Note that the result, is non-fuzzy as soon as, when the designation is ambiguous in the context, the fuzzy pattern-matching procedure indicates it and stops the evaluation. Another kind of execution will lead to list all the objects with non-zero membership grade in the fuzzy set BIG, with the corresponding grades, for the first example (Le Faivre 1974). A similar execution of the second instruction would yield the possibility distribution of the weights we can obtain!

Especially in robotics, in order to execute fuzzy instructions we have to interpret them in a non-fuzzy way. This "defuzzification" is now imposed by the confrontation with the real world. The execution may remain fuzzy if the real conditions of the execution are ignored.

Examples:
- ($\alpha$) do <u>some</u> steps on the left
- ($\beta$) increase by <u>x</u> the pressure (x is an identifier of the already computed fuzzy quantity)
- ($\gamma$) go-and-fetch the <u>greenish</u> <u>big</u> cub.

In examples ($\alpha$) and ($\gamma$) where the occurences must have discrete values (the numbers of cubs or steps are integers) a fuzzy instruction is defuzzified by using the already cited maximum meaningfulness principle : from the possibility distribution resulting from the evaluation of the fuzzy elements in the instruction, we keep among the really existing occurences those which correspond to a maximal possibility. In the example (β) we directly take a pressure value which corresponds to the maximum of the possibility distribution associated with the fuzzy quantity x, because the pressure can be continuously assigned. The second example of the introduction sketches a robotics-like sequence of fuzzy instructions. Note that in the execution of an instruction such that "go about 300 meters", a demon (in the sense of Artificial Intelligence) should be activated in order to indicate when begins and ends the area where it is more or less possible to find the narrow street the robot is looking for. (This area is defined by the support of the fuzzy set "about 300 meters" i.e. the set of distances with non-zero membership grades).

In robotics, after Zadeh's (1973) and Goguen's (1975) hints, Smith (1975) designed, in the framework of JASON project, a plan generator fitted to an uncertain environment and ill-stated problems. In the same spirit (Tanaka & Mizumoto 1975, Uragami et al. 1976) have dealt with the control of robots by means of fuzzy instructions. Shaket's system (1976) allows fuzzy designations of objects in a Winogradian world of blocks. Fuzzy hints are used by Gershman (1976) in order to guide a robot in a maze.

In some situations, results have to be defuzzified in order to be used. But in other situations fuzzy results can not only be accepted but even wanted. Indeed, these fuzzy results are a powerful method to represent and communicate information in a human-like way.

Examples of fuzzy results got after linguistic approximation:

- x is very large
- the temperature is rather small.

If the results are expressed in a procedural way, they can be viewed as fuzzy instructions. Such fuzzy instructions may constitute valuable plans for a robot. Using fuzzy designations is a way to sum up the useful information; the recording of fuzzy plans may be a powerful way of planning.

## III. FUZZY PATTERN MATCHING

In the following a brief account of a pattern matching system developed by Cayrol, Farreny and Prade (1980) is given. An example illustrates it.

The current pattern-matching procedures are based on a rigid conception of the similitude between the pattern and the datum under consideration. Comparing patterns and data, ordinary methods decide if the datum is acceptable by the pattern. Patterns and data are represented as nested lists of atoms. The similarity between an atom of the datum and an atom of the pattern is considered as a two-valued logical variable, even if this similarity is calculated through more complex functions than the identity. The global similarity between the pattern and a datum is computed as a logical combination of atomic similarities.

In knowledge-based systems it is very convenient to use words of a natural language as atoms.

This system is an attempt to make more flexible the associative access by taking into account the imprecisions due to the fuzziness of the words of natural language.

A word of a natural language may often designate a collection of objects, rather than a unique object; moreover this collection has generally no crisp boundary: it is more or less possible that an object of the collection is the object actually described. This fuzzy set can be viewed as the meaning of the word (Zadeh).

The membership function of the set is viewed as a possibility distribution.

In this approach a pattern is a tree-like structure whose leaves are "atomic pattern". A possibility distribution is associated with each of them. In the same way the data are represented as tree-like structures whose leaves are "atomic data", also associated with possibility distributions.

Two measures of the similarity between the atomic pattern and an homologous atomic datum (homologous in the sense of the morphism between the structures) are evaluated:

-The possibility (Zadeh) that what is designated in the datum is (at least approximately) equal to what is required by the pattern. If $\mu_P$ and $\mu_D$ denote, respectively, the possibility distributions attached to an atomic pattern and an atomic datum, the possibility measure is taken to be equal to:

$$\Pi(P_\circ R|D) = \sup_{u\in U, v\in U} \min(\mu_P(u), \mu_D(v), \mu_R(u,v))$$

where U is the universe of discourse, and $\mu_R$ is the membership function of a relation of (possibly approximate) equality. $\mu_R$ models a tolerance. In the following example $\mu_R(u,v) = \max(0, 1-|u-v|/\lambda)$ with $\lambda \geq 0$.

-The necessity $N(P \circ R|D)$ of the same event. This concept of necessity is a dual concept of the one of possibility in the sense that: $N(P \circ R|D) = 1 - \Pi(\overline{P \circ R}|D)$; the necessity of an event is equal to "the impossibility of the opposite event" which yields:

$$N(P \circ R|D) = \inf_{v \in U} \max (\sup_{u \in U} \min (\mu_P(u), \mu_R(u,v)), 1 - \mu_D(v)).$$

These atomic measures of possibility and necessity are aggregated in order to yield global measures between the whole pattern and the whole datum. This aggregation (making use of "min" operator) preserves the respective semantics of possibility and necessity of the measures.

Both similarity measures are computed between the pattern and each datum present in the data basis. The data which have the best similarity measures with respect to the pattern are yielded by the system. Several situations may happen:

-No datum sufficiently matches the pattern: the similarity measures are below some thresholds.

-One or several data match the pattern in the same way: there may be several data which have (at least approximately) the same possibility measure and the same necessity with respect to the pattern; then the data are said to be interchangeable.

-Several non-interchangeable data are yielded: two data may be sufficiently similar to the pattern, but one of them has the best possibility measure while the other has the best necessity measure. In this case, interchangeable data are gathered in a same cluster; clusters are incomparable from a choice point of view.

The system described above is implemented in LISP and is currently experimented. This system may be particularly useful in the framework of the communication in natural language with robots, where approximate descriptions of real world situations and approximatively specified rules are needed.

The following example deals with (more or less approximate) description of people. The data available in the considered data base are:

D1: (JOHN THIRTY:AGE TALL:HEIGHT AROUND-SIXTY:WEIGHT)
D2: (PETER ABOUT-FIFTY:AGE MEDIUM:HEIGHT OVER-SEVENTY:WEIGHT)
D3: (PAUL ABOUT-FORTY:AGE RATHER-TALL:HEIGHT ABOUT-FIFTY:WEIGHT)
D4: (TOM RATHER-YOUNG:AGE SMALL:HEIGHT FIFTY:WEIGHT)
D5: (RICHARD ABOUT-FORTY-FIVE:AGE TALL:HEIGHT ABOUT-SEVENTY-FIVE:
WEIGHT)
D6: (HENRY AROUND-FORTY:AGE VERY-SMALL:HEIGHT AROUND-SIXTY:WEIGHT)

D7: (ROBERT ABOUT-FORTY-FIVE:AGE MEDIUM-HEIGHT ABOUT-SEVENTY-FIVE: WEIGHT)

D8: (PATRICK MORE-OR-LESS-MEDIUM:AGE MEDIUM:HEIGHT ABOUT-SEVENTY-FIVE:WEIGHT)

D9: (DAVID AROUND-FORTY:AGE MEDIUM:HEIGHT SIXTY-NINE-AND-HALF:WEIGHT)

The pattern is P : (>X AROUND-FORTY:AGE MEDIUM:HEIGHT OVER-SEVENTY: WEIGHT)

The representation (by 4-tuples) used in the example are given in Appendix 1. With a tolerance equal to 0 ($\lambda = 0$) for each universe the pattern-matching procedure yields as acceptable (because their similarity measures are above the thresholds) the data D7, D8, D9. The data D7 and D8 are interchangeable: the difference between the homologous similarity measures is below 0.05. More precisely we have:

$\Pi(P|D7) = 1$ $\quad \Pi(P|D8) = 1$ $\quad \Pi(P|D9) = 0.75$

$N(P|D7) = 0.42$ $\quad N(P|D8) = 0.41$ $\quad N(P|D9) = 0.5$

We observe that D7 and D9 (or D8 and D9) are not interchangeable.

The effect of an increasing of the tolerance parameters of the universe of heights is detailed in Appendix 1. In this example, it does not lead to select more data. For sure it is not always the case.

Note that the results are coherent with our intuition. The system yields the data that <u>may</u> correspond to the pattern.

The selection procedure sketched in the introduction, once computed the global profile of each candidate, obviously corresponds to a fuzzy pattern-matching problem.

## IV. TOWARDS FUZZY DATA TYPES

A data type consists of:

-a set of objects, called its members

-a set of functions, called its primitive operations whose domains and ranges are specified. At least either the domain or the range of each primitive operation must include (possibly within some Cartesian product) the own set of members of the data type. The primitive operations operate on or produce members of the data type.

-a set of axioms which describe the way the primitive operations interact with one another, and perhaps with other functions.

The concept of data type has been extensively discussed by Liskov and Zilles, Guttag et al., Parnas et al..

Moreover, each data type has, as a member, an ideal element, namely REJECT which is the output of any operation when it has no genuine output of the expected sort (i.e. when some members of the domain of the operation are not mapped onto members of the range).

After this brief recall, we may try to introduce what is a fuzzy data type (see also Prade et al. 1978).

Naturally, there are several kinds of "fuzzy data types" more or less fuzzy, depending on where fuzziness lies.

First, only the members of the data-type may be fuzzy entities. But the primitive and derived operations are ordinary ones. The axioms are also of a classical kind. Thus, we have for instance the data-types "Fuzzy Set" (which is different from the data-type "Set") or "Fuzzy Number" (different from the data-type "Real Number" - some axioms which are true for real numbers are no more valid for fuzzy numbers, see Dubois & Prade 1980), but they are also data types in the ordinary sense. It corresponds to the non fuzzy manipulation of fuzzy quantities.

A little more fuzzy, is the concept of fuzzy rejection. It corresponds to the ordinary functions equipped with a fuzzy domain A (see Dubois and Prade 1980). It means that all the elements of X are not equally suitable as inputs of the function under consideration. Thus each x is rejected with a degree equal to $1 - \mu_A(x)$ (if $\mu_A(x) = 0$ the output of the function is the classical REJECT). We have, thus an example of fuzzy control!

Already intrinsically fuzzy are the data types where some primitive operations or functions are fuzzy. The members are or are not fuzzy. The axioms are classical ones: it is not because an operation is fuzzy that it interacts in a fuzzy manner with itself or others.

Lastly, the axioms of a data type may be themselves fuzzy, even if the operations or/and the members are not fuzzy. What is a fuzzy axiom? It is a fuzzy description of the way operations are allowed to interact with one another. The description is fuzzy because the allowed interaction is supposedly ill-known. For instance, we may say that the operation * is approximately commutative if A * B is approximately equal to B * A for all A and B where "approximately equal" is modelled by a fuzzy relation. * is approximately commutative in an another sense if A * B = B * A for <u>most</u> of the A and B. Fuzzy axioms are linguistic statements on the allowed interactions of the operations. Linguistic statements can be represented using possibility distributions.

Thus there are several kinds of fuzzy data types because, members, operations and even axioms may be fuzzy independently of each other.

Obviously, it is there just some hints. The theory remains to be done.

## V. CONCLUDING REMARKS

Fuzzy programming does not seem to be either a dream or a dubious speculation. Fuzzy procedures may be useful in many problems of Artificial Intelligence. A language such as LISP seems to be a good support to implement fuzzy procedures after defining some general facilities as fuzzy pattern matching or manipulation of fuzzy numbers.

ACKNOWLEDGEMENTS

The author wishes to thank Michel Cayrol, Henri Farreny, Louis Feraud for the role they played in the discussion of the ideas presented here.

## REFERENCES

Adamo, J. M., 1980, L.P.L. - A fuzzy programming language. 1. Syntactic aspects, Fuzzy Sets and Systems, Vol. 3, No. 2, pp. 151-179.

Adamo, J. M., 1980, L.P.L. - A fuzzy programming language. 2. Semantic aspects, Fuzzy Sets and Systems, Vol. 3. No. 3, pp. 261-289.

Adamo, J. M., 1981, Some applications of the L.P.L. language to combinatorial programming, Fuzzy Sets and Systems, Vol. 6, pp. 43-60.

Baldwin, J. F., 1980, An automated fuzzy reasoning algorithm, C.N.R.S. Round Table on Fuzzy Sets, Lyon, France, June 23-25.

Bonissone, P., 1978, A pattern recognition approach to the problem of linguistic approximation in systems analysis, Memo, UCB/ERL M78/57, Berkeley.

Cayrol, M., Farreny, H., and Prade, H., 1979, Fuzzy instructions: Execution/Production, in "Towards the use of fuzzy set theory in A.I.", L.S.I. Tech. Rep., Toulouse, France.

Cayrol, M., Farreny, H., and Prade, H., 1980, Fuzzy reasoning based on multivalent logics in the framework of production-rules systems, Proc. 10th IEEE Int. Symp. Multivalent Logic, Evanston, Illinois, pp. 143-148.

Cayrol, M., Farreny, H., and Prade, H., 1980, An advanced pattern-matching method taking into account the uncertainty in meaning, Int. Conf. Art. Int. & Information-Control Syst. of Robots, Smolenice near Bratislava, Czechoslovakia, June 30-July 4, in "Computers and Artificial Intelligence, Vol. 1, No. 1, pp. 47-61, 1982, Bratislava.

Chang, C. L., 1975, Interpretation and execution of fuzzy programs, in "Fuzzy sets and their applications to cognition and decision processes", Zadeh, Fu, Tanaka, Shimura, eds., pp. 191-218, Academic Press.

Chang, S. K., 1972, On the execution of fuzzy programs using finite state machines, IEEE Trans. Comput., Vol. 21, pp. 241-253.

Dubois, D. and Prade, H., 1980, Fuzzy sets and systems: theory and applications, Academic Press.

Dubuisson, A., 1979, Conception et réalisation du compilateur et de l'exécuteur du language LPL, Thesis 3rd cycle, Lyon-I, France.

Eshragh, F., and Mamdani, E. H., 1979, A general approach to linguistic approximation, Int. J. Man-Machine Studies, Vol. 11, pp. 501-519.

Fellinger, W. L., 1978, Specification for a fuzzy systems modelling language, Ph. D. Thesis, Oregon State, Univ., Corvallis.

Floyd, R., 1967, Nondeterministic algorithms, J. of A.C.M., Vol. 14, No. 4, pp. 636-644.

Freksa, C., 1980, L-FUZZY - an AI language with linguistic modification of patterns, AISB Conference, Amsterdam.

Gershman, A., 1976, Fuzzy sets methods for understanding vague hints for maze running, M. S. Thesis, Comp. Sci. Dept., U.C.L.A.

Goguen, J. A., 1975, On fuzzy robot planning, in "Fuzzy sets and their applications to cognition and decision processes", Zadeh, Fu, Tanaka,Shimura, eds., Academic Press.

Goguen, J. A., 1976, Robust programming languages and the principle of maximal meaningfulness, Milwaukee Symp. on Automatic Computation and Control, pp. 87-90.

Guttag, J. V., Horowitz, E., and Musser, D. R., 1978, Abstract data types and software validation, Com. of ACM, Vol. 21, No. 12, pp. 1048-1064.

Hamilton, M. and Zeldin, Z., 1976, Higher order software. A methodology for defining softwares, IEEE Trans. Soft. Eng., Vol. 2, No. 1, pp. 9-32.

Hinde, C. J., 1977, Algorithms embedded in fuzzy sets, Int. Computing Symp., Morlet et Ribbens, eds., pp. 381-387, North-Holland.

Jakubowski, R. and Kasprzak, A., 1973, Application of fuzzy programs to the design of machining technology, Bulletin of the Polish Academy of Sciences, Vol. 21, pp. 17-22.

Kling, R., 1974, Fuzzy-PLANNER: Reasoning with inexact concepts in a procedural problem-solving language, J. of Cybernetics, Vol. 4, No. 2, pp. 105-122.

Le Faivre, R. A., 1974, The representation of fuzzy knowledge, J. of Cybernetics, Vol. 4, No. 2, pp. 57-66.

Le Faivre, R. A., 1974, Fuzzy problem-solving, Ph.D. Thesis, Univ. Wisconsin.

Liskov, B. H. and Zilles, S. N., 1975, Specifications techniques for data abstractions, IEEE Trans. Soft. Eng., Vol. 1, No. 1, pp. 7-19.

Mamdani, E. H., 1975, F.L.C.S.: A control system for fuzzy logic, Queen Mary Coll., London.

Mandic, N. J. and Mamdani, E. H., 1980, A fuzzy linguistic calculator, 4th European Cong. on Operations Research, July 22-25, Cambridge, England.

Noguchi, K., Umano, M., Mizumoto, Tanaka, K., 1976, Implementation of fuzzy artificial intelligence language FLOU, Tech. Rep. Autom. Language IECE, Osaka.

Parnas, D. L., Shore, J. E., and Weiss, D., 1976, Abstract data types defined as classes of variables, Proc. Conf. on Data, Salt Lake, Sigplan Notices, Vol. 8, No. 2, pp. 149-154.

Prade, H. and Vaina, L., 1980, What 'Fuzzy H.O.S.' may mean, Proc. 4th IEEE Computer Software & Applications Conference, Oct. 27-31, pp. 850-857, Chicago.

Santos, E., 1977, Fuzzy and probabilistic programs, in "Fuzzy automata and decision processes", Gupta, Saridis, Gaines, eds., pp. 133-147, North Holland.

Shaket, E., 1976, Fuzzy semantics for a natural-like language defined over a world of blocks, M.S. Thesis, Memo no 4, A. I. Comp. Sci. Dept., U.C.L.A.

Smith, M. H., et al., 1975, The system design of Jason, a computer controlled mobile robot, IEEE S.M.C. Conf., San Francisco, Sept.

Tanaka, K. and Mizumoto, M., 1975, Fuzzy programs and their execution, in "Fuzzy Sets and Their Applications to Cognition and Decision Processes", L.A. Zadeh, K.S. Fu, K. Tanaka, and M. Shimura, eds., pp. 41-76, Academic Press.

Umano, M., Mizumoto, M. and Tanaka, K., 1978, FSTDS-system: a fuzzy set manipulation system, Information Science, Vol. 14, pp. 115-159.

Uragami, M., Mizumoto, M., and Tanaka, K., 1976, Fuzzy robot controls, J. Cyber, Vol. 6, pp. 39-64.

Zadeh, L. A., 1968, Fuzzy algorithms, Information and Control, Vol. 12, pp. 94-102.

Zadeh, L. A., 1973, Outline of a new approach to the analysis of complex systems and decision processes, IEEE Trans. S.M.C., Vol. 3, pp. 28-44.

Zadeh, L. A., 1978, PRUF - A meaning representation language for natural languages, Int. J. for Man-Machine Studies, Vol. 10, pp. 395-460.

## APPENDIX 1: NUMERICAL DATA OF THE EXAMPLE

1) <u>Universe : age</u> ; tolerance : $\lambda = 0$

| | Fuzzy labels : | $\Pi(P\|D)$ | $N(P\|D)$ |
|---|---|---|---|
| Pattern : | AROUND-FORTY: (35,45,5,5) | | |
| data | THIRTY: (30,30,0,0) | 0 | 0 |
| | ABOUT-FIFTY; (48,52,2,2) | 0.57 | 0 |
| | ABOUT-FORTY: (38,42,2,2) | 1 | 1 |
| | RATHER-YOUND: (25,30,3,6) | 0.53 | 0 |
| | ABOUT-FORTY-FIVE: (43,47,2,2) | 1 | 0.42 |
| | AROUND-FORTY: (35,45,5,5) | 1 | 0.5 |
| | MORE-OR-LESS-MEDIUM: (35,45,7,7) | 1 | 0.41 |

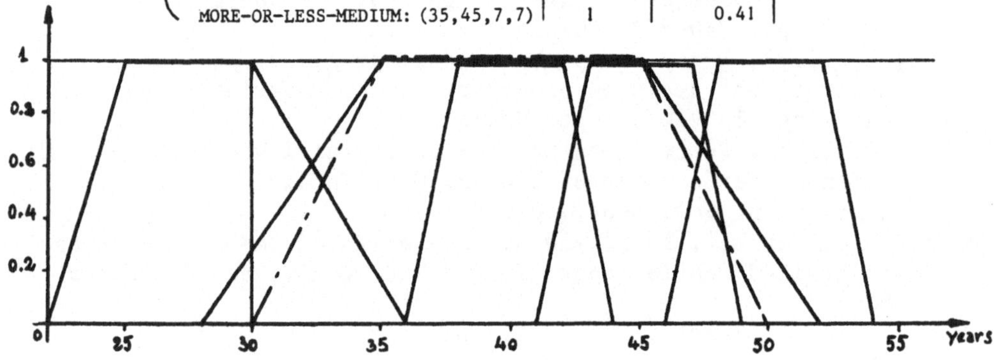


2) <u>Universe : height</u>

| | Fuzzy labels | tolerance : $\lambda = 0$ $\Pi(P\|D)$ | $N(P\|D)$ | tolerance : $\lambda = 5$ $\Pi(P\|D)$ | $N(P\|D)$ |
|---|---|---|---|---|---|
| Pattern | MEDIUM: (165,175,2,2) | | | | |
| | TALL: (175,210,10,15) | 1 | 0 | 1 | 0 |
| | MEDIUM; (165,175,2,2) | 1 | 0.5 | 1 | 0.77 |
| Data | RATHER-TALL: (170,180,10,10) | 1 | 0 | 1 | 0.11 |
| | SMALL: (155,160,5,5) | 0.28 | 0 | 0.58 | 0 |
| | VERY-SMALL: (150,155,5,5) | 0 | 0 | 0.15 | 0 |

(Continued)

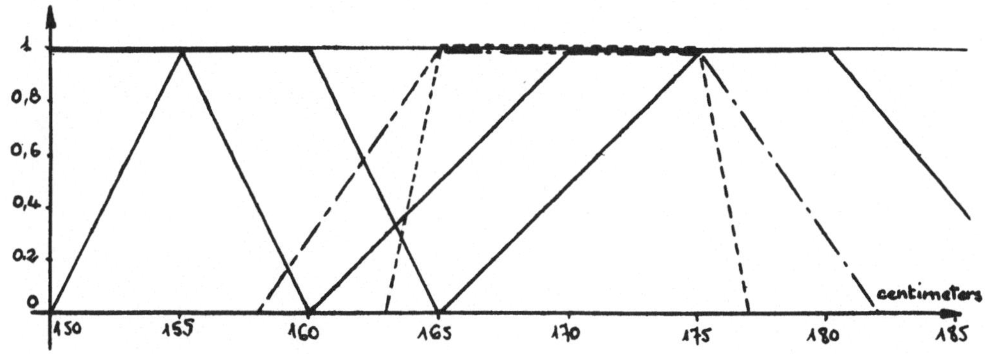


3) Universe : weight

tolerance : λ = 0

| | | Π(P\|D) | N(P\|D) |
|---|---|---|---|
| Pattern | OVER-SEVENTY: (70,120,2,0) | | |
| Data | AROUND-SIXTY: (55,65,3,3) | 0 | 0 |
| | ABOUT-FIFTY: (47,53,2,2) | 0 | 0 |
| | FIFTY: (50,50,0,0) | 0 | 0 |
| | ABOUT-SEVENTY-FIVE: (72,78,4,4) | 1 | 0.67 |
| | SIXTY-NINE-AND-HALF: (69.5 , 69.5 , 0,0) | 0.75 | 0.75 |

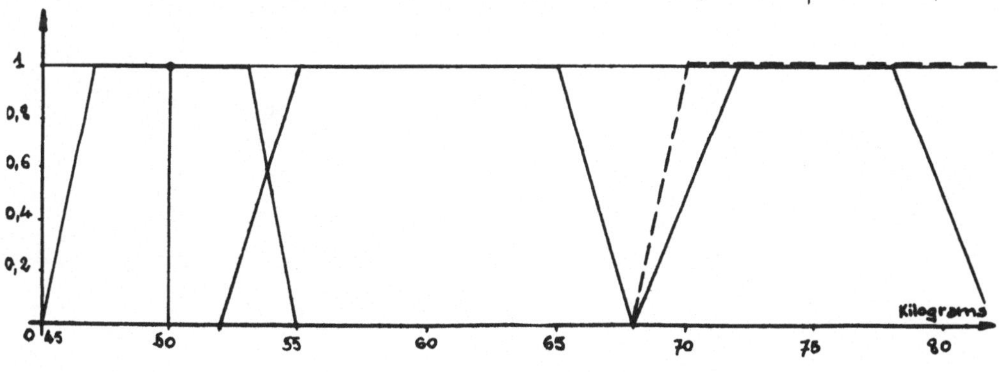

# A FUZZY, HEURISTIC, INTERACTIVE APPROACH TO THE OPTIMAL NETWORK PROBLEM

Didier Dubois*

Department d'Etudes et de Recherches en Automatique
CERT
avenue Edouard Belin BP N° 4025
31055 TOULOUSE CEDEX - FRANCE

## INTRODUCTION

Among the various problems that have to be solved in the transportation planning field, that of designing optimally a network has focused the attention of many researchers. The network is formally represented as a graph and we must find a subset of the maximal set of links, which are for instance streets or roads. The problem belongs to the combinatorial field, and its optimal solving is impossible for practical-sized sets of data; we are forced to use heuristic methods which are not completely satisfactory by their very nature.

Usual approaches involve three questionable assumptions:

- Interaction between users and transportation system is well-known, i.e. for a given system, we can forecast the number of users.
- Appropriate easy-to-compute criteria can be defined for the choice of a good network.

---

*This paper is a summary of the author's thesis, which was done at the Centre d'Etudes et de Recherches de Toulouse (France), Department d'Etudes et de Recherches en Automatique, under the sponsorship of the Ecole Nationale Superieure de l'Aeronautique et de l'Espace (Toulouse France). The author was supported by a scholarship of the Societe des Amis de l'Ensae (Paris).

- Numerical data are precise, which is not the case in socioeconomic fields (for instance the trip matrix, travel times).

This paper intends to present a methodology which tries to resolve these difficulties, and provides a procedure for approximate network generation, which builds distinct alternatives, to be evaluated more precisely in a further step.

After a brief survey of existing methods, the methodology is described and discussed. Approximate models, taking into account the partial lack of knowledge, will be preferred to delusive precise models. Necessity for man-machine interaction and local criteria will be emphasized. As an illustration of this methodology, an interactive, heuristic, fuzzy procedure is described for bus transportation network design in a town. It explicitly uses tools from Zadeh's fuzzy sets theory [33] [34].

## I. EXISTING METHODS

### 1) Formal Statement of the Problem

The optimal network problem is to find a good tradeoff between a sparse network where little investment is required, but which does not satisfy the users, and the maximal network, which is too expensive, but provides excellent service.

We are given a set N of zones represented by nodes to be joined, a set $\hat{A}$ of links - possible direct paths joining zones - and the number of trips between zones - i.e. the trip matrix.

A binary vector $X = (x_1, \ldots, x_\ell)$ can be defined such that

$\ell = |\hat{A}|$ = total number of links $\alpha_k$, $k = 1, \ell$

$x_k = 0 \iff \alpha_k \notin A \subseteq \hat{A} \quad k = 1, \ell$

$x_k = 1 \iff a_k \in A \subseteq \hat{A} \quad k = 1, \ell$

X is specific of a network $G = (N, A)$; A is a subset of $\hat{A}$. Given the investment cost $P_k$ of each link $\alpha_k \in \hat{A}$, the total investment cost is $P = \sum_{k=1}^{\ell} P_k \cdot x_k$ To evaluate the difficulty of moving through the network, total travel time is often used. This is given by $T = \sum_{ij \in N^2} D_{ij}\, t_{ij}(A)$ where

$D_{ij}$ is the ij - term of the trip matrix

$t_{ij}(A)$ is the shortest time to go from i to j when the network is (N,A).

In fact, this evaluation only considers the negative aspects of travel - i.e. losing time, and neglects their attractiveness - to reach places of interest. There are two ways of taking it into account:

- o to assign to each origin - destination pair the user's profit in monetary units [1]. This is not very realistic since this profit involves individual unknown factors (see [39]).
- o to use accessibility indices which model the feed-back effect of the network on the trip matrix (see [27,42]). However those indices are specific of one zone compared with others. So, there is no easily justified way to define a global criterion; moreover its mathematical expression is more complex than the one of total travel time.

Using the latter, three statements of the static problem can be found in the literature.

ø1: Minimizing total travel time under an investment constraint $P<P_T$ ([26],[31]).

ø2: Minimizing total investment cost under a travel-time constraint $T<T_0$ ([35],[39]).

ø3: Minimizing the sum T+P ([4],[22]).

ø1 is the most satisfactory because it only assumes a given investment budget not to be overpassed, which sounds realistic. On the other hand, ø2 constraint $T<T_0$ is not known a priori, but arbitrarily selected. It can be evaluated only through the model. Moreover, when minimizing T, we are not sure of getting rid of local defects (too long travel times in some parts of the network), which are filtered by this average criterion. Stating only $T<T_0$, this tendency may be enhanced because $T_0$ is not necessarily minimum in that case. Lastly, in ø3, travel times are assumed to be converted to monetary units. This is a tricky operation, since travel costs may change according to the people. Thus conversion factors are not homogeneously distributed through the network. Implicitly, relative weights are assigned to T and P, without knowledge of the corresponding trade-off.

2) Discussion of Methods

a) Optimal methods. The only optimal method is implicit

enumeration of feasible solutions; we do not intend to describe it here, owing to the large number of papers dealing with this topic. For example Branch and Bound or Branch and Backtrack algorithms were proposed by Ochoa-Rosso [26], Scott [31], Boyce, Fahri, Weischedel [5], Hoang Hai Hoc [15], Los [22] and many others.

Computational experience has shown these methods usually would not succeed in solving the problem whenever the networks had more than ten nodes, whether because of too high computational times, or too large memory requirements. A first cause of that failure is the expensive necessity of recomputing shortest paths in the network each time a link is dropped or added to the network. But basically, the intractability depends upon the NP - completeness of the optimal network problem. It means that this problem belongs to a class for which polynomial bounded growth deterministic algorithms have never been found. Specialists generally assume such algorithms do not exist. The equivalence of NP - complete problems was proven (cf Karp [19], Sahni [20]). Their common feature is the necessity, when solving them, of making choices whose a priori optimality is dubious, and which will eventually be contradicted. The optimal network problem was proven NP - complete only recently [18].

b) Heuristic methods. These methods are fitted to realistic-sized sets of data and provide a feasible solution which is hoped close to the optimal one. Usually, the procedures that can be found in the literature deal with statement ø3 of the problem (cf Billheimer [4] Avila [1], Los [22]).

On the other hand, the author developed three routines adapted to statement ø1 described in [2] or [7], together with computational experience. These are all algorithms of the greedy type (see Nemhauser et al. [21]) and seem to provide good results: a weak increase of total travel time T, short computation times (between 2 and 4 minutes on IBM 360/44 for 53 node networks), together with low memory requirements.

Heuristic methods are commonly thought of as makeshifts; this point of view is unfortunate in the case of network generation for two main reasons:

- As the network size increases, the optimum, in the sense of T, T+P, and any average-type global criterion, degenerates into a group of equivalent solutions whose evaluations are very close to each other. The more complex the network, the more insensitive the criterion, when a link is dropped or added.

- Owing to the uncertainties on the data, the system structure, and the criteria formulation, the optimal solution of the mathematical problem may not correspond with the optimal solution in the real world.

However, the main reproach that can be raised against Ø1, Ø2, Ø3, statements of the problem, is their lack of realism even when heuristic methods are provided to reduce intractability. The classical operations research approach (optimizing one criterion under linear constraints) is to some extent irrelevant to transportation network design. Computationally feasible algorithms, taking into account the complexity of the real problem, together with our inability to assimilate it precisely, are needed.

## II. A METHODOLOGY

To make a step in that direction, a methodology is now proposed; it can be considered more generally as an attempt to deal with complex systems, especially those in which man is involved. Roughly outlined it consists of three points.

1. the necessity of interactive procedures.
2. the relevance of Zadeh's approach using fuzzy sets theory to model incomplete available information about humanistic systems.
3. the necessity of local evaluations and decision making.

### 1) Interactive Programming

The idea of an interactive method for the optimal network problem was first proposed and realized by Stairs [32].

Since results of heuristic algorithms applied to large networks show the existence of numerous solutions within a small range of values of T, T+P, accessibility indices etc., those criteria are insufficient to justify any choice of a network. Indeed, in such a situation, those methods build a solution by deleting or adding links at random; decisions are not checked. This lack of flexibility can be reduced by allowing the interaction with a human operator, who will be needed each time a poorly motivated choice is to be carried out by the computer.

Two main advantages can be pointed out. First, human intuition in the combinatorial field is sometimes faster than systematic exploration (see for instance Jeroslow [17]). Second, an interactive approach is useful to take into account various aspects of the network, naturally aggregated through the operator's vision of the problem. A man, experienced in real networks, is able to guide a procedure towards realistic solutions, i.e. good trade-off between investment costs, accessibilities of zones, travel times, plus practical feasibility.

A transportation company manager will prefer several approximately evaluated network alternatives to just one so called optimal or underoptimal solution which may violate some non-modelized constraint. Computers are reliable for summarizing the information, more than making ultimate global choices. Using an interactive program it seems possible to generate a class of equivalent networks, through interactive modifications of decisions at a given level of the tree of solutions.

2) Dealing with Incomplete Information

a) Introduction. For a problem such as transportation network synthesis, it is basically impossible to have precise information about a big system which has not worked yet. Good cost efficiency and attractiveness of a network is the consequences of a lot of factors such as urbanization, demography, fares, location of bus stops, and so on, the influence of which cannot be quantified exactly. Usual operation research methods do not consider this lack of knowledge about the future system.

A straight-forward way of doing it is sensitivity analysis. However, except in the case of continuous linear programming, sensitivity analysis is very cumbersome since various simulations runs are performed on closely related data. The only attempt to use the proximity of data sets, in order to derive a solution of a combinatorial problem from that of another already solved one, has been realized in the framework of branch and bound and cutting plane algorithms for integer linear programming (cf Nauss [25], Geoffrion [13]) - i.e. academic problems, usually.

Another classical approach to uncertainty in systems science is probability theory. The major drawback of such an approach is an increasing complexity of the model which renders already difficult problems even more untractable. Let us discuss an example.

Consider a bus system in a town. A very significant part of the data regarding the system are the possible speeds of the buses on each street. Such speeds depend upon the traffic flow and reflect the congestion of the network. Physical measurements can be carried out and histograms of speeds can be derived, hence probability densities for travel times on all streets. Strictly speaking, we are concerned with the probability, for any bus, to make a given trip within a given time. Probability theory asserts that a travel time density on a path can be calculated by adding the random variables that model travel times on each street of the path. It is well known that such additions are very time-consuming to perform, on a practical level, except if densities are Gaussian or the number of streets on the path is sufficient so as to apply the Central Limit theorem. Obviously both assumptions do not sound realistic, and probabilistic path length computations remain out of reach. It is thus delusive to think of solving a probabilistic optimal network problem.

Besides, probability theory models only random events for which sufficient statistical data are available. Obviously other kind of uncertainties do exist, which refer to human subjectivity, for instance. In humanistic systems science, part of the information is subjective, and can be expressed as beliefs ascertaining events. Let A and B be two disjoint events, subsets of a universe U. Benote b(A) and b(B) belief degrees for assertions $x \varepsilon A$ and $x \varepsilon B$ respectively. Without any statistical basis, it is unjustified to claim that $b(A \cup B) = b(A) + b(B)$, so that the set-function b can be anything but a probability measure.

Even in the realm of objective uncertainty, probability is not the only thinkable concept. Possibility -the degree to which an event is possible or not- is also worth considering, as an alternative to probability. Although probability degrees are more informative than possibility degrees, the latter can be considered even when statistical data are not sufficient; moreover fuzzy set theory, which is the topic of the next section, provides a mathematical tool for describing the concept of possibility (Zadeh [41]).

b) <u>Fuzzy sets theory</u>. Fuzzy sets theory was introduced by Zadeh [33] in 1965, in reaction to the complexity of models in research fields such as systems theory, operations research, artificial intelligence, econometrics. These models, often based upon formal analogies with the laws of physics and the principles of mechanics, emphasized too much quantitative aspects. However, to make, or simply to define good measurements for humanistic systems is usually very difficult. All the quantified information remains partially out of reach. Available evaluations are mostly opinions given by trained individuals - here a transportation specialist - who state them rather linguistically than numerically, using "names of fuzzy sets" according to Zadeh [24].

A fuzzy set is supposed to model a set without sharp boundary by assigning to each element a grade of membership which lies between 0 and 1 by convention. To associate a fuzzy set A on a reference set U with a variable x taking its values in U is to assign to each element u of U a weight $\pi_x(u)$ which expresses the degree of possibility of assigning the value u to the variable x. $\pi_x$ is called a possibility distribution and is such that $\forall\, u \in U$, $\pi_x(u) = \mu_A(u)$, grade of membership of u in A. In other words, if x is a real variable, the "possibilistic" modelling of uncertainty is interpreted as allowing x to range an interval whose bounds are soft, expressing tolerances, preferences, or simply physical possibility in terms of a weight distribution.

Informally a fuzzy set can model individual perception of

phenomena and especially its ability to summarize information - i.e. to keep in mind the main features of a phenomenon. Using fuzzy sets theory we can take into account the uncertainty which lies in socioeconomical evaluations; we also generalize usual tolerance intervals to locally valued ones [10].

NB for a rigorous and more exhaustive presentation of fuzzy sets theory see [20], [14], [43].

c) Fuzziness in some decision making problems. The aim of a fuzzy model can be to infer the level of uncertainty for results (or outputs) knowing that of data (or inputs) and structure. Anyway, the model is unable to provide conclusions more valid than hypotheses, unless it receives information from outside, so that "entropy" may fall [6]. This information is usually provided by an expert; this points out the link between fuzziness and interaction. A fuzzy algorithm is a generalization of heuristic methods which are nothing but an approximate translation of rules of thumb into binary instructions.

In the following we consider algorithms which manipulate fuzzy sets but do not have a fuzzy structure. We must define the most valid manipulations losing as little information as possible. Those are either induction through fuzzy relations or algebraic operations on fuzzy sets of the real line. They extend usual operations between real numbers, by means of an extension principle [12]. Lastly they can be logical operations such as intersection, union etc. of fuzzy sets.

Easy practical computation for algebraic operations between fuzzy sets is possible using analytically defined membership functions. It was shown (Dubois and Prade [10]) that providing a certain homogeneity on the type of functions, one "fuzzy operation" was equivalent to three classical ones. This result was applied to shortest paths algorithms (Dubois and Prade [8]), and is used in the network synthesis problem.

Using fuzzy concepts in such a problem provides a means to deal with the poor knowledge of data such as interzonal trips, amount of population and employments, usual traffic congestion on links. So, it is possible to evaluate the validity of a decision: as soon as several network alterations have equal possibility values of being the best one, the operator makes its own choice. If this indeterminacy is due to the lack of precision of data and criteria, he will have to look for better inputs and to clarify his preference pattern of objectives. If it is a real indeterminacy, he will make a choice according to his own view of the network.

NB Fuzzy set theory is often confused with Probability theory although each has distinct features (see [24]).

### 3) The Importance of Local Criteria

Notice first that such a global criterion as the total travel time T, in the case of the optimal network problem, evaluates a solution without taking into account its structure. It gives only average information about the speed of trips. Using only such a criterion, we implicitly assume that the existence of isolated zones can be counterbalanced by that of fast travels between other zones. This is a probabilistic-like network evaluation, links being considered alike and their characteristics added.

This way of adding elementary features to get a global estimate is questionable especially for some humanistic systems where a certain level of local good behavior can be demanded; whenever a part of the system fails the whole system is deteriorated. So local criteria must be used, or at least global ones in which local defects are not masked owing to an averaging effect. Note that the "min" and "max" operators of the fuzzy sets theory do satisfy this condition; they are not the only ones of course (see [44]).

In transportation networks, local features are very important; thus it seems useful to divide the global choice of a network alternative into partial decisions, coordinated between each other, about subsystems. This implies a first step where data are analyzed, and the various parts of the maximal network are considered and ranked according to actual and potential trip demands, a set of candidate actions is thus built. A second step is needed to choose among the possible actions, in an interactive way.

## III. A GENERATION PROCEDURE FOR TRANSPORTATION NETWORK SYSTEMS

### 1) Assumptions, Statement and Objectives

We consider the set of data described in I,1), i.e. a maximal network, a trip matrix and investment costs. Each zone i is characterized by population $p_i$ and a number of employments $e_i$, considered respectively as emission and attraction coefficients. A precise knowledge is assumed about the length of $d_\alpha$ of links $\alpha$, but speeds $v_\alpha$ on those links together with investment cost $P_\alpha$ are only approximately known. The maximal amount of money allowed for building the network is $P_T$.

We do not want to determine an optimal solution, but rather several alternatives having analogous performances, and possibly contrasted features. The aim is to provide easy exchanges between zones, through the network, for actual as well as potential users.

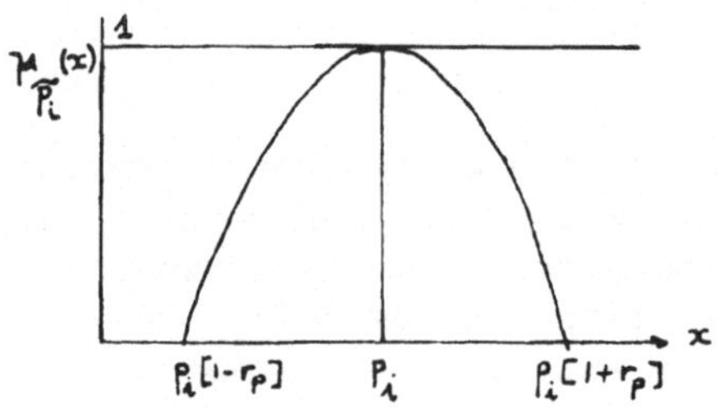


Fig. 1

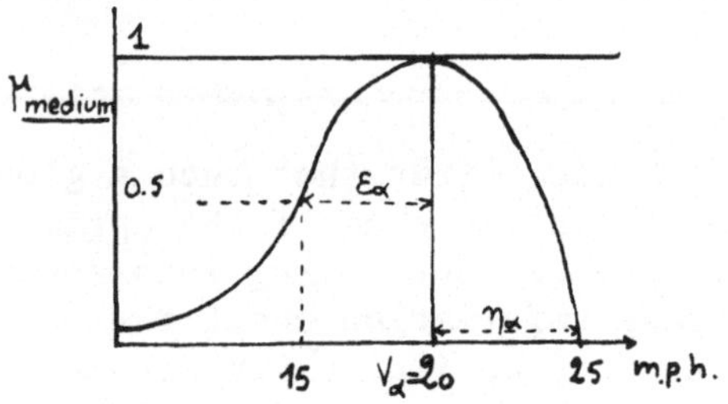


Fig. 2. Medium Speed

Each quantity $p_i$, $e_i$, $D_{ij}$, $P_\alpha$ is not precisely defined, and relative tolerances $r_p$, $r_e$, $r_d$, $r_c$, are assumed; but since the transition between membership and non membership to a tolerance interval is gradual rather than abrupt, we use fuzzy sets to represent for instance "approximately $p_i$". We write $\tilde{p}_i = (p_i, r_p \cdot p_i)$ whose membership function is

$$\mu_{\tilde{p}_i}(x) = \max\left(0, 1 - \left(\frac{x-p_i}{r_p p_i}\right)^2\right) \qquad \text{(cf Fig. 1)}$$

Speeds are assumed to belong to a set of defined types (e.g. very slow, slow, medium, fast, very fast) which are fuzzy sets as in Fig. 2. In the case study on which the procedure was tested, fuzzy speeds reflected the actual congestion of streets which buses could use. Here congestion due to private vehicles was assumed to be approximately known, and the bus fleet too small to alter it. We dealt with the assignment of mass transit users only. Membership function $\mu_{\tilde{v}_\alpha}(x)$ is defined as

$$\mu_{\tilde{v}_\alpha}(x) = \max\left[0, 1 - \left(\frac{x-v_\alpha}{\eta_\alpha}\right)^2\right] \triangleq R\left(\frac{x-v_\alpha}{\eta_\alpha}\right); \; x \geq v_\alpha$$

$$= \frac{1}{1+\left(\frac{x-v_\alpha}{\varepsilon_\alpha}\right)^2} \triangleq L\left(\frac{v_\alpha-x}{\varepsilon_\alpha}\right); \; x \leq v_\alpha$$

and we write

$$\tilde{v}_\alpha = (v_\alpha, \varepsilon_\alpha, \eta_\alpha)_{LR} \left[\begin{array}{l} v_\alpha \text{ is called mean value} \\ \varepsilon_\alpha, \eta_\alpha \text{ are left and right spreads} \end{array}\right.$$

$\max(0,1-x^2)$ and $1/(1+x^2)$ are called respectively "right" and "left references" (cf [10]).

On Figure 2, $v_\alpha = 20$; $\eta_\alpha = \varepsilon_\alpha = 5$. Two assumptions underlie the shape of $\mu_{\tilde{v}_\alpha}(x)$:

- there is a strict upper bound to speed on link $\alpha$ (given by $v_\alpha + \eta_\alpha$)

- owing to possible link congestion, the possibility of reaching very low speeds always exists.

Membership functions can be derived from statistical data (cf [43], [45]).

Other data can be quality values for the links. If we consider them as membership values, the network becomes a fuzzy graph [38].

The reader must notice that we do not deal with problems of construction staging and timing for the network. As a consequence the procedure will be mainly fitted to short range alterations of transportation networks. As established at the end of II, the procedure will involve two steps

1 Find a candidate set of paths
2 Choose among the paths to build a network.

N.B. In a preliminary step, shortest paths in average travel times are computed.

2) Step 1: Finding the Candidate Set

The main part of this step consists in a search for candidate paths; they are the fastest routes for biggest actual or potential demands for travel. As in Avila [1], paths, and not links, are taken as decision variables. For any origin destination pair, both the level of traffic and the best paths are determined. Shortest paths in average travel times (one per O-D pair) are assumed to be already known in the following.

a) Traffic estimation. Level of traffic is computed through a sequence of fuzzy logical tests. We define first, in a very straightforward manner, the fuzzy sets high actual demand (D), high amount of population, high amount of employment, short travel time (ST).

The membership function $\mu_D(Dij)$ for high actual demand is defined as:

$\mu_D(Dij) = 1;\ Dij > \overline{D}$

$$\mu_D(Dij) = \frac{Dij - \underline{D}}{\overline{D} - \underline{D}} \quad \underline{D} \leq Dij \leq \overline{D}$$

$$\mu_D(Dij) = 0 \quad Dij < \underline{D}$$

$\overline{D}$ and $\underline{D}$ are parameters chosen according to the town. $\overline{D}$ is the level of actual demand above which traffic is considered as important. Demands whose level is below $\underline{D}$ can be individually neglected, but will be assigned since they may combine.

High amount of population, high amount of employment, are built on the same principle, and the parameters are $(\underline{p},\overline{p})$, $(\underline{e},\overline{e})$, $(\underline{t},\overline{t})$. Short travel time is different from the others:

$t \leq \underline{t}$: the user will not be wronged by his travel:

$$\mu_{ST}(t) = 1$$

$$\underline{t} \leq t \leq \overline{t}: \quad \mu_{ST}(t) = \frac{\overline{t} - t}{\overline{t} - \underline{t}}$$

$t \geq \overline{t}$: the user will not travel: $\mu_{ST}(t) = 0$

there can be several classes of $(\underline{t},\overline{t})$ according to the origin and destination of the trips.

Informally a large demand for trips will be a large actual demand (evaluated by $D_{ij}$) or a large potential demand.

A large potential demand will be a strong attraction from one zone to the other together with a short travel time. Such a potential demand is useful, since the $D_{ij}$'s are not fitted to the maximal network analyzed in this step. $D_{ij}$ is measured on the existing network. However, the actual trip matrix is also considered, for we deal with short-range alterations of the network.

The attraction of one zone with respect to another can be estimated by, for instance, their respective level of employment and population; the membership function $\mu_\theta$ of large demand can be computed as $\mu_\theta^k(ij) =$

$$\max\{\mu_D(Dij),\ \mu_t(t_{ij}^k)\cdot\max[\min(\mu_p(p_i),\ \mu_e(e_j)),\ \min(\mu_p(p_j),\ \mu_e(e_i))]\}$$

where $t_{ij}^k$ is the average travel time on the $k^{th}$ route between i and and j. "max" and "min" are the usual operators for the union and intersection of fuzzy sets (cf [32]). In the formula, the use of $\mu_t(t_{ij}^k)$ as a multiplication is made to allow a progressive alteration of attraction between zones by the travel time. The above expression is a numerical translation(*) of the informal definition of a large demand: (high actual demand) or ((short travel time) between i and j) and (high amount of population in i (or j) and high amount of employment in j (or i))).

In this model the demand is network-dependent since its level is a function of travel times. Thus it is an implicit form of accessibility; the above formula may also remind the reader of gravity distribution models.

b) Shortest paths. The best paths in the maximal network are found through a partial enumeration technique. For any origin-destination pair ij for which, according to the average travel time on the shortest path, there is a sufficient demand for trips, all the shortest paths between i and j crossing a given zone are generated. Those paths are made of two adjacent shortest paths. It is proved in [7] that the non-existence of these paths means that the shortest path is a lot faster than others, at least when travel times on links have the same order of magnitude.

Note that the average travel time must be understood as $t^\alpha = d_\alpha / v_\alpha$. As a matter of fact, travel times are fuzzy, and

$$\tilde{t}_\alpha = d_\alpha / \tilde{v}_\alpha \simeq (t^\alpha, \underline{t}_\alpha, \overline{t}_\alpha)_{RL}$$

This latter notation is the same as for $\tilde{v}_\alpha$; using the approximate formula for the inverse of a fuzzy number, it was proved [8] that mean and spreads can be taken as:

$$t^\alpha = \frac{d_\alpha}{v_\alpha};\quad \underline{t}_\alpha = \frac{\eta_\alpha d_\alpha}{v_\alpha^2};\quad \overline{t}_\alpha = \frac{\varepsilon_\alpha d_\alpha}{v_\alpha^2} \quad (**)$$

The main point is that the above formula does not depend upon the functional forms of membership functions, provided that the references are the same for all $\tilde{t}^\alpha$. This contrasts with stochastic approaches [36] where, instead of a fuzzy number, $\tilde{t}^\alpha$ is a random variable. $\tilde{t}_\alpha$ must then be assumed normal for the sake of computational simplicity. Travel time $\tilde{t}^k_{ij}$ on path $p^k_{ij}$ between i and j is:

$$\tilde{t}^k_{ij} = \sum_{\alpha \text{ on } p^k_{ij}} \tilde{t}^\alpha = \left(\sum t^\alpha, \sum \underline{t}_\alpha, \sum \overline{t}_\alpha\right)_{\alpha \text{ on } p^k_{ij}}, \ RL$$

using the exact formula for the addition of fuzzy numbers ([10]).

(*) according to Zadeh's fuzzy logic [40].

(**) references are those of $\tilde{v}_\alpha$ but exchanged ([10]).

For each path $p^k_{ij}$ the truth value of the proposition "$p^k_{ij}$ is the shortest path between i and j" is computed, i.e.:

$$v^k_{ij} = \max_{x,y:x\leq y} \min (\mu^k_{ij}(x), \mu_{ij}(y))$$

where

$\mu^k_{ij}$ and $\mu_{ij}$ are the membership functions of $\tilde{t}^k_{ij}$ and $\tilde{t}_{ij}$

$\tilde{t}_{ij}$ is the fuzzy travel time of the path having the shortest average travel time.

$v^k_{ij}$ is the degree of consistency of $\mu^k_{ij}$ and $\mu_{ij}$ (see [40]). $v^k_{ij}$ is easily computed (see [11]), and only the paths such that $v^k_{ij} \geq v_{min}$ are kept. $v_{min}$ is a parameter of the procedure, and is used to control the selectivity of Phase 1.

c) Traffic Assignment. A multipath assignment is then performed, where the amount of the flow on $p^k_{ij}$ is proportional to $\frac{v^k_{ij} - v_{min}}{1 - v_{min}}$. This "fuzzy assignment" assumes that the user chooses the shortest path, but his knowledge of possible speeds on links is approximate. It is supposed that as soon as $v^k_{ij} \geq v_{min}$, there will be some users thinking of $p^k_{ij}$ as the shortest path between i-j. This assignment is stable: a slight modification of length of links, or speeds does not significantly alter the distribution of the flows on the routes. When the procedure finds only one path between two nodes related to a large demand, the links of the path are permanently included in the future generated networks. When more than one path is available, all are kept in the memory together with $v^k_{ij}$, and $\mu_\theta(ij)$. Note that the memory requirement is very moderate for keeping path $p^k_{ij}$: only the origin node, the destination node, and an intermediate node $\ell$ are needed, such that $p^k_{ij}$ is the union of the shortest paths between i and $\ell$ and between $\ell$ and j. This kind of assignment is similar to probabilistic assignment [37].

We call "cluster" a set of paths having the same origin and destination. Within a cluster, paths are ordered in decreasing value of speed possibility ($v^k_{ij}$). Clusters are themselves ordered in decreasing order of maximal demand ($\mu^k_\theta(ij)$).

A flow chart of Step 1, including traffic estimation, speed evaluation, fuzzy assignment, and final ordering of paths, is given below

NB the "$\mu_\theta(ij) \geq \mu_{min}$" test is used to discriminate high-traffic-single-path O-D pairs from medium-low traffic single path ones. The former are automatically added to the final network whatever it will be; the latter are put into the candidate set and will possibly be added during step 2.

The fuzzy assignment is then used to detect links mainly carrying local traffic; those links can be added or dropped without altering traffic distribution and are independent subsystems.

It is also used to detect links carrying mainly connecting traffic; insufficient demands may combine on those links, to induce a significant flow. In this case the corresponding links are considered as real O-D pairs.

<u>N.B.</u> It should be clear that all the terms of the trip matrix are actually assigned to at least one path. No trip is neglected.

3) <u>Step 2: Decision Analysis</u>

a) <u>Criteria</u>. At the end of step 1 a set of routes, clustered according to their extremities, is available, the clusters are sorted in decreasing order of traffic magnitude; on the other hand, a partial network has been built made of single high-traffic routes.

The problem is now to choose one route out of each cluster. Let i and j be the end nodes of a given cluster; each route k in it is characterized by an intermediate node $\lambda_k$ which is at the

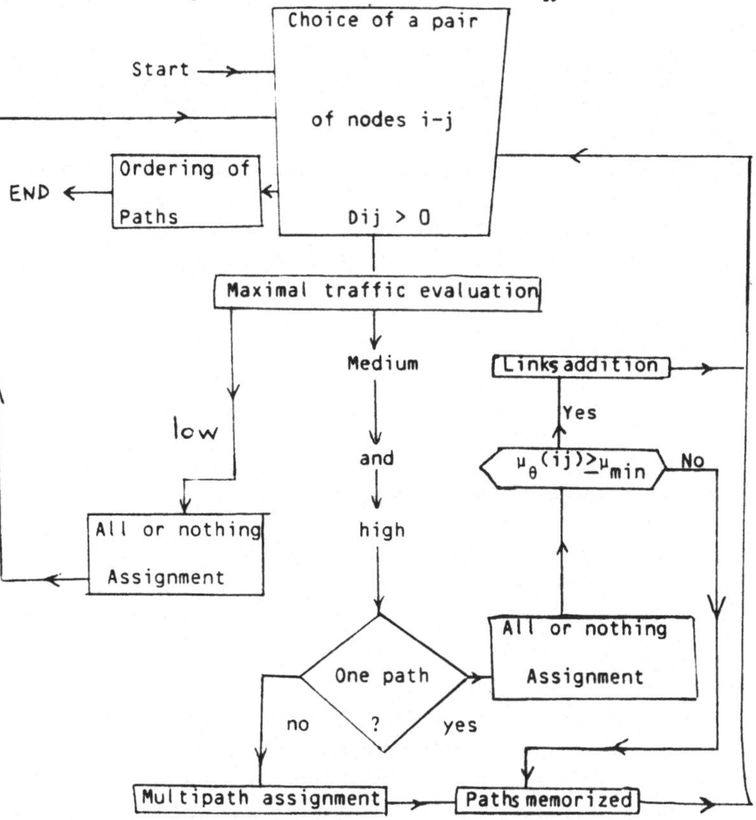


Fig. 3. Flowchart No. 1 (Step1)

junction of shortest paths $i-\lambda_k$ and $\lambda_k-j$. Note that $\mu_\theta^k(i-j)$ possibly differs from $\mu_\theta^{k'}(i-j)$, if $k \neq k'$ and the potential demand overpasses the actual one.

Five criteria will be considered to evaluate routes, three of them are gathered under the name of "socioeconomic":

C1 The speed of a route, compared with others in the cluster.

C2 The investment cost of the route when added to the partial network.

C3 The total time saved by the addition of the route, for potential or actual trips. We consider more particularly:

C31 The amount of time saved by actual demands.

C32 The amount of time saved between populations.

C33 The amount of time saved between work places and populations.

Criteria C1 and C2 are local because their values depend only upon the considered path. C3 tries to evaluate the influence of the path on the remainder of the partial network. Other criteria could be considered, for instance the quality of a path. It could involve the number of lanes, the type and degree of damage of the pavement and so on. Using quality indices $q_\alpha$ for links $\alpha$, as membership values of a fuzzy graph [39] (cf. III 1), we could set the quality $q^k$ of a path as that of the worse link:

$$q^k = \min_{\alpha \text{ on path } k} q_\alpha .$$

The expression of the criteria for route k are

*$C1(k) = 1 - v_{ij}^k$ where $v_{ij}^k$ is the truth value of the proposition: "route k is the shortest path" (see III2b)).

$C1(k) = 0$ means k is the fastest route.

*$C2(k) = \sum_i P_i$ for all links $\alpha_i$ belonging to route k and not present in the partial network.

For C3, rudimentary but easily interpreted indices were preferred to sophisticated accessibilities, owing to the lack of precision of the data. Let $\tau$ be the minimal increase in travel time that can be perceived by the users: when this time increases by more than $\tau$, users complain and utilization begins to fall.

*C31(k) is the proportion of actual users whose travel time in the partial netwr rk overpasses the shortest mean travel time in the maximal (=best) network, by at least $\tau$, when route k is added.

NB. $\tau$ can change according to the standard of living of users; it then depends on the origin and destination zones; and can be estimated through polls. There is no objection to considering $\tau$ as a fuzzy number [10].

*C32(k) is a proportion of employments, weighed by the population that can reach them.

*C33(k) is a proportion of populations, weighed by the amount of population than can reach them.

C31(k), C32(k), C33(k) are built in the same way. Those three indices express a percentage of travels whose time can be diminished, and evaluate to what extent route k is fitted to the partial network.

NB: shortest paths in the graph made of route k and partial network are not recomputed using usual shortest path algorithms, which is expensive, but through a fast procedure generalizing the one of Loubal for link addition [23] to route addition (see details in [7]).

Recall that most of the data used, to compute the values of the criteria are only approximately known, so that these values are fuzzy numbers obtained by the fast algebraic methods already proposed in [10], from the spreads of speeds, population, employment, actual demand and investment costs.

b) Multiple aspect aggregation. The proposed aggregation scheme was first suggested by Jain [16] and improved in [9]. The decision maker, supposed to be a transportation specialist, must give three sets of weights ($\tilde{\Psi}_1$ $\tilde{\Psi}_2$ $\tilde{\Psi}_3$), ($\tilde{\omega}_1-$, $\tilde{\omega}_2-$, $\tilde{\omega}_3-$), ($\tilde{\omega}_1+$, $\tilde{\omega}_2+$, $\tilde{\omega}_3+$).

The first set gives the respective priority of actual demand, employment-population and population-population potential demand. $\tilde{\Psi}_1$, $\tilde{\Psi}_2$, $\tilde{\Psi}_3$ are a numerical fuzzy translation of linguistic estimations such as "important", "very important", "not important" etc., and are fuzzy numbers of [0,1]. ($\tilde{\omega}_1-$, $\tilde{\omega}_2-$, $\tilde{\omega}_3-$) are given to sort C1, C2, C3 in order of priority, when the considered traffic is low. The last set is used for the same purpose but when the considered traffic is high. As a matter of fact the relative priority of route speed and investment cost will change according to the amount of considered traffic. The higher the traffic the more crucial the speed. For each route k a set of weights ($\tilde{\omega}_1$, $\tilde{\omega}_2$, $\tilde{\omega}_3$) is then computed through an interpolation, according to the value of $\mu_\theta^k(ij)$; the aggregation is then performed according to the scheme below:

$$C3(k) = \sum_{i=1}^{3} \tilde{\psi}_i \ \theta \ C31(i)$$

$$\tilde{C}(k) = \sum_{i=1}^{3} \tilde{\omega}_i \ \theta \ Ci(k)$$

where the criteria values are normalized, together with the weights. $\Sigma$ and $\theta$ means extended sum and product of fuzzy numbers (see [10]). $\tilde{C}(k)$ is thus a fuzzy number, which is the result of a fuzzy linear aggregation. The closer to 0 is $\tilde{C}(k)$, the better is route k.

c) Decision-making. Each route of a cluster is evaluated by the fuzzy value $\tilde{C}(k)$ whose shape looks like in the figure below. C(k) is called the mean value and has membership value 1. Suppose that the routes are sorted according to their mean value, so that

$C(k_1) \leq C(k_2)$ --- $\leq C(k_p)$ if p routes are considered.

For simplicity assume p=2. The truth value of the proposition "$\tilde{C}(k_1) \leq \tilde{C}(k_2)$" is 1; the one of the proposition "$\tilde{C}(k_2) \leq \tilde{C}(k_1)$" is $b_{12} = \mu_{\tilde{C}(k_1)}(a_{12})$, the height of the fuzzy set $\tilde{C}(k_1) \cap \tilde{C}(k_2)$, where $\cap$ means intersection. (cf [11] and fig. 4).

In the procedure two cases are considered

- $b_{12}$ is smaller than a threshold $\theta$: Route $k_1$ is automatically chosen to be added to the partial graph. (figure 4a)).

- $b_{12}$ is greater than $\theta$. The procedure calls for the decision maker; if he considers that the values $\tilde{C}(k_1)$, $\tilde{C}(k_2)$ are precise enough, he chooses either $k_1$ or $k_2$ from external considerations (fig. 4b)). If he considers the values are not precise enough to allow a comparison, he will have to revise his preference scheme (to give more precise weights, or different ones), or improve the quality of the data (fig. 4c)).

d) Outline of Step 2. The algorithm is composed of a sequence of cluster evaluations. For each cluster a decision is made as described above and a route is added to the partial network. The procedure stops when the monetary envelop can be overpassed by the investment costs. A network is thus generated. Other networks, having similar performance characteristics. but eventually distinct features can be built

- Either replacing the last added path by others which were still in the candidate set

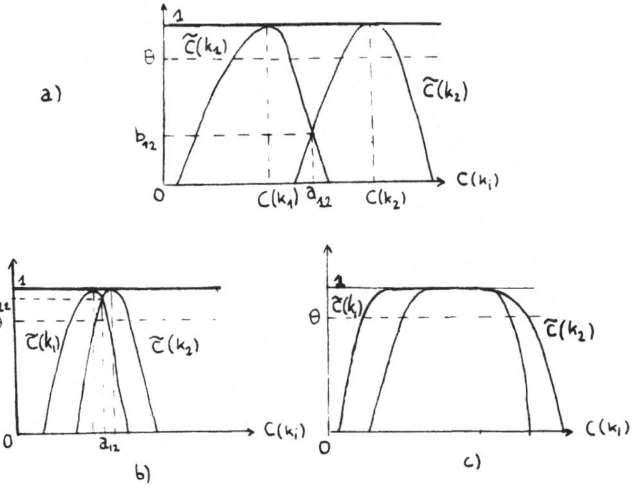

Fig. 4

- or by modifying the choices of the decision maker, each time two candidate paths are equivalent.

Step two is summarized on flowchart no. 2.

NB The procedure could be modified easily to output the k best paths, whenever they exist in the candidate set, at each trial of Phase 2, so as to allow a complete external decision making.

## IV. RESULTS - COMMENTS

A Fortran program has been written to implement the above procedure; until now complete runs were only performed on small size sets of data (10 zones, 25 links); several networks were obtained interactively. Parts of the program have been tested in a case study (100 nodes, 160 links) for the Toulouse transportation system (see [2]). The selectivity of phase 1 in the definition of a candidate set can be modified thanks to threshold parameters. Storage requirements are modest, around 15K words for 50 zones and 100 links, 40K words for 100 zones and 200 links on IRIS 80.

Presently the program is only a prototype. However, it is easy to forecast that computation time increases only polynomially(*) with the size of the network: owing to decomposition techniques,

(*) ($O(n^3)$ for initial shortest path search, $O(n^2)$ for the assignment and path addition, where n is the number of nodes)

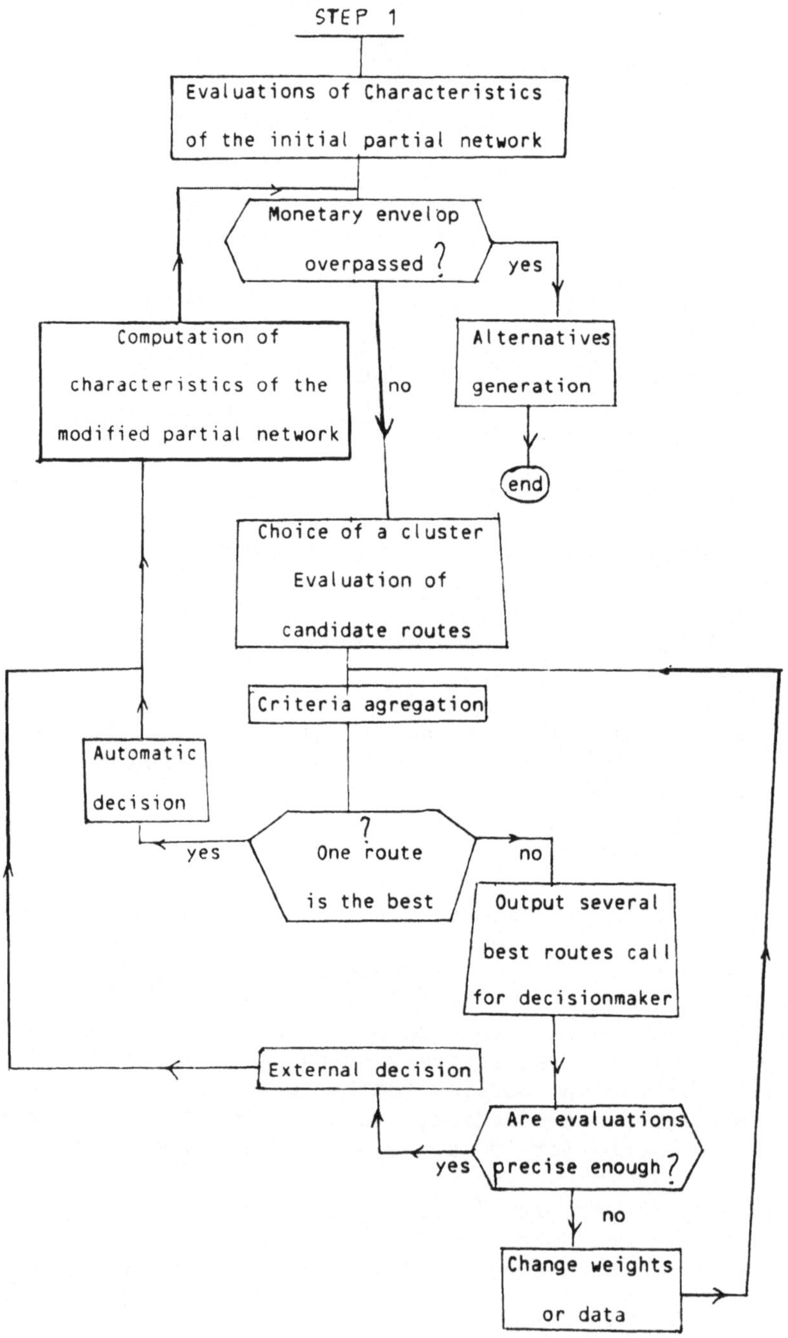


Fig. 5. Flowchart No. 2 (Step 2)

partial evaluations are fast. The whole procedure is obviously much more rapid than any branch and bound algorithm, and the problem it solves is much less academic than the classic optimal network problem.

Improvements of the above procedure can be considered. Firstly the construction of the fuzzy logical rules for demand evaluation can be modified to take into account the experience that a transportation specialist has acquired about the behavior of mass-transit users in a specific town. However, the idea of a logical estimation of potential traffic in a network seems worth keeping anyway. Secondly, such a procedure can really be efficient only if pertinent outputs are provided to the operator so as to facilitate its decisions. This problem belongs no more to the transportation planning field, but to the one of man-machine interaction. The aim is to suppress ambiguities, and to fit the program to human language.

Any implemented version of the procedure is not necessarily fitted to any set of data for any town. Data processing and choice of criteria can be specific; moreover some structural constraints may exist, which determine certain features of the network. On the other hand the methodology, i.e. the division of global choice into partial decision, the use of concepts from fuzzy sets theory, and man-machine interaction in general, can be applied to other synthesis problems in the field of humanistic systems analysis. For instance the design of bus routes can be realized in the framework of the same approach (see [7] ch IV).

## CONCLUSION

This paper suggests that classical operations research methods are not always fitted to the determination of alternative projects in a transportation planning study, and more generally the design of humanistic systems. In a complex process, whose behavior rules and evaluation are ill-known, constraints and criteria become fuzzy.

To provide an aid to the imagination of planners, it seems interesting to develop fast procedures which, in an interactive way, build several alternative projects that are approximately evaluated. Sophisticated global models are somewhat utopic, since they attempt to eliminate the uncertainty which is always pervasive in decisions concerning humanistic systems. Owing to its irreducibility, it seems more useful to evaluate the level of uncertainty so as to measure the legitimacy of choices.

The procedure described above is a step in that direction. Man-machine interaction allows us to reduce the combinatoric aspect of the network generation problem while providing realistic solutions. Decomposition of global choice, and the use of local criteria make this interaction more easy.

Lastly, the incorporation of concepts and tools from fuzzy sets theory into the model permits the algebraic manipulation of ill-known quantities (represented by sets instead of single precise values) in order to evaluate the legitimacy of decisions. Another application of fuzzy sets theory to operations research, using a similar approach can be found in Prade [28] [29] in the framework of scheduling problems.

## REFERENCES

1. J. L. A. Avila, "Optimal design of transportation network with fluctuating demands: A case in multicommodity network flows". Stanford University, Calif. 1972.
2. G. Bel, D. Dubois, and M. Llibre, A set of methods in transportation network synthesis and analysis, J. Opl. Res. Soc., 30, 797-808, 1979.
3. R. Bellman and L. A. Zadeh, Local and Fuzzy Logics, Memo. ERL M504 Electronics Research Lab., University of California, Berkeley, 1976.
4. J. W. Billheimer, Network design of fixed and variable elements. Transportation science, Vol. 7, No. 1, 1973, pp. 49-74.
5. D. E. Boyce, A. Fahri, and R. Weishedel, "Optimal network problem: a branch and bound algorithm." Environment and Planning, Vol. 5, 1973, pp. 519-533.
6. A. De Luca and R. M. Capocelli, Fuzzy sets and decision theory. Inf. and Cont. 23, 1973, pp. 446-473.
7. D. Dubois, Quelques outils Methodologiques pour la conception de reseaux de transport. These Doct. Ing. ENSAE/CERT-DERA Toulouse, France, 1977.
8. D. Dubois and H. Prade, Algorithmes de plus court chemins pour traiter des donnees floues, R.A.I.R.O. (1978, No. 2).
9. D. Dubois and H. Prade, A comment on "Tolerance analysis using fuzzysets" and "A procedure for multiple aspect decision making using fuzzy sets by R. Jain". in Int. J. on Systems Science, Vol. 9, No. 3, 1978, pp. 357-360.
10. D. Dubois and H. Prade, "Operations on fuzzy numbers", Int. J. on Systems Science, 1978, Vol. 9, No. 6, 613-626.
11. D. Dubois and H. Prade, "Systems of linear fuzzy constraints", Fuzzy Sets and Systems, 3, 37-48, 1980.
12. B. R. Gaines, Foundations of fuzzy reasoning, Int. J. Man Machine Studies, 8, 1976, pp. 623-668, also in [14].
13. A. M. Geoffrion and R. Nauss, "Parametric and post-optimality analysis in integer linear programming", Working paper, No. 246, 1976, UCLA.
14. M. M. Gupta, G. Saridis, and B. R. Gaines, Fuzzy Automata and

Decision Processes, North-Holland, 1977.
15. Hoang Hai Hoc, A computational approach to the selection of an optimal network, Management Science, 19, 1973, No. 5, pp. 488-498.
16. R. Jain, A procedure for multiple aspect decision making using fuzzy sets, Int. J. on Systems Science, 8, 1977, Vol. 1, pp. 1-7.
17. Jeroslow, "A one constraint 149-variable zero one program requiring 9 million years for solving using branch and bound method. Techn. Rep. Carnegie Mellon Univ., Pittsburgh, 1973.
18. D. S. Johnson, J. K. Lenstra, and A. G. H. Rinnoy Khan., "The complexity of the network design problem," Math. Cent. Dept. of Opns. Res., Amsterdam, 1977.
19. R. M. Karp, "On the computational complexity of combinatorial problems", Networks, 5, 1975, pp. 45-68.
20. A. Kaufmann, "Introduction a la theorie des sous ensembles flous et a ses applications." Tomes 1,2,3,4, 1973-1976, Masson.
21. G. L. Nemhauser, L. E. Trotter, and M. J. Magazine, "When the greedy solution solves the knapsack problem." Opns. Res. 23, 2, 1975, 207-217.
22. M. Loss, Optimal network design without congestion: some computational results. Univ. of Montreal Centre de Recherches sur les transports #40, 1976.
23. Loubal, "A network evaluation procedure." Bay Area Transportation Commission, 1966.
24. Nahmias, "Fuzzy Variables." presented at ORSA TIMS Meeting, Miami, 1976, also "Fuzzy Sets and Systems, Vol. 1, No. 2, 1978, pp. 97-111.
25. R. M. Nauss, Parametric Integer Programming, Ph.D. Dissertation UCLA. Working paper No. 226, 1975.
26. Ochoa-Rosso, "Optimum project addition in urban transportation network via descriptive traffic assignment models. D.C.E./M.I.T., Cambridge, Mass., 1968.
27. Poulit, "Urbanisme et transport: le critere d'accessibilite et de development urbain", S.E.T.R.A. Division Urbaine, 1970.
28. H. Prade, Ordonnancement et temps reel. These doct. ing. ENSAE/CERT-DERA, Toulouse, France, 1977.
29. H. Prade, "Exemple d'approche heuristique, interactive, floue pout un probleme d'ordonnancement." Communication presented at AFCET Congress "Modelisation et Maitrise des Systemes" Versailles. Nov., 1977. Editions Hommes et Techniques. vol. 2, pp. 347-355. "Using Fuzzy Sets Theory in a Scheduling Problem: A Case Study", "Fuzzy Sets and Systems", 2, 153-165, 1979.
30. S. Sahni, "Computationally related problems", SIAM J. on Computers, 3, 1976, No. 4.
31. A. J. Scott, The optimal network problem: some computational

procedures Transportation Research, Vol. 3, 1969, pp. 201-210.
32. S. Stairs, "Selecting an optimal traffic network", Jr. of Transport, Economics, and Policy, 1968, pp. 218-231.
33. L. A. Zadeh, Fuzzy sets, Inf. and Cont. 8, 1965, pp. 338-353.
34. L. A. Zadeh, "Outline of a new approach to the analysis of complex systems and decision processes", IEEE Trans. on System, Man and Cybernetics, 3, pp. 28-44.
35. Y. P. Chan, "Optimal travel time reduction in a transport network", Res. Rep. R-69-39, M.I.T.-D.C.E., Cambridge, Mass., 1969.
36. C. Daganzo and Y. Sheffi, "On stochastic models of traffic assignment," Trans. Sci., 11 (3) 1977, pp. 253-274.
37. R. B. Dial, A probabilistic multipath assignment model which obviates path enumeration, Trans. Res. 5, 1971, 83-111.
38. A. Rosenfeld, "Fuzzy graphs," in L. A. Zadeh, K. S. Fu, K. Tanaka, and M. Shimura, Eds. Fuzzy Sets and their Applications to Cognitive and Decision Sciences, Academic Press, New York, 1975, pp. 77-95.
39. Steenbrink, "Optimization of Transport Networks", Wiley, 1974.
40. L. A. Zadeh, PRUF - A meaning representation language for natural language. Memo ERL M 77 61, Univ. of Calif., Berkeley, 1977.
41. L. A. Zadeh, "Fuzzy sets as a basis for a theory of possibility", "Fuzzy Sets and Systems" I, n°I, pp. 3-28, 1978.
42. J. G. Koenig, Indicators of urban accessibility: theory and applications, Transportation, 9, n°2, 1980, 145-172.
43. D. Dubois and H. Prade, Fuzzy Sets and Systems: Theory and Applications, Academic Press, 1980.
44. D. Dubois and H. Prade, Criteria aggregation and ranking of alternatives in the framework of fuzzy set theory, to appear in "Fuzzy set and decision analysis" (H. J. Zimmermann and L. A. Zadeh, eds.) TIMS Series in the management sciences, 1982, North Holland.
45. D. Dubois and H. Prade, Unfair coins and necessity measures submitted to Fuzzy Sets and Systems.

# APPLICATION OF FUZZY SET THEORY TO ECONOMICS

Guo quan Chen*, Samuel C. Lee*, and Eden S. H. Yu+

School of Electrical Engineering
and Computer Science*
University of Oklahoma
Norman, Oklahoma

College of Business Administration+
University of Oklahoma
Norman, Oklahoma

## I. INTRODUCTION

It has been widely recognized by economists that economic behavior generally involves both elements of stochasticness and/or fuzziness. The objects which economic theory deals with are replete with all sorts of fuzzy emotions, perceptions and processes. While a rich literature on economic behavior under a stochastic environment has developed during the past decade,[1] it is notable that virtually no systematic attempt has been made to-date to investigate economic behavior under fuzziness.

A powerful tool of economic analysis is utility theory. However, the indispensable fuzzy aspects of a human mind in the area of utility and preference remains to be explored. The primary purpose of this Chapter is to investigate certain properties of a fuzzy utility function. Unlike the traditional approach, utility indicators will be expressed by a fuzzy set in lieu of real numbers. A utility function becomes a mapping from a n-nary commodity space to the set of fuzzy subsets of utility space -- a one-dimensional Euclidean space. A secondary purpose of this Chapter is to analyze consumer behavior characterized by maximizing fuzzy utility subject to a budget constraint.

---

[1]See, for example, Batra (1975) and the references contained herein.

For the convenience of readers unfamiliar with fuzzy set, basic concepts and operations of fuzzy set used in our subsequent analysis are summarized in Section II. Section III develops the notions of fuzzy preferences and fuzzy utility functions. Section IV investigates several interesting properties of fuzzy utility functions. Section V is devoted to the maximization of fuzzy utility function and the optimum consumption equilibrium. Concluding remarks and implications are included in Section VI.

## II. FUZZY SETS - BASIC CONCEPTS AND OPERATIONS

The theory of fuzzy set was originated with the seminal work by Zadeh (1965). The theory was refined and further developed by Kaufman (1975), Kandel and Lee (1979) and Dubois and Prade (1980), among many others.

To begin with, let X be the universe of discourse, whose generic elements are denoted by $\chi$.

A <u>fuzzy set</u> A in X is characterized by a <u>membership function</u> $\mu_A(\chi)$ which is associated with each point in X a real number in the interval [0,1], with the value of $\mu_A(\chi)$ at $\chi$ representing the grade of membership of $\chi$ in A. (See Figs. 1 and 2.) Thus, the closer the value of $\mu_A(\chi)$ to unity, the higher the grade of membership of $\chi$ in A. When A is an ordinary set, its membership function can take on only two values 0 and 1, with $\mu_A(\chi) = 1$ or 0 according as $\chi$ does or does not belong to A. $\mu_A(\chi)$ is referred to as the characteristic function of the set A.

A fuzzy set A is completely characterized by the set of pairs

$$A = \{(\chi, \mu_A(\chi)), \chi \varepsilon X\}$$

A more convenient notation is as follows, when X is a finite set

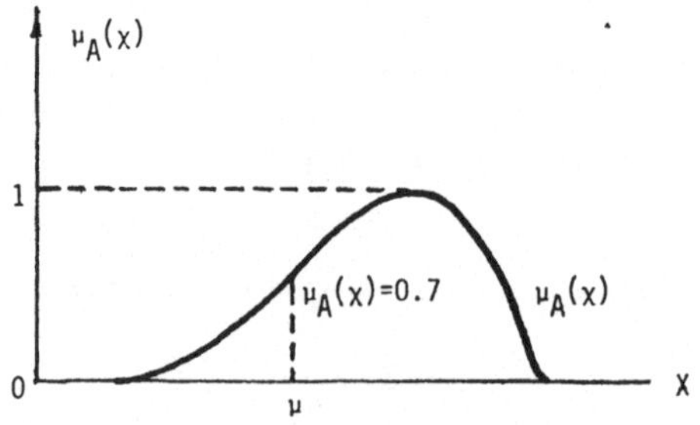


Fig. 1. A one-dimensional fuzzy set and its membership function.

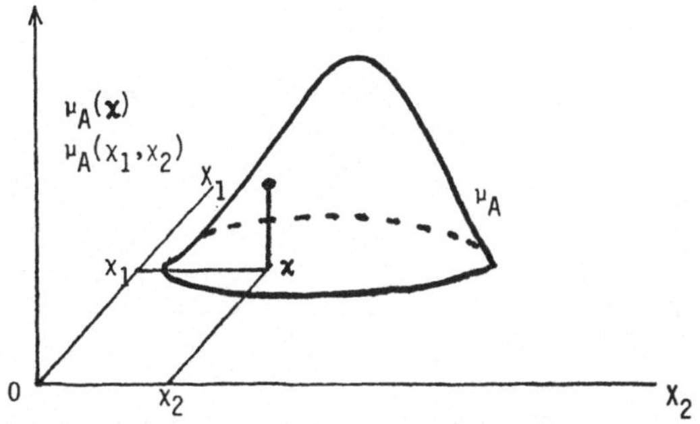


Fig. 2. A two-dimensional fuzzy set and its membership function.

$\{\chi_1,\ldots,\chi_n\}$, a fuzzy set on X is expressed as

$$A = \mu_A(\chi_1)/\chi_1 + \ldots + \mu_A(\chi_n)/\chi_n = \sum_{i=1}^{n} \mu_A(\chi_i)/\chi_i$$

where + and $\Sigma$ represent the union operation.

The <u>support</u> of a fuzzy set A (See Fig. 3) is an ordinary subset of X:

$$\text{Supp } A = \{\chi \varepsilon X,\ \mu_A(\chi) > 0\}$$

A is said to be normalized if $\exists \chi \varepsilon X,\ \mu_A(\chi) = 1$.

A fuzzy set is <u>empty</u> if and only if its membership function is identically zero on X.

Two fuzzy sets A and B are <u>equal</u>, if and only if $\mu_A(\chi) = \mu_B(\chi)$ for all $\chi$ in X.

A is <u>contained</u> in B if and only if $\mu_A(\chi) \leq \mu_B(\chi)$. In symbols

$$A \subset B \longleftrightarrow \mu_A \leq \mu_B.$$

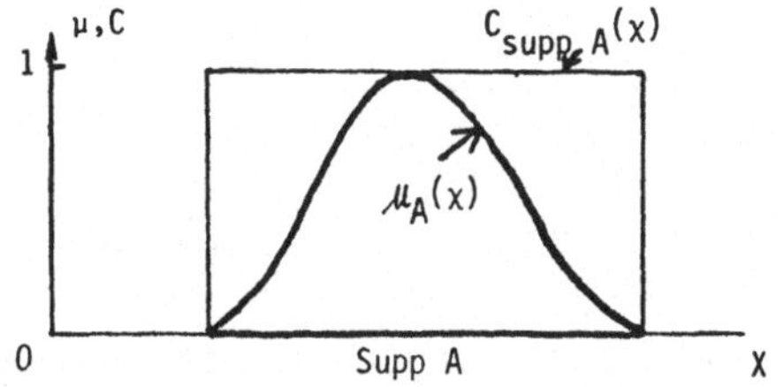


Fig. 3. Support of fuzzy set A and its characteristic function.

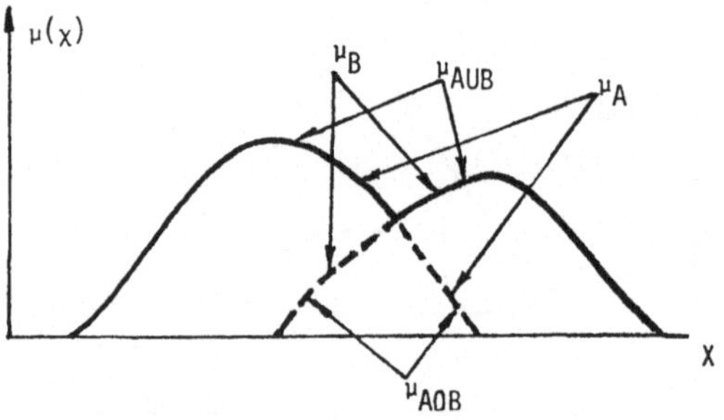


Fig. 4. The union and intersection of two fuzzy sets.

The union $\cup$ and intersection $\cap$ of two fuzzy sets A and B are: (See Fig. 4.)

$$\forall \chi \varepsilon X,\ \mu_{A\cup B}(\chi) = \max\ (\mu_A(\chi),\ \mu_B(\chi))$$

$$= \mu_A(\chi) \vee \mu_B(\chi)$$

$$\forall \chi \varepsilon X,\ \mu_{A\cap B}(\chi) = \min\ (\mu_A(\chi),\ \mu_B(\chi))$$

$$= \mu_A(\chi) \wedge \mu_B(\chi)$$

where $\mu_{A\cup B}$ and $\mu_{A\cap B}$ are the membership functions of $A\cup B$ and $A\cap B$, respectively. In the case where the membership function of the fuzzy set B is a constant $\beta$, we may express the intersection of the two fuzzy sets $B\cap A$ by $\beta A$.

An $\alpha$-cut of a fuzzy set A (See Fig. 5) is an ordinary set of such elements with membership grade greater than or equal to a threshold $\alpha$, $0 \leq \alpha \leq 1$.

$$A_\alpha = \{\chi \varepsilon X,\ \mu_A(\chi) \geq \alpha\}$$

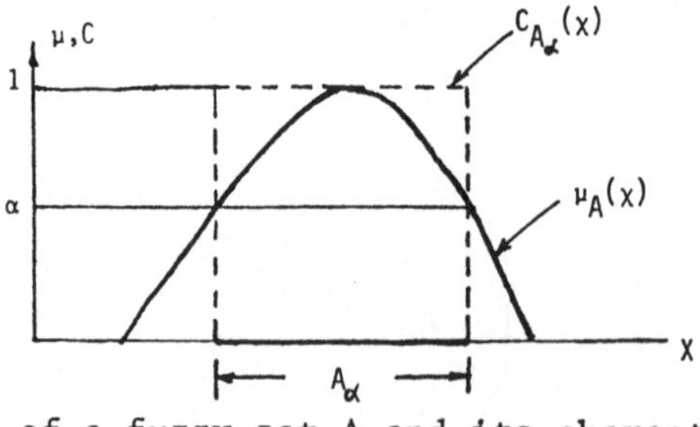


Fig. 5. An $\alpha$-cut of a fuzzy set A and its characteristic function $C_{A_\alpha}(\chi)$.

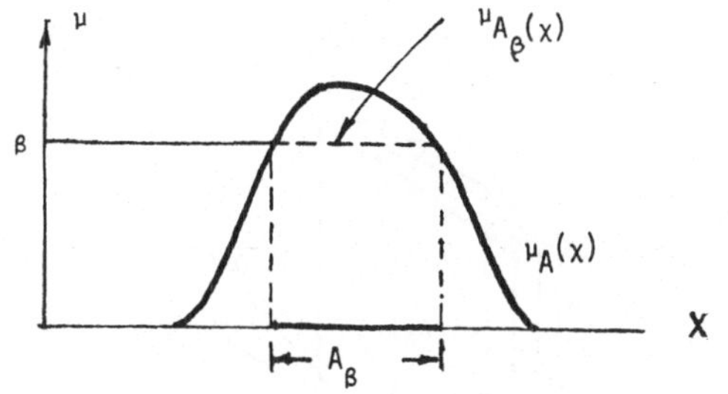


Fig. 6. A β-cut of a fuzzy set A and its membership function $\mu_{A_\beta}(\chi)$.

A β-cut of a fuzzy set A (See Fig. 6) is a fuzzy set that $A_\beta = \{\beta/\chi \mid \chi \in X,\ \mu_A(\chi) \geq \beta\}$.

Let $X_1,\ldots,X_n$ be n universes. An n-ary fuzzy relation R in $X_1 \times \ldots \times X_n$ is a fuzzy set on $X_1 \times \ldots \times X_n$.

The projection of a fuzzy relation R (See Fig. 7) on $X_{i_1} \times \ldots \times X_{i_k}$, where $(i_1,\ldots,i_k)$ is a subsequence of $(1,2,\ldots,n)$, is a relation of $X_{i_1} \times \ldots \times X_{i_k}$ defined by

$$P_{X_{i_1} \times \ldots \times X_{i_k}} R = \sum_{X_{i_1} \times \ldots \times X_{i_k}} \ \sup_{x_{j_1},\ldots,x_{j_\ell}} \mu_R(\chi_1,\ldots,\chi_n)/(\chi_{i_1},\ldots,\chi_{i_k})$$

where $j_1,\ldots,j_\ell$ is the subsequence complementary to $(i_1,\ldots,i_k)$

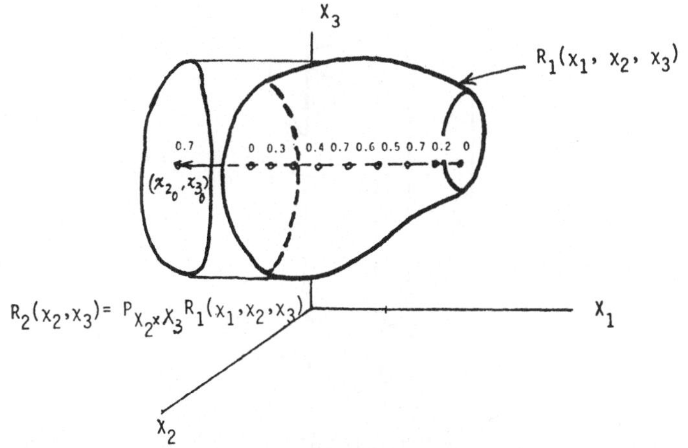


Fig. 7. Fuzzy relation and its projection.

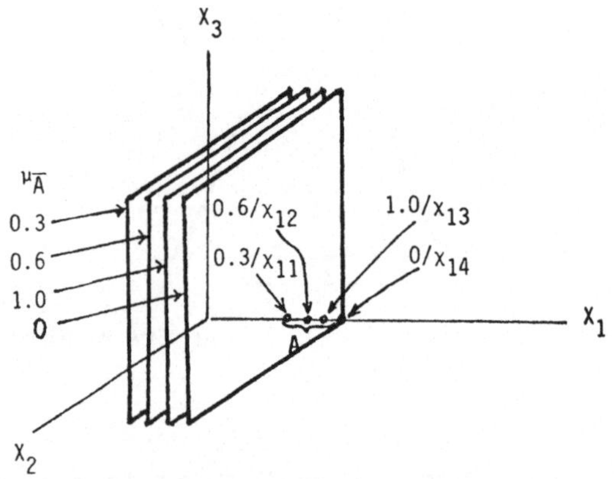


Fig. 8. Expansion of a one-dimensional fuzzy set into a three-dimensional fuzzy set through cylindrical extension.

in $(1,\ldots,n)$.

If A is a fuzzy set in $X_{i_1} \times\ldots\times X_{i_k}$, then its <u>cylindrical extension</u> in $X_1 \times\ldots\times X_n$ is a fuzzy set $\bar{A}$ in $X_1 \times\ldots\times X_n$ defined by

$$\bar{A} = \sum_{X_1 \times\ldots\times X_n} \mu_A(x_{i_1},\ldots,x_{i_k})/(x_1,\ldots,x_n),\ k < n.$$

Figures 8 and 9 give two examples of cylindrical extensions. In Fig. 8 a 1-dimensional fuzzy set

$$A = 0.3/x_{11} + 0.6/x_{12} + 1.0/x_{13} + 0/x_{14}$$

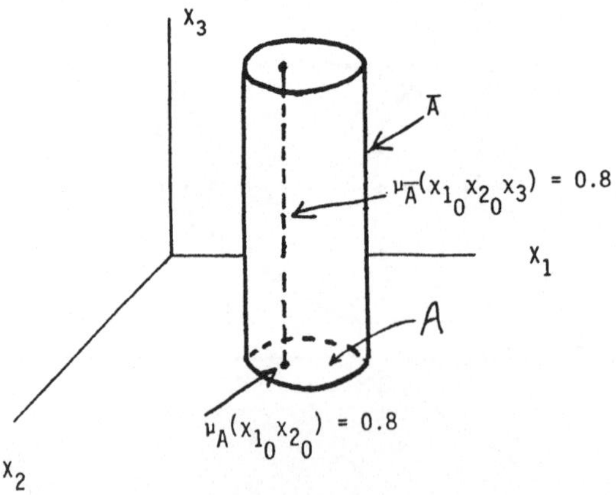


Fig. 9. Expansion of a two-dimensional fuzzy set into three-dimensional fuzzy set through cylindrical extension.

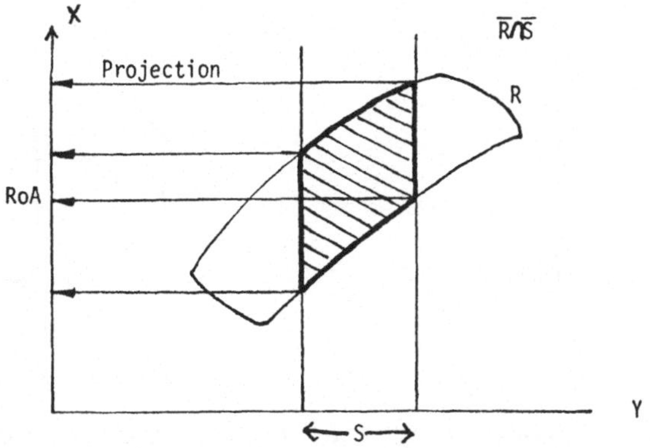


Fig. 10. Composition of fuzzy relations R and S.

becomes a 3-dimensional fuzzy set $\overline{A}$ via a cylindrical extension operation. Note that $\overline{A}$ is a fuzzy set in $X_1 \times X_2 \times X_3$, $\mu_{\overline{A}}(\chi_1,\chi_2,\chi_3)$ is independent of $\chi_2$ and $\chi_3$. In Fig. 9, a 2-dimensional fuzzy set A in $X_1 \times X_2$ is extended into a 3-dimensional fuzzy set $\overline{A}$ in $X_1 \times X_2 \times X_3$, with its $\mu_{\overline{A}}(\chi_1,\chi_2,\chi_3)$ independent of $\chi_3$.

A conditional relation of fuzzy relation $R(\chi_1,\ldots,\chi_n)$ is a fuzzy relation in which some of the coordinates remain constant.

Let R be a fuzzy relation in X x Y and S be a fuzzy relation in Y x Z, the <u>composition</u> of R and S denoted RoS is defined as

$$RoS = P_{X \times Z}(\overline{R} \cap \overline{S})$$

$$\mu_{RoS}(\chi,z) = \bigvee_y [\mu_R(\chi,y) \wedge \mu_S(y,z)]$$

$$= \max_y [\min(\mu_R(\chi,y) \wedge \mu_S(y,z))]$$

where $\overline{R}$ and $\overline{S}$ are cylindrical extension in X x Y x Z. Figure 10 demonstrates this operation with a simplfied example.

A fuzzy set A is <u>convex</u> iff all of its α-cut sets are convex. An equivalent definition of convexity is: A is convex iff

$$\forall\chi_1 \varepsilon X,\ \forall\chi_2 \varepsilon X,\ \forall\lambda \varepsilon [0,1],$$

$$\mu_A(\lambda\chi_1 + (1-\lambda)\chi_2) \geq \min(\mu_A(\chi_1),\ \mu_A(\chi_2))$$

Note that this definition does not necessarily imply that $\mu_A$ is a convex function of $\chi$. (See Fig. 11.)

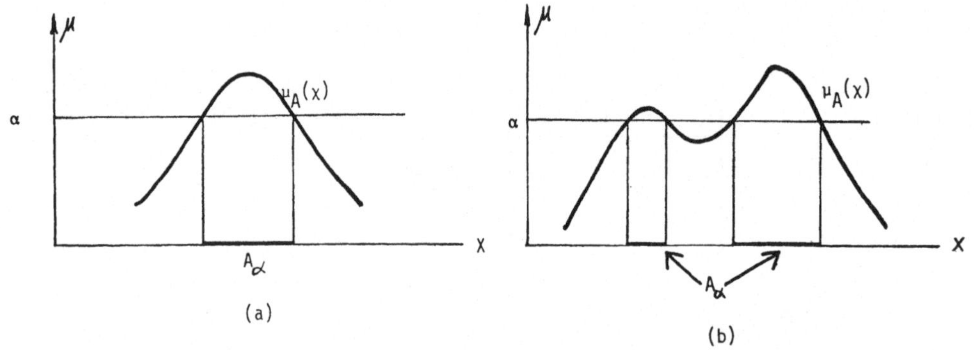


Fig. 11. (a) Convex fuzzy set and its convex α-cut.
(b) Non-convex fuzzy set and its non-convex α-cut.

A fuzzy number is a convex normalized fuzzy set A defined on the real line E such that

a) $\exists\, \chi \in E \quad \mu_A(\chi_0) = 1$

b) $\mu_A$ is piecewise continuous.

A fuzzy variable is a variable which takes on fuzzy numbers as its values.

If E is restricted to a real interval, one can express adjectives in natural language as fuzzy sets in this interval. For example, assume the universe of discourse is [0,100], one can express old, young, very old, not very old and not very young, etc., as fuzzy numbers in [0,100] interval as depicted in Fig. 12.

## III. SOME BASIC CONCEPTS OF FUZZY ECONOMICS

We define commodity bundle as

$$\chi = (\chi_1, \ldots, \chi_n).$$

With $\chi_i$ denoting the amount of the $i^{th}$ good $(i=1,\ldots,n)$.

By P(X) we denote the collection of ordinary subsets of X.

Define fuzzy commodity bundle as a fuzzy subset A in the commodity space X such that

$$A = \mu_1/a_1 + \ldots + \mu_i/a_i + \ldots$$

where $\mu_1, \ldots, \mu_i, \ldots$ denote the membership grades of commodity

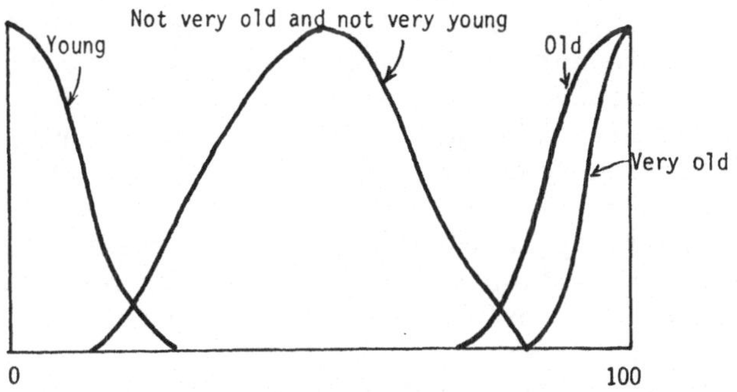


Fig. 12. Age described by fuzzy numbers.

bundles $a_1,\ldots,a_i,\ldots$ in A, respectively, / the separator and + the union operation.

Fuzzy commodity bundles are pervasive and evident in the real world. For example, a "good" library is a fuzzy commodity bundle, because it consists of a "decent" collection of books and an "effective" management system, which are fuzzy statements.

By F(X) we denote the collection of all fuzzy subsets in X.

The addition of two fuzzy commodity bundles, A and B and the scalar multiplication of a fuzzy commodity bundle are

$$A \dot{+} B = \sum_i [(\mu_A(a_i) \wedge \mu_B(b_i))/(a_i \dot{+} b_i)]$$

$$\lambda A = \Sigma(\mu_i/\lambda a_i)$$

where $\Sigma$ denotes the continual union operation and $\dot{+}$ denotes the arithmetic addition operation.

It is well-known that theories of consumer behavior and welfare economics can be constructed upon the properties of consumer preferences. See Rader (1972, Ch. 5 and 6). We write $\chi R y$ to indicate that "the consumer thinks the bundle $\chi$ is at least as preferred as the bundle y." It is generally assumed that R has some or even all of the properties regarding completeness, reflexitivity, transitivity, continuity, local nonsatiation, strong monotonicity and convexity. These properties are not necessarily unreasonable and they are desirable in deriving utility functions. If R is a regular ordering, i.e., the upper contour sets $R(\chi)$ are closed, preferences are representable by utility functions.

Consumers often reveal their preferences in a fuzzy manner[2]; preferences are indeed loaded with fuzziness. There is a need to reexamine the traditional utility theory for better describing consumer behaviors. Although we are not concerned with the question of representing fuzzy preferences by fuzzy utility indicators in this study, simple experimental procedures can be constructed to solicit information from a consumer about his/her degree of satisfaction toward various combinations of commodities. Appropriate fuzzy numbers are assigned to represent the consumer's expressions, in everyday language[3], about their subjective responses. With a sufficient collection of these fuzzy numbers, crude fuzzy utility function can be derived for the entire commodity space.

We define fuzzy utility indicator as a fuzzy number, i.e., a convex, normalized fuzzy subset, in utility space (one-dimensional Euclidean space)U. The collection of fuzzy (ordinary) subset of U is denoted by F(U)(P(U)). For practical purpose, U may be restricted to the [0,1] interval.

Define a fuzzy utility function as a mapping from X in F(U)

$$f: X \rightarrow F(U)$$

where f associates each commodity bundle in the commodity space X with a fuzzy utility indicator in the utility space U. It follows that the locus of f in the product space X x U, called consumption space, is a fuzzy subset of X x U, or equivalently a fuzzy relation $R_f$ between X and U.

Figure 13 depicts one type of fuzzy utility function. In this figure, every nonfuzzy commodity bundle $\chi$ in the commodity space X corresponds a fuzzy subset M in U with the exception in which the zero commodity bundle corresponds to a determinate zero utility indicator. Let $R_f^{-1}(u,\chi) = R_f(\chi,u)$ be the inverse of $R_f$. We define an indifference set of commodity bundles associated with a fuzzy utility indicator M, $M \varepsilon F(U)$, as a fuzzy subset $I_m$ in the commodity space X such that

$$I_m = M o R_f^{-1} = P_x(\overline{M} \cap R_f^{-1})$$

[2]Statements in our daily life like "I would rather prefer compact cars with dark blue color," "I like that nice restaurant very much," or "A calculator is quite useful for students in an econometrics class," are full of fuzziness.

[3]For example, a consumer may be asked to reveal his preferences in terms of the order of satisfaction: very satisfactory, quite satisfactory, barely satisfactory, not satisfactory, or absolutely not satisfactory, etc.

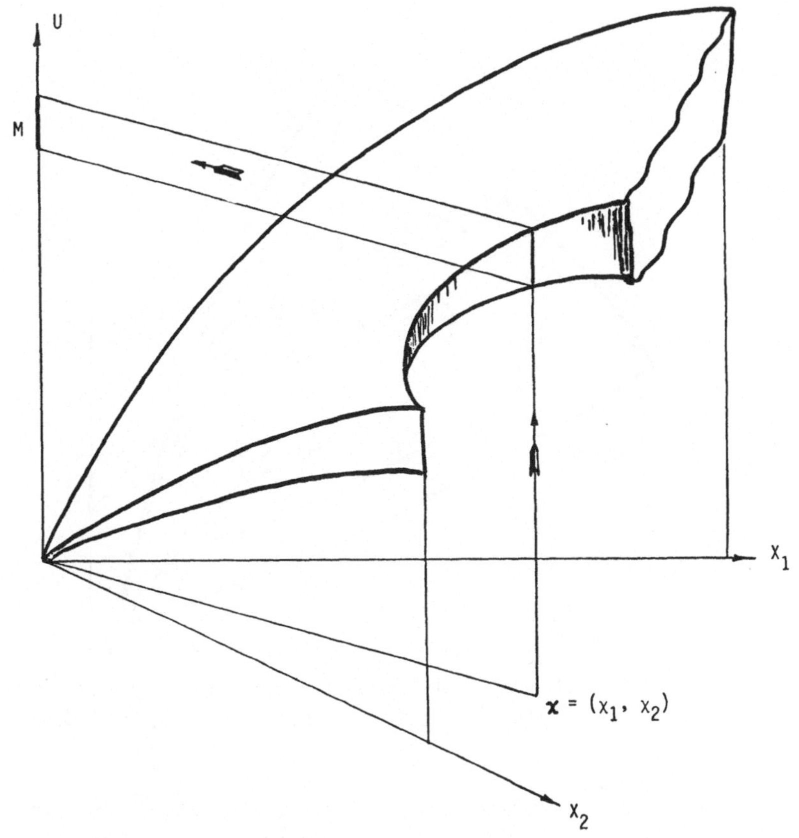


Fig. 13. One type of fuzzy utility function.

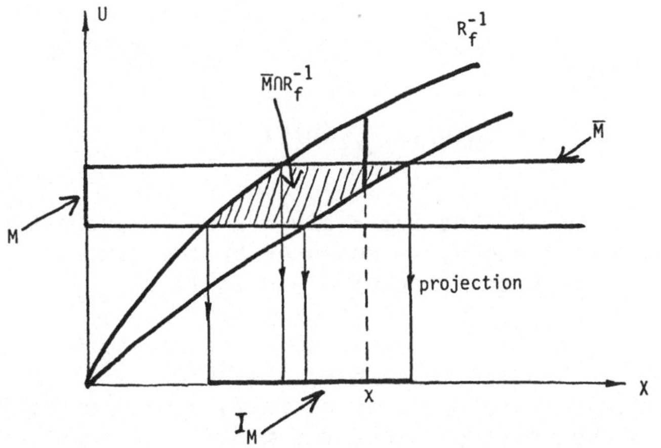


Fig. 14. A one-dimensional indifference set of commodity bundles.

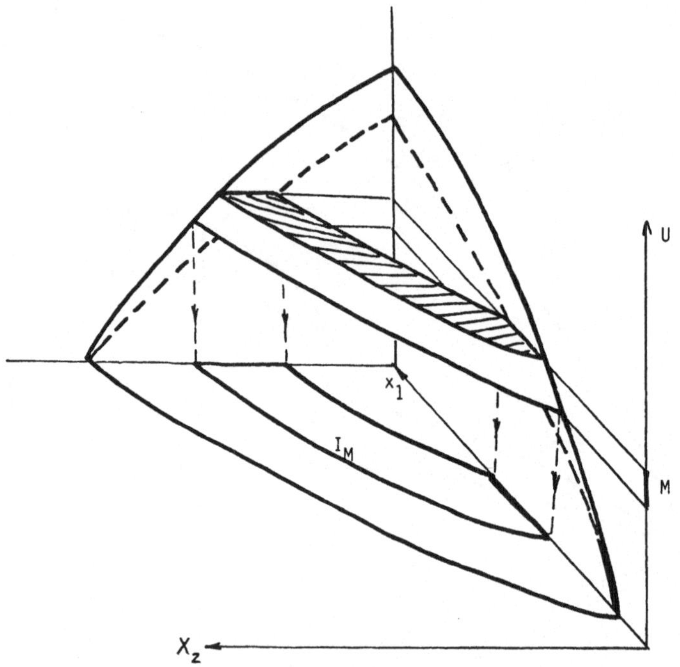


Fig. 15. A two-dimensional indifference set of commodity bundles $I_M$ with the "same" (in the fuzzy context) utility M.

where $P_X$ denotes the operation of projection to X, $\overline{M}$ is the cylindrical extension of M in X x U. Obviously, an indifference set is a fuzzy commodity bundle in the commodity space.

Consider a 1-dimensional indifference set of commodity bundles. (See Fig. 14.) Since $I_M$ is an indifference set, all elements of $I_M$ have the "same" utility indicator M. The degree of the sameness between M and $f(\chi)$, the fuzzy utility indicator of $\chi$, describes the membership grade for the commodity bundle $\chi$ in $I_M$. According to the definition of $I_M$

$$\mu_{I_M}(\chi) = \operatorname*{Max}_{u} \operatorname{Min}[\mu_M(u), \mu_{f(\chi)}(u)].$$

This definition means that for a given $\chi$ the closeness between two fuzzy set, $f(\chi)$ and M, is measured by the maximum of membership grades of the intersected set of M and $f(\chi)$.

In Fig. 15, $I_M$ is an example of a 2-dimensional indifference set of commodity bundles which have the "same" utility indicator M. When the utility indicator is an ordinary number u, rather than a fuzzy set, in the utility space, the indifference set of a commodity bundle is still a fuzzy subset in X, and

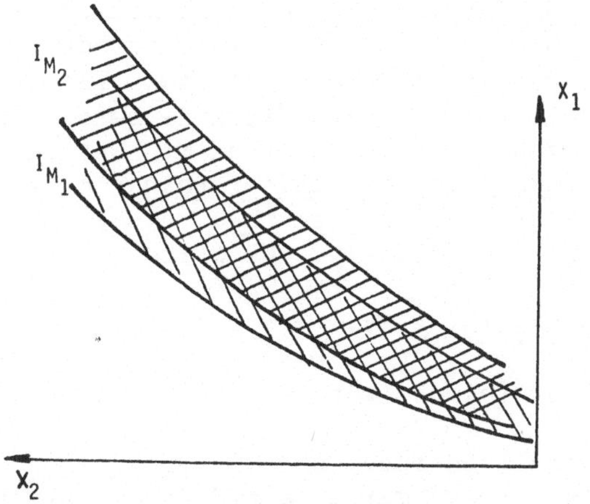


Fig. 16. Two indifference sets of commodity bundles associated with difference fuzzy utility indicators $M_1$ and $M_2$.

$$I_u = P_X(R_f^{-1}(u,\chi))$$

which is the projection of the conditional relation of $R_f^{-1}$ to X.

Proposition 1: If $N \subset M$, $N,M \in F(U)$, then $I_N \subset I_M$

Proof: $I_M = P_X(\overline{M} \cap R_f^{-1})$

$$I_N = P_X(\overline{N} \cap R_f^{-1})$$

$$\mu_{I_M} = \bigvee_u (\mu_M(u) \wedge \mu_{R_f^{-1}}(u,\chi))$$

$$\mu_{I_N} = \bigvee_u (\mu_N(u) \wedge \mu_{R_f^{-1}}(u,\chi)) .$$

Since $N \subset M$ thus, $\mu_M(u) \geq \mu_N(u)$. Then $\mathrm{Min}(\mu_M(u), \mu_{R_f^{-1}}(u,\chi)) \geq \mathrm{Min}(\mu_N(u), \mu_{R_f^{-1}}(u,\chi))$ for each $\chi$. And

$$\operatorname*{Max}_u \mathrm{Min}(\mu_M(u), \mu_{R_f^{-1}}(u,x)) \geq \operatorname*{Max}_u \mathrm{Min}(\mu_N(u), \mu_{R_f^{-1}}(u,x))$$

thus $I_N \subset I_M$ Q.E.D.

It may be noted that when $N = \{1/u\} \subset M$, $I_u \subset I_M$. Hence, the indifference set of a determinate utility indicator is included in the indifference set of a fuzzy utility indicator provided that the determinate utility indicator is included in the fuzzy counterpart. In general, two indifference sets are covered with each other to some extent. (See Fig. 16.)

Define a <u>weak preference set</u> $R_M$ of commodity bundles as the collection of the commodity bundles which the consumer "at least as prefer as" the fuzzy indifference set $I_M$. The weak preference set $R_M$ is the union of all the indifference sets with fuzzy utility indicators equal or greater than M. Note that a fuzzy set K is <u>greater than</u>, <u>equal to</u>, or <u>smaller than</u> another fuzzy set L if all of the α-cut of K is <u>greater than</u>, <u>equal to</u>, or <u>smaller than</u> the α-cut of L, respectively. We write

$$R_M = \bigcup_i \{I_i = M_i \circ R_f^{-1} | M_i \geq M\}.$$

An example of weak preference set is depicted in Fig. 17.

Let M be a normalized fuzzy set. A <u>fuzzy step function</u> of M, denoted by 1(M), is a fuzzy set obtained from M by letting $\mu_M(\chi) = 1$ for all $\chi > \chi_0$. $\chi_0$ is the point where the unity membership grade is first time reached. We have:

Proposition 2: The union of all fuzzy subsets N in U such that N greater than M is a fuzzy step function of M.

Proof: Firstly, since all N's are greater than or equal to M, all membership grades of the union set before the membership grade of M reaches to 1 will be impossible to be greater than that of M, otherwise it will violate the assumption that $N \geq M$. Secondly, we can always find a fuzzy set greater than M and its membership grade equals to 1 at every point after the membership grade of M reaches to 1. Thus, the membership grades of the union set will maintain 1 thereafter. (See Fig. 18.)

Define the fuzzy upper blunt partition set of a fuzzy relation $R(\chi,u)$ partitioned by $M \in F(U)$ as the fuzzy relation

$$_M R = R \cap \overline{1(M)},$$

with the understanding that $\overline{1(M)}$ is the cylindrical extension of 1(M) in X x U. When M is replaced by a nonfuzzy number $u \in U$, we obtain an <u>upper sharp partition</u> set of R, denoted by $_u R$,

$$_u R = R \cap \overline{1(u)}.$$

Reducing the membership grade of 1(u) from 1 to β we obtain a <u>β-upper sharp partition</u> set of R denoted by $_\beta R$

$$_\beta R = R \cap \overline{\beta 1(u)}$$

$_M R$, $_u R$, and $_\beta R$ are fuzzy relations in the consumption space X x U

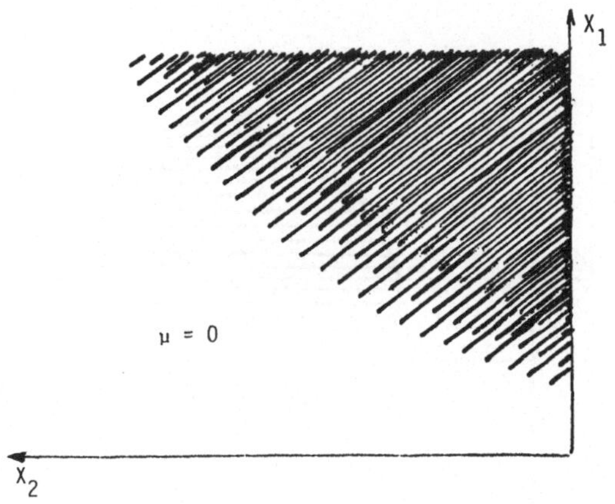


Fig. 17. A two-dimensional weak preference set.

Figures 19, 20, and 21 depict ${}_MR$, ${}_uR$, and ${}_\beta R$, respectively, where commodity spaces are 1-dimensional spaces.

In Fig. 19, a fuzzy relation $R(u,\chi)$ is partitioned to three sections by a fuzzy upper blunt partition. In the section shadowed by dotted lines, the membership grades are determined by the minimum of $\mu_{R(u,\chi)}$ and $\mu_M$; the membership grades in the section shadowed by real lines are maintained unchanged and the membership grade of the blank section are replaced by zero. The fuzzy upper blunt partition set ${}_MR$ consists of the first two sections.

An upper sharp partition simply divides a fuzzy relation into two sections. The membership grades of the upper section remain unchanged with that of another section replaced by zero. (See Fig. 20).

The only difference between the β-upper sharp partition and the upper sharp partition is that the membership grades of the upper section are determined by the minimum of ordinary membership grades and β. (See Fig. 21.)

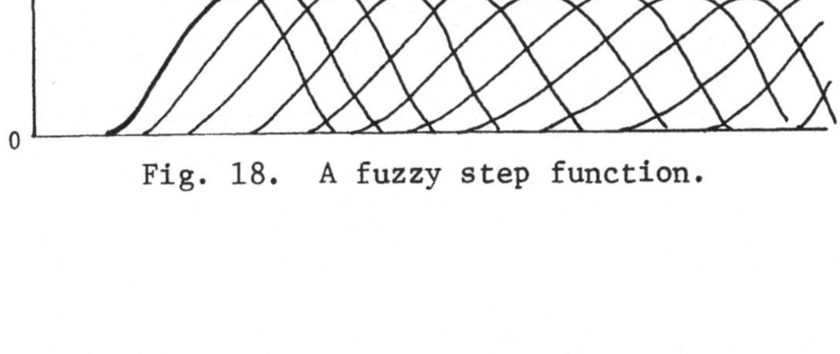

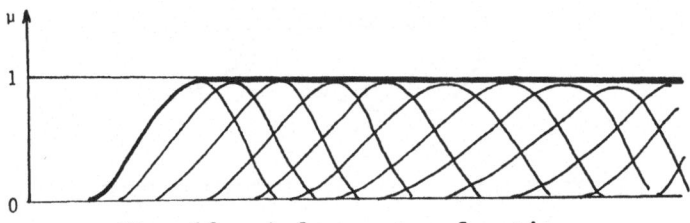

Fig. 18. A fuzzy step function.

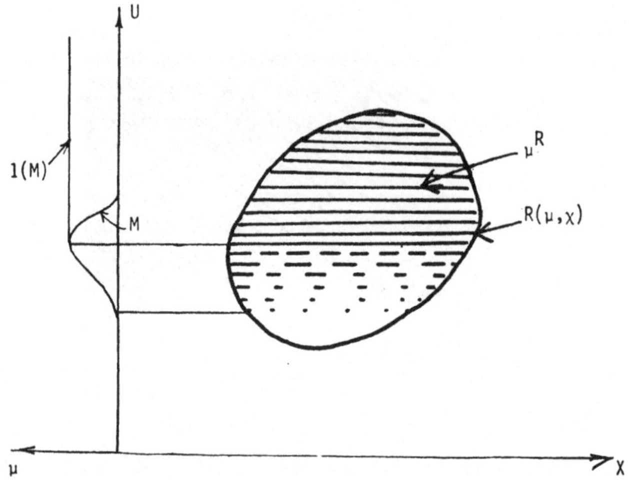


Fig. 19. A fuzzy upper blunt partition set.

Proposition 3: The weak preference set with the utility indicator M, denoted by $R_M$, is equal to the project of the upper blunt partition set ${}_M R_f^{-1}$ to the commodity space X.

$$R_M = \bigcup_i \{I_i = M_i o R_f^{-1} | M_i \geq M\} = 1(M) o R_f^{-1} = P_X({}_M R_f^{-1}).$$

This proposition means that we prefer to make only one composition operation after the union operation of all $M_i \geq M$ rather than a union operation of many indifference sets, resulting from a series of composition operations of $M_i$ and $R_f^{-1}$.

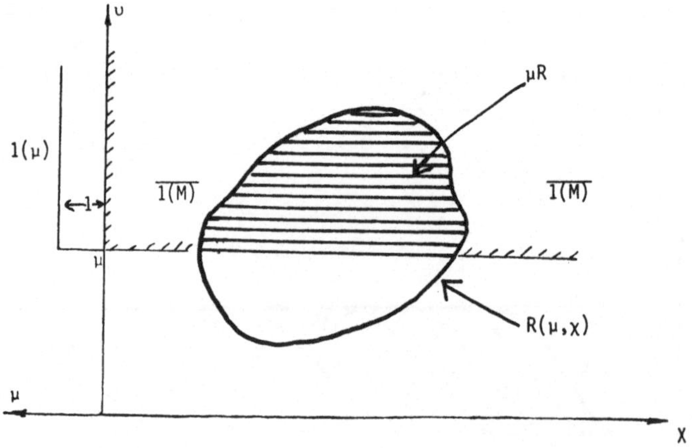


Fig. 20. An upper sharp partition.

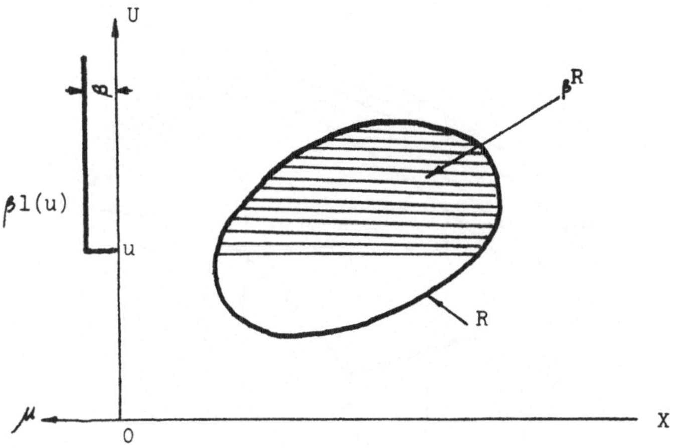


Fig. 21. A β-upper partition.

Proof: From the definition of weak preference set and by using the associative law of the composition operation, we have

$$R_M = \bigcup_i \{I_i = M_i o R_f^{-1} | M_i \geq M\}$$

$$= (\sum_i M_i) o R_f^{-1} \ ; \quad M_i \geq M$$

$$= 1(M) o R_f^{-1}$$

$$= P_x({}_M R_f^{-1})$$

similarly,

$$R_u = P_x({}_u R_f^{-1})$$

$$R_\beta = P_x({}_\beta R_f^{-1})$$

Q.E.D.

If $R_M$ is called weak M - preference set, $R_u$ and $R_\beta$ can be called weak u-preference set and weak β - preference set, respectively.

Proposition 4: If B $(=\{\beta/u | \forall u \geq u_0\}) \subset$ N $(=\{1/u | \forall u \geq u_0\} \subset$ M, then $R_\beta \subset R_u \subset R_M$.

Proof: See Proposition 1 and consider that if $A_1 \subset B_1, \ldots, A_n \subset B_n$ then $U(A_1, \ldots, A_n) \subset U(B_1, \ldots, B_n)$. Q.E.D.

## IV. PROPERTIES OF FUZZY UTILITY FUNCTIONS

In this section, we prove some theorems regarding the properties of a fuzzy utility function. The properties shed

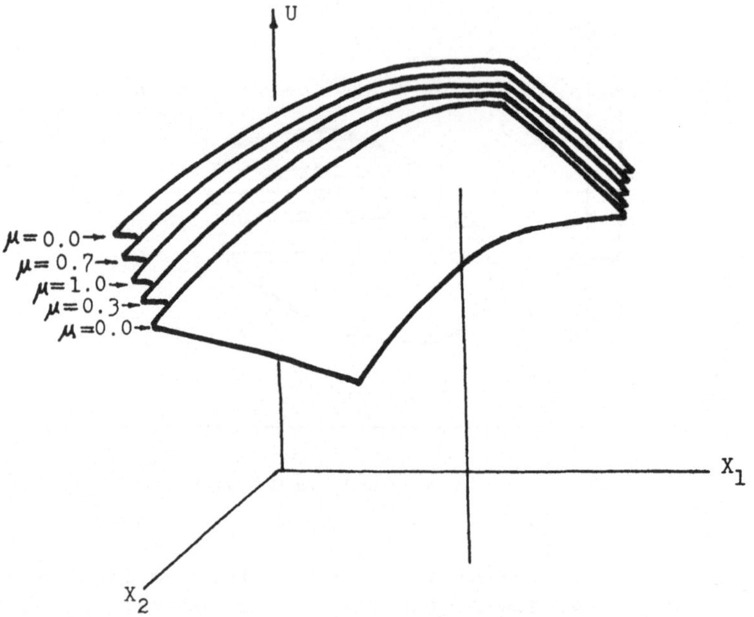


Fig. 22. Part of fuzzy utility function and its iso-membership grade surfaces.

light on the conditions for obtaining consumption equilibriums.

For expositional purpose, we confine the analysis to a simple case of two commodities, labor service and wheat. Let $X_1$ represent the amount of labor supplied by a household and $X_2$ the amount of wheat available to the household.

Theorem 1: If every fuzzy utility indicator of a fuzzy

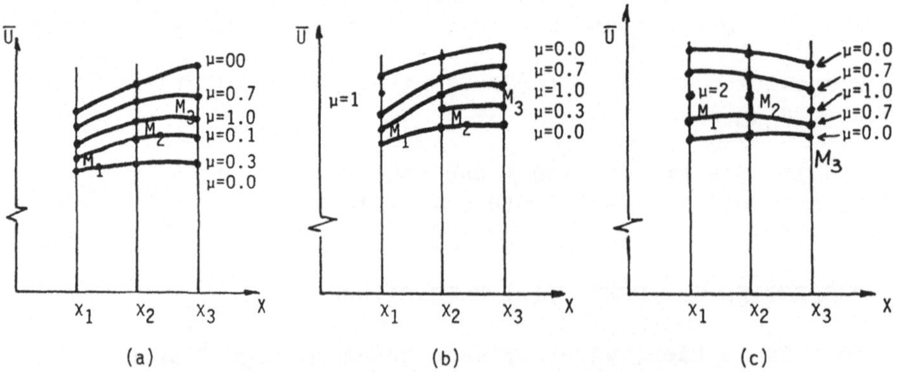


Fig. 23. Convexity and connectivity.

utility relation is convex and normalized, the iso-membership grade surfaces (See Fig. 22) of the fuzzy utility relation are connective and never intersect each other.

Proof: This theorem is apparent directly from the definition of convexity of fuzzy set. Several examples of iso-membership grade surfaces (here reduced to lines) for convex and normalized fuzzy utility indicators are shown in Fig. 23(a). Figure 23(b) shows a case where one of the fuzzy utility indicators, $M_1$, is non-convex, and Fig. 23(c) shows another case where one of them, $M_2$, is non-normalized. These cases lead to the existence of intersections between the iso-membership grade surfaces and/or the lack of connectiveness. Q.E.D.

The lack of an intersection between the iso-membership grade surfaces in fuzzy utility relation further leads to the lack of intersections among the contours of α-cut in the fuzzy indifference set.

Remark: If the marginal utilities of every iso-membership grade surface of a fuzzy utility relation is diminishing, i.e., $\partial^2 f/\partial^2 x_i > 0$, then every contour of the α-cut of indifference set is concave. (See Fig. 24.) While diminishing marginal utilities or diminishing marginal rate of substitutions implies concavity of

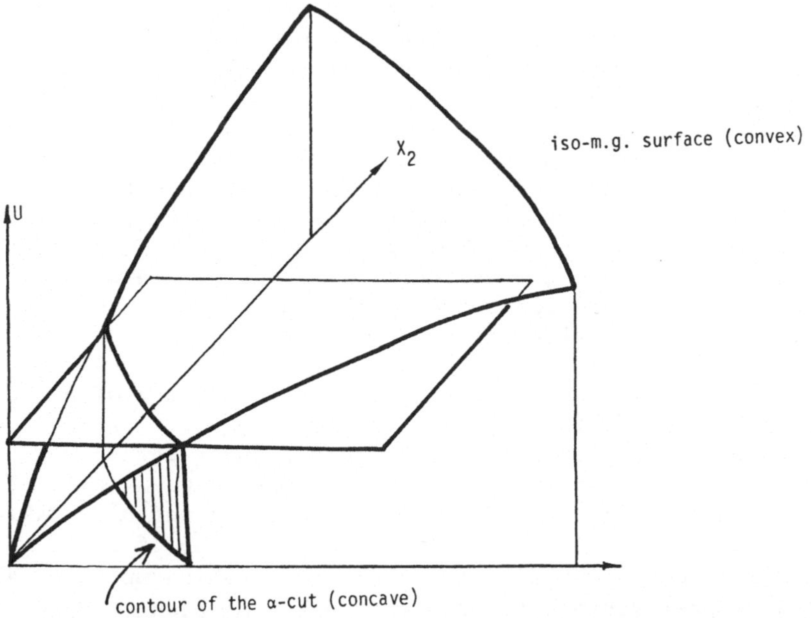


Fig. 24. Iso-membership grade surface and contour of the α-cut of an indifference set.

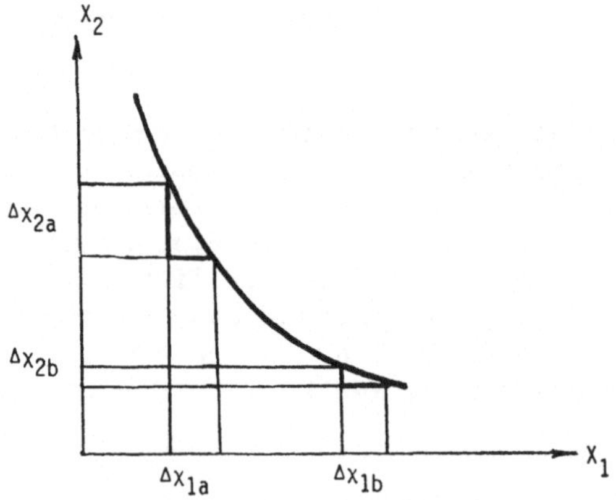


Fig. 25. Diminishing marginal rate of substitution.

the indifference curve, diminishing marginal utility here implies convexity of iso-membership grade surface of a fuzzy utility relation. Figure 25 is used for the explanation of diminishing marginal rate of substitutions. Note that following the increase of $\chi_1$, the same increment $\Delta_{\chi_1}$ corresponds to a smaller increment $\Delta_{\chi_2}$.

Theorem 2: If every fuzzy utility indicator of a fuzzy utility relation $R_f$ is convex and normalized, and if there is a diminishing marginal utilitie for every iso-membership grade surface of $R_f$ then the weak M- preference set $R_M$, the weak u-preference set $R_u$ and the weak $\beta$-preference set $R_\beta$ are all convex.

Proof: Since $R_u$ and $R_\beta$ are special cases of $R_M$, it suffices to prove the convexity of $R_M$. Let us delineate the procedure of constructing the weak preference set $R_M$ first. According to Proposition 3, for obtaining the weak preference set, all we have to do is to make the composition operation between 1(M) and $R_f^{-1}$. The operations include the cylindrical extension operation of 1(M), the intersection operation between 1(M) and $R_f^{-1}$ and the projection operation of intersection set to commodity space X.

In Fig. 26, we construct the section labelled F of a fuzzy utility relation $R_f^{-1}$ in any plane vertical to $X_2$. Cutting F by a line with equal utility level u, we obtain fuzzy set A on $X_1$. Assume that $u_1 \varepsilon 1(M)$ and its membership grade is $\mu_{1(M)}(u_1)$. The intersection of $\mu_A(\chi_1)$ and $\mu_{1(M)}(u_1)$ is a fuzzy set A'. Taking the union of all A' for all $u_1$ in the support set of 1(M), we obtain all commodity bundles and their membership function values of the weak preference set below this vertical plane. Repeating this process for all $\chi_2 \varepsilon X_2$, we get the entire weak preference set.

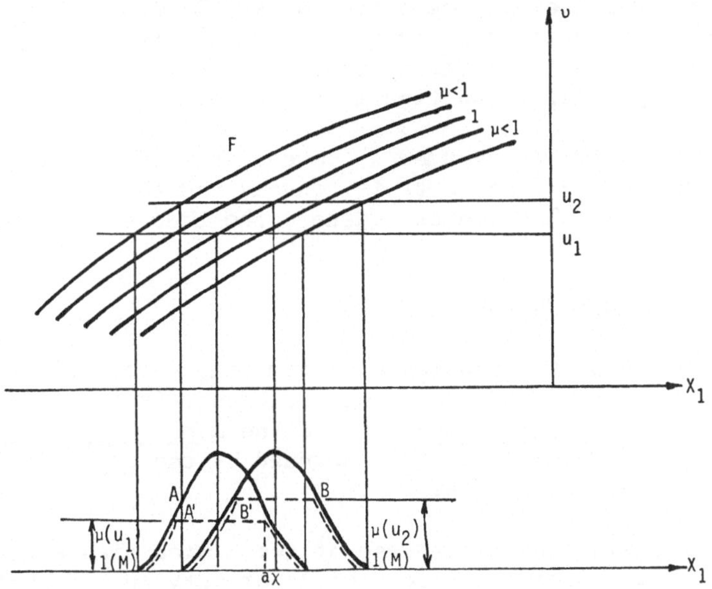


Fig. 26. The procedure of constructing the weak preference set.

Now, let us focus our attention on point a. For $\forall\chi\,|\,\chi \geq a$, we can always find an $u_2 > u_1$, for which

$$\mu_{R_f^{-1}(\chi,u_2)}(\chi) = 1.$$

Due to the convexity property of 1(M), we have $\mu_{1(M)}(u_2) \geq \mu_{1(M)}(u_1)$. Thus

$$\mu_{1(M)}(u_2) \wedge \mu_{R_f^{-1}(\chi,u_2)}(\chi) = \mu_{1(M)}(u_2) \geq \mu_{1(M)}(u_1) \geq \mu_A(\chi).$$

It means that we can always find a fuzzy set B' which is the component of weak preference set and which covers one specific point in the lagging edge of A'. The union operation in the process of constructing weak preference set will delete this specific point in A'. The lagging edge of A' is a part of the lagging edge of A. Another part of the lagging edge of A has already lost its effect when the intersection operation between A and $\mu_{1(M)}(u_1)$ was performed. Thus, the entire lagging edge of fuzzy set A fails to make any contribution to the construction of the weak preference set.

Since this is true for every u, the membership grade of all the points in the consumption space under the surface in which the membership grade equals to 1 may be replaced by 1 without affecting the evaluation of the weak preference set.

From this replacement, we obtain a new $F_f^{-1}$ denoted by $R_f^{-1}$. It is convex if every utility indicator of the determinate commodity bundle is convex and the marginal utility of every iso-membership surface of fuzzy utility function is diminishing.

Recall that l(M) is assumed to be convex. Since the intersection of two convex fuzzy sets is convex, and the project of a convex fuzzy set is also convex. Hence, the theorem follows. Q.E.D.

## V. OPTIMUM CONSUMPTION EQUILIBRIUM

The solution to the problem of maximizing fuzzy utility function subject to a traditional budget constraints yields an optimum consumption equilibrium. We define a nonfuzzy budget set as a subset in the commodity space bounded by the price line and the consumer's initial endowments.

Consider again the simple case of two commodities e.g., labor service and wheat. Let the prices of labor service and wheat be denoted by $p_1$ and $p_2$, respectively. Thus, $p_1\chi_1 + p_2\chi_2$ represents the amount, net of income earned by supplying labor services the consumers spend on wheat. Assume that the consumer has an initial holdings of wheat $\chi_1^0$ and faces an exogeneously determined set of prices. The consumer maximizes his fuzzy utility from consuming wheat and providing labor services subject to the budget constraint, $p_1\chi_1 + p_2\chi_2 = p_1\chi_1^0$. The nonfuzzy budget line (or price line) is the boundary of the nonfuzzy budget set.

What we face here is to seek the "tangent point" between a fuzzy set and a straight line.

Define a u- consumption equilibrium point as a point $\hat{\chi} = (\chi_1,\chi_2)$ at which the budget line is tangent to the edge of the convex $\hat{\alpha}$-cut set of $R_u$. Note that the grade of membership of a u-consumption equilibrium (with respect to the fuzzy consumption equilibrium set) is given by $\hat{\alpha}$. (See Fig. 27.) Obviously, another value u will induce another weak preference set $R_u$ and the point, at which the convex edge of $\hat{\alpha}$-cut set of $R_u$ is tangent to budget line, and its membership grade $\hat{\alpha}$ will be changed at the same time. The fuzzy u-consumption equilibrium set is defined by

$$E_u = \sum_u \hat{\alpha}_i/\hat{\chi}_i$$

where every u, $u_i$, is the value of utility which is associated with the consumption equilibrium point $\hat{\alpha}_i/\hat{\chi}_i$. $E_u$ is a fuzzy subset of X on the budget line.

We define β- consumption equilibrium point as a point

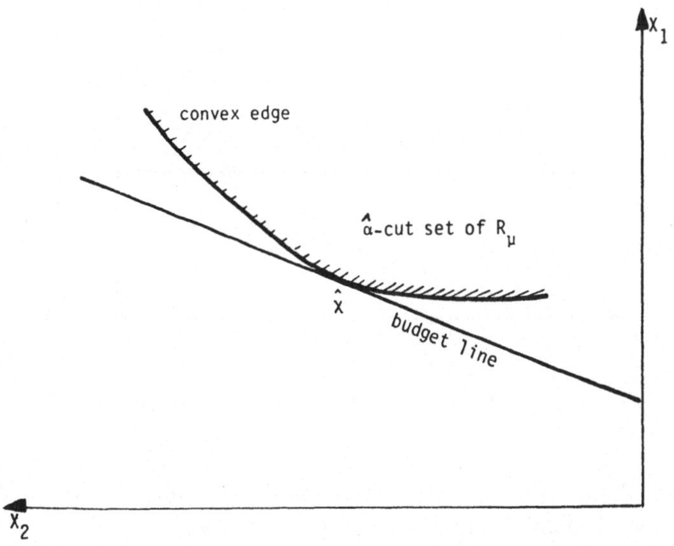


Fig. 27. Definition of the μ-consumption equilibrium point.

$\check{X} = (X_1, X_2)$ at which the budget line becomes tangent to the edge of convex $\check{\alpha}$-cut set of $R_\beta$. Its grade of membership equals to $\check{\alpha}$, $\check{\alpha} = \beta\hat{\alpha}$.

Define M-consumption equilibrium set as a fuzzy set, denoted by $E_M$, of X on the budget line.

$$E_M = \sum_M \sum_\beta \check{\alpha}_{ij}/\check{X}_{ij}$$

where $\check{\alpha}_{ij}/\check{X}_{ij}$ is the β-consumption equilibrium point for the $\beta_i$-cut of $1(M_j)$

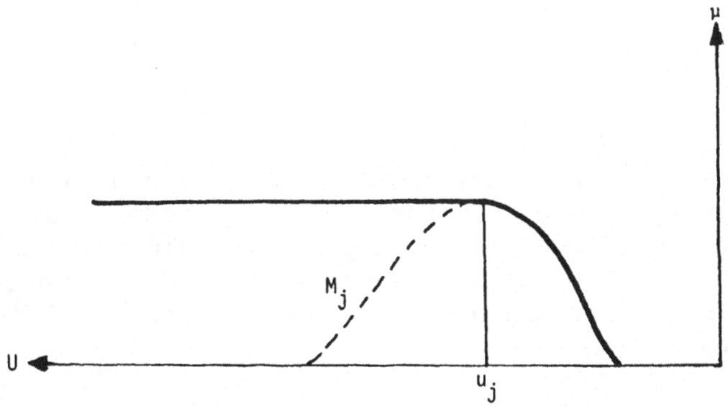


Fig. 28. Two partition sets.

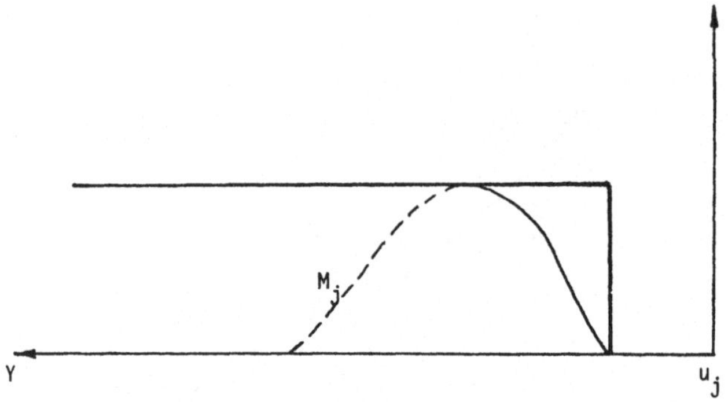


Fig. 29. Two other partition sets.

Theorem 2 warrants the uniqueness of χ and χ for every M (associated with one β) and u, respectively.

Theorem 3: The fuzzy u- consumption equilibrium set is equal to the fuzzy M- consumption equilibrium set, i.e., $E_u = E_M$.

Proof: There are two parts in this proof.

I. Consider two partition sets as described in Fig. 28. It is obvious that $E_{u_j} \subset E_{M_j}$ because $E_{u_j}$ is a component of $E_{M_j}$ with $\beta = 1$. It is known that if $A_1 \subset B_1, A_2 \subset B_2, \ldots, A_n \subset B_n$, then $A_1 \cup \ldots \cup A_n \subset B_1 \cup \ldots \cup B_n$. Thus

$$\sum_{u>u_j} E_{u_j} \subset \sum_{M>M_j} E_{M_j}$$

and

$$\sum_{u>o} E_{u_j} \subset \sum_{M>0} E_{M_j}$$

or

$$E_u \subset E_M \; .$$

II. Consider two other possible partition sets as described in Fig. 29 in which $1(u_j)$ is the support of 1(M). Obviously, the membership grade of every β consumption equilibrium point for every β-cut of 1(M) is equal or smaller than that of u-consumption equilibrium point corresponding to the α-cut set of 1(M), where $\alpha = \beta$. Hence

$$\sum_{M>M_j} E_{M_j} \subset \sum_{u>u_j} E_{u_j}$$

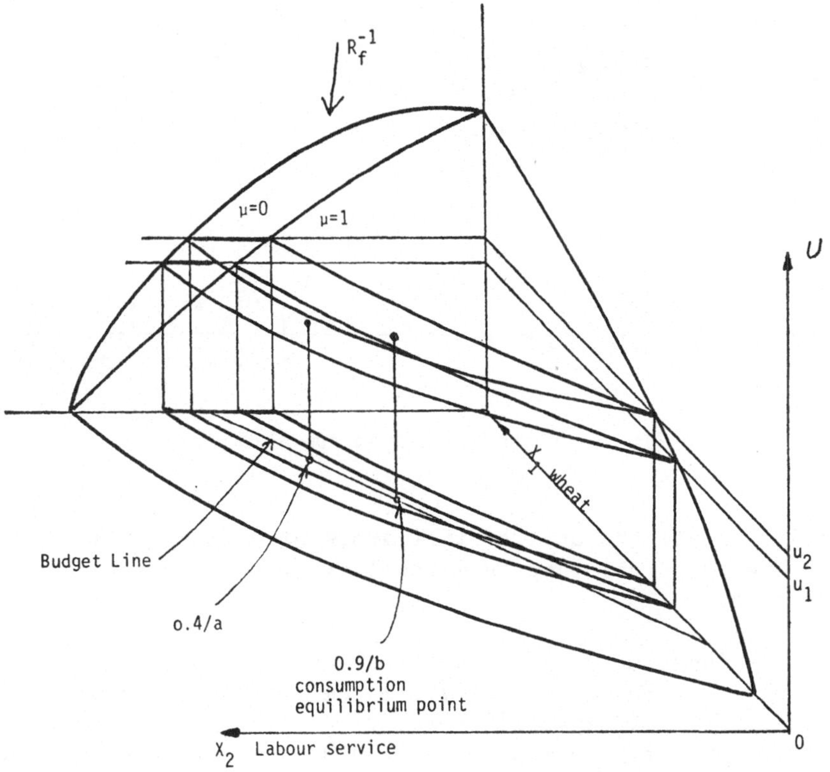


Fig. 30. The existence of multiple consumption equilibriums.

where $M_j$ ranges over F(U), $u_j$ ranges over U we have

$$E_M \subset E_u$$

thus $E_M = E_u$, since $E_M$ and $E_n$ include each other. Q.E.D.

It is of interest to note the existence of multiple consumption equilibriums. Consider, for example, a fuzzy utility function defined over nonfuzzy wheat and fuzzy labor services. Such a fuzzy utility function is depicted in Fig. 30. With the exceptions of the corner points, all contours of various α-cut of weak u preference set for every u are nonintersecting owing to the convexity of utility indicators. Assuming marginal utility is diminishing, the contours of equal α-cut of various weak u preference sets are also nonintersecting. However, the contours of different levels α-cut in different u may interest each other. It follows that the points of tangency between the budget lines and the contours of the α-cut in different u cannot overlap.

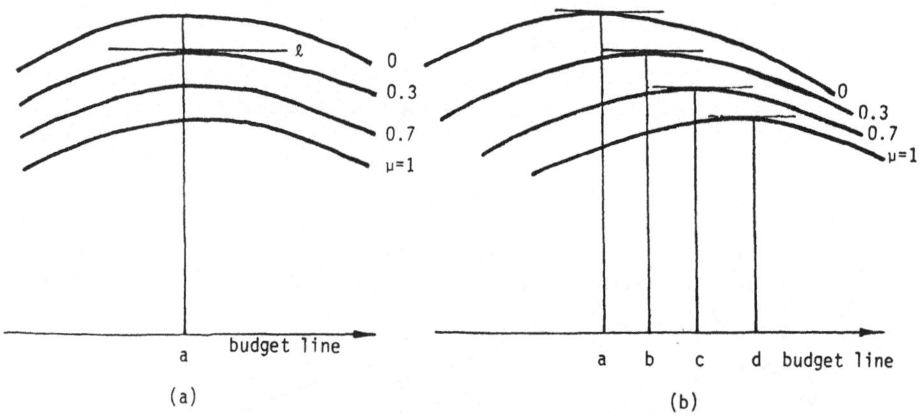


Fig. 31. Two kinds of optimum equilibrium.

Theorem 4: If all the utility indicators of a fuzzy utility function are comparable, optimum equilibrium set reduces to a singular point.

Proof: Construct a plane vertical to X and passing through the budget line. We can identify all the intersections between this plane and the fuzzy utility relation. (See Fig. 31 a.) Suppose the contour of $\alpha_1$-cut with $\alpha_1$ = 0.3 for example, is tangent, at point a, to the straight line ℓ, which runs in parallel with the budget line. The tangency point a is a consumption equilibrium with membership grade $\alpha_1$. Because of the convexity of $R_f^{-1}$, the

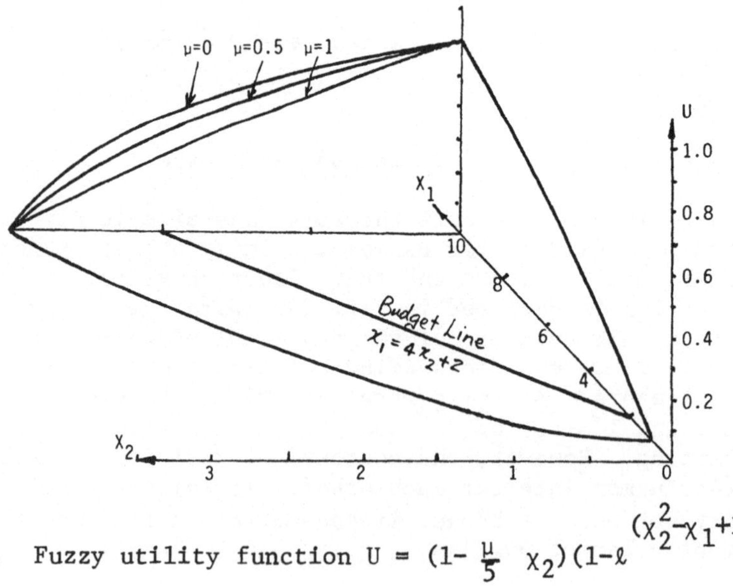


Fig. 32. Fuzzy utility function $U = (1-\frac{\mu}{5}\; X_2)(1-\ell^{(X_2^2-X_1+1)/10})$.

membership grades of all the other points on the budget line for the same utility level must be smaller than $\alpha_1$. In this case, where all of the utility indicators are comparable, the membership grade of the point a is maximum at any of the other utility levels. The case in which the utility indicators are not comparable is depicted in Fig. 31 b.

Theorem 5: From the above discussions, it follows immediately that the uniqueness of the consumption equilibrium requires the fulfillment of the following three conditions: (1) utility indicators are convex; (2) utility indicators are comparable; and (3) marginal utility of every iso-membership grade surface of a fuzzy utility function is diminishing.

Let us consider an analytically representable fuzzy utility function described by

$$U = (1 - \frac{k}{5}\chi_2)(1 - e^{(\chi_2^2 - \chi_1 + 1)/10})$$

where $\chi_1$ represents one kind of commodity and $\chi_2$ another kind. U represents the fuzzy utility indicators of commodity bundles $(\chi_1, \chi_2)$. Parameter k, $0 \leq k \leq 2$, relates the membership grade $\mu$, $0 \leq \mu \leq 1$, via the following equations

$$\begin{cases} \mu = k, \ 0 \leq k \leq 1 \\ \mu = 2 - k, \ 1 < k \leq 2 \end{cases}$$

If $\chi_1$ represents a certain amount of wheat and $\chi_2$ the hours of labor service required for exchanging the $\chi_1$ wheat, then U is the degree of satisfaction of the person after this trade. Obviously, for any

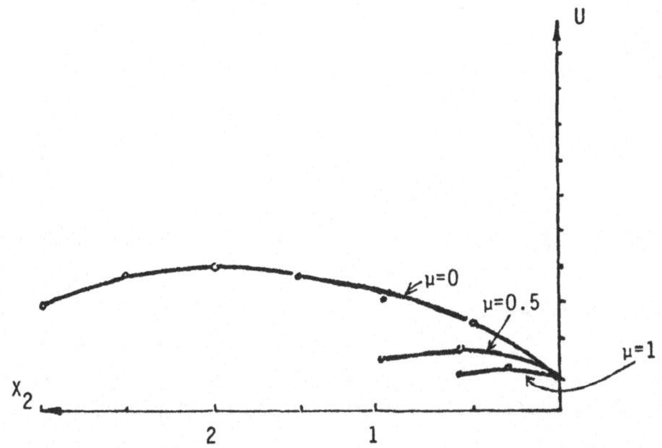


Fig. 33. Relation between U and $X_2$ on the budget line.

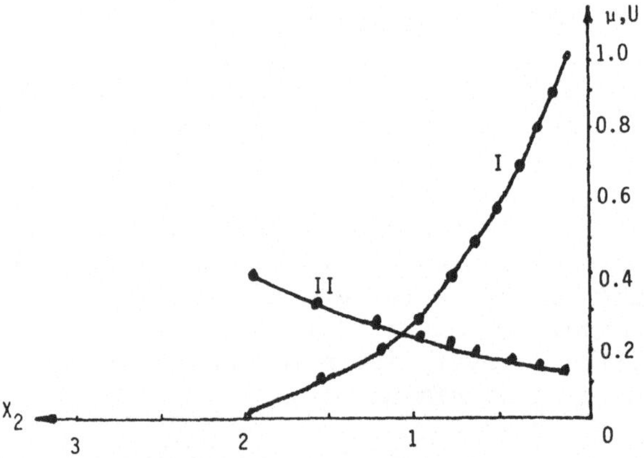


Fig. 34. I equilibrium point and its μ.
II utility on equilibrium point.

given $\chi_1$ and $\chi_2$, a fuzzy set $U_1 \varepsilon F(U)$ can be obtained. When $\chi_2 = 0$, U becomes independent of k (therefore, also μ); since we assumed that the consumer in question was precise in the estimation of utility of wheat but fuzzy in that of labor.

It was pointed out in Theorem 2 that the lagging edges of all fuzzy utility indicators failed to make any contribution to the construction of the weak preference set, hence, also the consumption equilibrium. Thus, it is sufficient to consider the range $0 \leq k \leq 1$. In Fig. 32 some iso-membership surfaces of fuzzy utility function within this range are depicted in company with the budget line $\chi_1 = 4\chi_2 + 2$. Figure 33 shows the intersection between the iso-membership surface and the budget plane. The maxima of these curves determine the consumption equilibrium points. From Fig. 34, it is seen that the membership grade of a point in the consumption equilibrium set and the correspondent utility indicator run in the opposite directions. We express consumption equilibrium points only by one of their components $\chi_2$, since $\chi_2$ is given, $\chi_1$ can be obtained from the budget line.

## VI. SUMMARY

In the preceeding analysis, several interesting properties of a fuzzy utility function have been established. Given diminishing marginal utility of every iso-membership grade surface of a fuzzy utility relation, the optimum consumption equilibrium is a fuzzy set on the budget line in the commodity space, provided that every fuzzy utility indicator of a fuzzy utility relation is convex and normalized. It has been pointed out that the utility level associated with a consumption equilibrium increases as the membership grade of this equilibrium decreases. Since the membership grade implies the degree of confidence, the consumption equilibrium with the highest confidence level is at the point where the membership grade equals one.

When utility function is fuzzy, multiple consumption equilibriums emerge. The resulting demand relation becomes fuzzy in that a given price is associated with various quantities of a commodity. The quantity purchased depends on both the prevailing price and the membership grade; the consumption equilibrium set reduces to a singular point when all the utility indicators of a fuzzy utility function are comparable.

REFERENCES

Batra, R. N., Pure Theory of International Trade Under Uncertainty, Halstead Press, New York, 1975.

Blin, J. M., et al., "Pattern Recognition in Micro-Economics," Journal of Cybernetics, 3, 4, pp. 17-27, 1974.

Chang, S. S. L., "Application of Fuzzy Set Theory to Economics," Kybernetics, vol. 6, pp. 203-207, 1977.

Dubois, D. and Prade, H., Fuzzy Sets and Systems: Theory and Applications, Academic Press, New York, 1980.

Kandel, A. and Lee, S. C., Fuzzy Switching and Automata, Crane Russak, New York, 1979.

Kaufmann, A., Introduction to the Theory of Fuzzy Subsets, Vol. 1, Academic Press, New York, 1975.

Rader, T., Theory of Microeconomics, Academic Press, New York, 1972.

Taranu, C., "The Economic Efficiency - a Fuzzy Concept," Modern Trends in Cybernetics and Systems, pp. 163-173, 1974.

Zadeh, L. A., "Fuzzy Sets," Information and Control, 8, pp. 338-353, 1965.

Guo quan Chen on leave from: Department of Radio-Electronics, The University of Science and Technology of China, Hefei, Anhwei, China.

# USE OF FUZZY LOGIC FOR IMPLEMENTING RULE-BASED CONTROL OF INDUSTRIAL PROCESSES

E. H. Mamdani, J. J. Østergaard and E. Lembessis

Department of Electrical and Electronic Engineering
Queen Mary College
University of London
Mile End Road
London E1 4NS, England

## 1. INTRODUCTION TO RULE-BASED CONTROLLERS

Rule-based methods have been investigated for process control applications since mid-1970's where the emphasis has been on the use of fuzzy logic for implementing the linguistic control rules. The method is now being used commercially for the control of cement kilns, using linguistic rules of the type shown in Fig. 1. The rules are a collection of situation-action pairs expressed as IF ... THEN ... statements and implemented as logical implications. Thus a set of rules, R, is given as

$$R = \{R^1, R^2, \ldots, R^n\}$$
$$= \{s^1 \rightarrow a^1, s^2 \rightarrow a^2, \ldots, s^n \rightarrow a^n\} .$$

This forms the a priori information expressing what to do under a set of hypothetical situations. In general, both the situation $s^j$ and the action $a^j$ are expressed in terms of values of a finite set of variables. Thus a situation $s^j$ is a logical product or a conjunct of a finite set of instances: $s^{j1} \times s^{j2} \times \ldots \times s^{jp}$. This indicates that $s^j$ is a product of p variable values. For exmaple: if BZ temperature is LOW and BE temperature is HIGH. Here BZ and BE are the variables with LOW and HIGH at their values, given linguistically. Fuzzy logic is particularly useful for expressing these linguistic values as fuzzy subsets of the universe of discourse which is the complete range of measurements expressed by the variables. Notice that superscripts are used to distinguish the rules in a collection.

| Case | Condition | Action to be taken | Reason |
|---|---|---|---|
| 10 | BZ OK | a. Increase I.D. fan speed | To raise back-end temperature and increase oxygen percentage for action 'b' |
| | OX low | | |
| | BE low | b. Increase fuel rate | To maintain burning zone temperature |
| 11 | BZ OK | a. Decrease fuel rate slightly | To raise percentage of oxygen |
| | OX low | | |
| | BE OK | | |
| 12 | BZ OK | a. Reduce fuel rate | To increase percentage of oxygen for action 'b' |
| | OX low | b. Reduce I.D. fan speed | To lower back-end temperature and maintain burning zone temperature |
| | BE high | | |
| 13 | BZ OK | a. Increase I.D. fan speed | To raise back-end temperature |
| | OX OK | b. Increase fuel rate | To maintain burning zone temperature |
| | BE low | | |
| 14 | BZ OK | NONE. However, do not get overconfident and keep all conditions under close observation. | |
| | OX OK | | |
| | BE OK | | |
| 15 | BZ OK | When oxygen is in upper part of range | |
| | OX OK | a. Reduce I.D. fan speed | To reduce back-end temperature |
| | BE high | When oxygen is in lower part of range | |
| | | b. Reduce fuel rate | To raise oxygen percentage for action 'c' |
| | | c. Reduce I.D. fan speed | To lower back-end temperature and maintain burning zone temperature |
| 16 | BZ OK | a. Increase I.D. fan speed | To raise back-end temperature |
| | OX high | b. Increase fuel rate | To maintain burning zone temperature and reduce percentage of oxygen |
| | BE low | | |
| 17 | BZ OK | a. Reduce I.D. fan speed slightly | To lower percentage of oxygen |
| | OX high | | |
| | BE OK | | |

Fig. 1. Extract from the manual for cement kiln operators.

The action side of a rule $a^j$ is, in general, a union of r variable values: $a^{j1} + a^{j2} + \ldots + a^{jr}$. Thus the set of rules becomes

$$R = \{ \bigcup_{j=1}^{n} s^j \rightarrow a^j \}$$

$$= \{ \bigcup_{j=1}^{n} s^{j1} \times s^{j2} \times \ldots \times s^{jp} \rightarrow a^{j1} + a^{j2} + \ldots + a^{jr} \}$$

By using the notion of distributivity we can write

$$R = \{ \bigcup_{j=1}^{n} s^{j1} \times s^{j2} \times \ldots \times s^{jp} \rightarrow a^{j1}, \ \bigcup_{j=1}^{n} s^{j1} \times s^{j2} \times \ldots \times s^{jp} \rightarrow a^{j2},$$

$$\ldots, \ \bigcup_{j=1}^{n} s^{j1} \times s^{j2} \times \ldots \times s^{jp} \rightarrow a^{jr} \}$$

$$= \{ \bigcup_{k=1}^{r} \ \bigcup_{j=1}^{n} s^{j1} \times s^{j2} \times \ldots \times s^{jp} \rightarrow a^{jk} \}$$

$$= \{ RB^1, RB^2, \ldots, RB^r \} .$$

The above represents the rule structure for the general case of a multi-input multi-output system. R consists of r subsets of rules (rule blocks, RB), one for each output. Each rule block consists of n rules in which the situation has p variables and the action only one variable. Thus for a multi-output system the multi-output controller can be decomposed to a number of single-output controllers.

An actual system situation is denoted by subscripts, so $s_i \rightarrow a_i$ denotes the system state at instant i. The set i may be infinite and gives the time trajectory of the system. Given the actual situation $s_i$, the controller algorithm has to infer the action $a_i$ by using the law of modus ponens applied to each rule in turn

$$\frac{\begin{matrix} s_i \\ R \end{matrix}}{a_i} \equiv \bigcup_{j=1}^{n} \left\{ \frac{\begin{matrix} s_i \\ s^j \rightarrow a^j \end{matrix}}{a_i^j} \right\} \tag{1}$$

and

$$a_i = \bigcup_{j=1}^{n} a_i^j .$$

This is a rule by rule matching of the actual state $s_i$ with the situation side of each rule $s^j$. In fuzzy logic this match leads to a varying degree of fulfillment (DOF) between 0 and 1. Then for each rule j the action $a^j$ is modified (weighed down) to the extent of the degree of fulfillment achieved for that rule to give $a_i^j$. that is, $a_i^j$ is a modified version of $a^j$ where

$$\text{if DOF} = 1 \text{ then } a_i^j = 0 \quad \text{and}$$

$$\text{if DOF} = 1 \text{ then } a_i^j = a^j .$$

Finally, the action to be taken by the controller $a_i$ is simply a fuzzy union of the individual actions $a_i^j$.

Figure 2 shows the controller described applied to a real process. The output interface shown measures and expresses $s_i$ in the language of fuzzy sets. This interface includes

1) Scaling factors used to map the range of values of the controlled variables into predefined universes of discourse.
2) A quantisation procedure to assist in the above mapping if discrete membership functions are used and therefore discrete universes of discourse.
3) An estimator of the rate of change of the controlled variables. This rate of change is treated by the controller as another input variable.

The input interface includes

1) A procedure to convert the fuzzy information supplied by the controller into one unique control action. There are several ways of achieving this: "mean of maxima" procedure, "modified

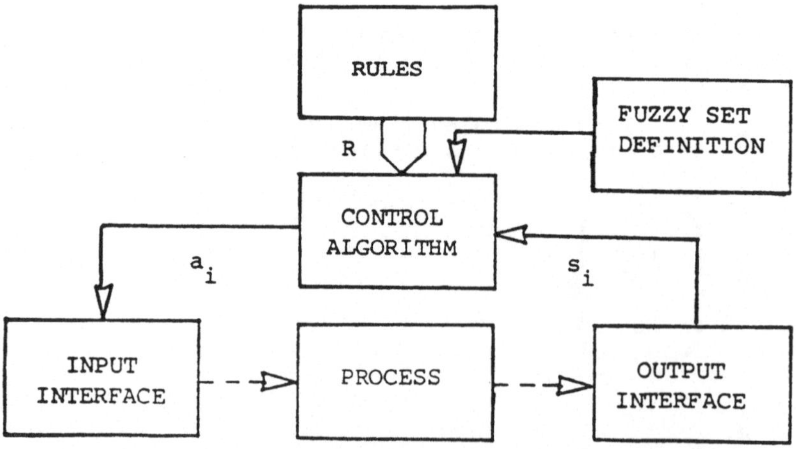


Fig. 2. Block diagram of the control system.

mean of maxima" procedure, "centre of gravity" procedure. The first and last will be described later.

2) Scaling factors to map the action values into real process input values.
3) An integrator, which means that the controller output is used as a change in output; the final control action being an aggregation of these changes.

Figure 2 also shows provision being made for the definition of the fuzzy sets, thus their membership functions can be continuous, discontinuous or discrete. Two examples of fuzzy control algorithms will follow.

(a) An algorithm used in cement kiln control [2] which uses continuous membership functions and the "centre of gravity" procedure to obtain control action, and
(b) A more widely used algorithm [3] which uses discrete membership functions and the "mean of maxima" procedure.

The control rules are given in terms of a finite set of variables and also each rule is expressed in a language containing finite linguistic values, thus the state space of the system can be viewed as a collection of finite points on which a rule may or may not be present (for example see Fig. 3). If a rule is present at its given point then it will carry one of a finite set of action values. An absent rule is simply a rule that carries an action value equal to zero. Expressing a situation fuzzily allows the influence of a rule to spread to neighboring points. Thus each collection of rules creates a unique "field pattern" in the state space to act on the system state. Now, suppose that the system state at instant i falls on a point on the state space where there exists a rule $s^j \rightarrow a^j$. The action calculated by the controller for that rule will then be $a_i^j = a^j$ (since DOF = 1). The final control decision will depend on the implementation of the input interface. If the "mean of maxima" procedure is used then the control action will be determined solely by this one rule. If the "centre of gravity" procedure is used then neighboring rules will also affect the decision slightly. However, if the system state at instant i falls on a point on the state space where there is no rule, then the neighboring rules will contribute to the decision, the ones in the immediate neighborhood will be of most importance to the decision. Actually, if the "mean of maxima" procedure is used, the rules in the immediate neighborhood will be the only ones contributing to the final control decision.

The system state space trajectory can be described by the sequence

$$s_1 \rightarrow a_1 \cdots \rangle\ s_2 \rightarrow a_2 \cdots \rangle\ s_3 \rightarrow a_3 \cdots\ s_i \rightarrow a_i \cdots \rangle$$

The dotted arrows indicate that in general $a_{i-1}$ does not give rise

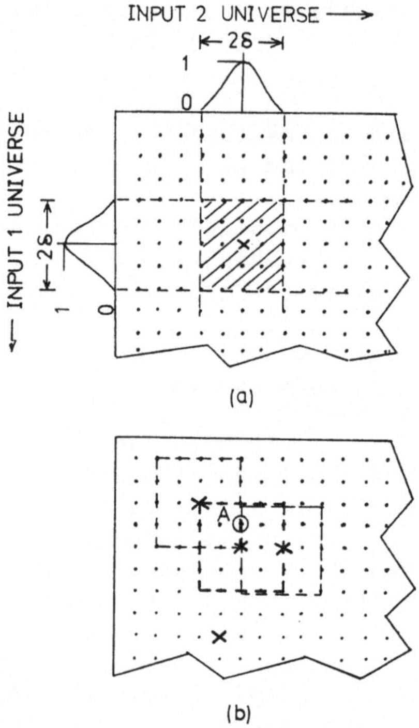


Fig. 3. State-space for controller with two inputs/one output
(a) area of influence of single rule
(b) three rules influencing decision when system state is at A.

to $s_i$ but could be a whole sequence of past actions $a_{i-k}$, $a_{i-(k+1)}$, ... . In systems where this happens it is almost impossible to discover linguistic rules. However, learning procedures can be designed in which the desired action can be discovered iteratively. Such a learning of self-organizing controller is described in Section 3.

## 2. SIMPLE CONTROLLER DESIGNS

### 2.1 Cement Kiln Control Algorithm

The control strategy consists of n rules of the form: IF $s^j$ THEN $a^j$, $j = [1,n]$. The condition $s^j$, as well as the control action $a^j$ are established by means of the following fuzzy primary terms.

LP, MP, SP, ZP = large, medium, small, zero positive
LN, MN, SN, ZN = large, medium, small, zero negative
PO = positive
ZE = zero
NE = negative
HI = high
OK = okay
LO = low

In addition the fuzzy logic operators NOT, AND, OR are used to construct composite terms of $s^j$, e.g., u = LP and v = OK.

For convenience it is assumed that all variables are scaled quantities in the interval [-1,1]. Hence, by letting u denote an arbitrary value in that interval, the linguistic terms LP, MP etc. are represented by the membership functions which associate with each value u a number $\mu_{LP}(u)$, $\mu_{MP}(u)$ etc. in the interval [0,1] which represents the grade of membership of u in LP, MP etc.

Figure 4 shows some of the applied membership functions. By using u = LP and u = OK as examples the fuzzy logic operators NOT, AND, OR, are defined by

$$\mu_{NOT\ LP}(u) = 1-\mu_{LP}(u)$$

$$\mu_{LP\ AND\ OK}(u) = \min(\mu_{LP}(u),\mu_{OK}(u))$$

$$\mu_{LP\ OR\ OK}(u) = \max(\mu_{LP}(u),\mu_{OK}(u)) \ .$$

Evaluation of the primary terms u = LP, u = MP etc. entering the condition $s^j$ results in the membership function values $\mu_{LP}(u)$, $\mu_{MP}(u)$ etc., which are scalars that express to what extent u is large positive, medium positive etc. Thus, evaluation of the composite linguistic term $s^j$ yields a scalar $s_i^j$ which represents the

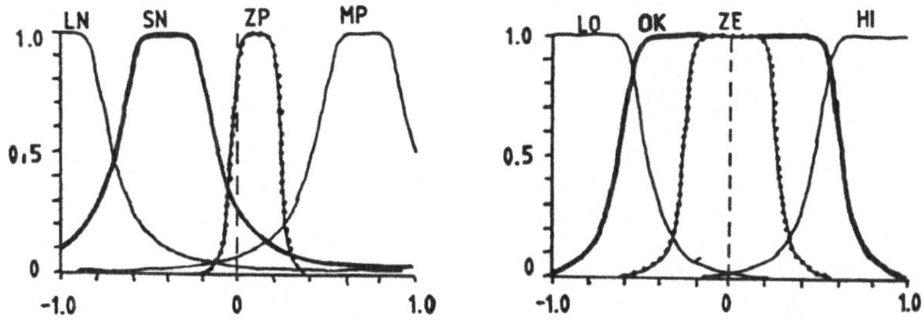


Fig. 4. Definition of some fuzzy primary terms.

degree of fulfillment (DOF) of control rule j. On the other hand, a control action $a^j$ expressed for example by the primary term SN is represented by a vector $(a^j)_k$ consisting of M discrete values of the membership function $\mu_{SN}$. Therefore, the result of control rule j is given by the vector

$$(a_i^j)_k = s_i^j \cdot (a^j)_k \equiv DOF \cdot (a^j)_k, \; k = [1,M] \; .$$

Comparing the above with eqn. 1 we see that the inference has been implemented by a multiplication operation.

The overall fuzzy control action $(a_i)_k$ resulting from n control rules is given by the following vector

$$(a_i)_k = \bigcup_{j=1}^{n} (a_i^j)_k \; , \quad k = [1,M] \; .$$

As an illustration of the fuzzy logic computations outlined above, consider the 9 rules for fuel adjustment of a cement kiln (ΔFUEL), for a kiln drive torque change (ΔTQUE) and a certain free lime content in the kiln (FCAO), see Fig. 5. Each row in the figure represents one rule. Thus, the first rule is

If ΔTQUE is zero and FCAO is low THEN ΔFUEL is medium negative.

Consider an actual value of torque change, ΔTQUE = -1.2%/hour, and a free lime content FCAO = +0.54%. The points of intersection between the value -1.2%/hour and the curves in the first column show that -1.2%/hour is ZERO, NEGATIVE and POSITIVE to the degrees 1.0, 0.29 and 0.0. Likewise, the second column shows that a free lime value equal to 0.54% is LOW, OK and HIGH to the degrees 0.98, 0.63 and 0.0. Thus, the DOF's for the 9 rules are

$DOF_1$ = min(1.0, 0.98) = 0.98

$DOF_2$ = min(1.0, 0.63) = 0.63

$DOF_3$ = min(1.0, 0.0) = 0.0

$DOF_4$ = min(0.29, 0.98) = 0.29

$DOF_5$ = min(0.29, 0.63) = 0.29

$DOF_6$ = min(0.29, 0.0) = 0.0

$DOF_7$ = min(0.0, 0.98) = 0.0

$DOF_8$ = min(0.0, 0.63) = 0.0

$DOF_9$ = min(0.0, 0.0) = 0.0

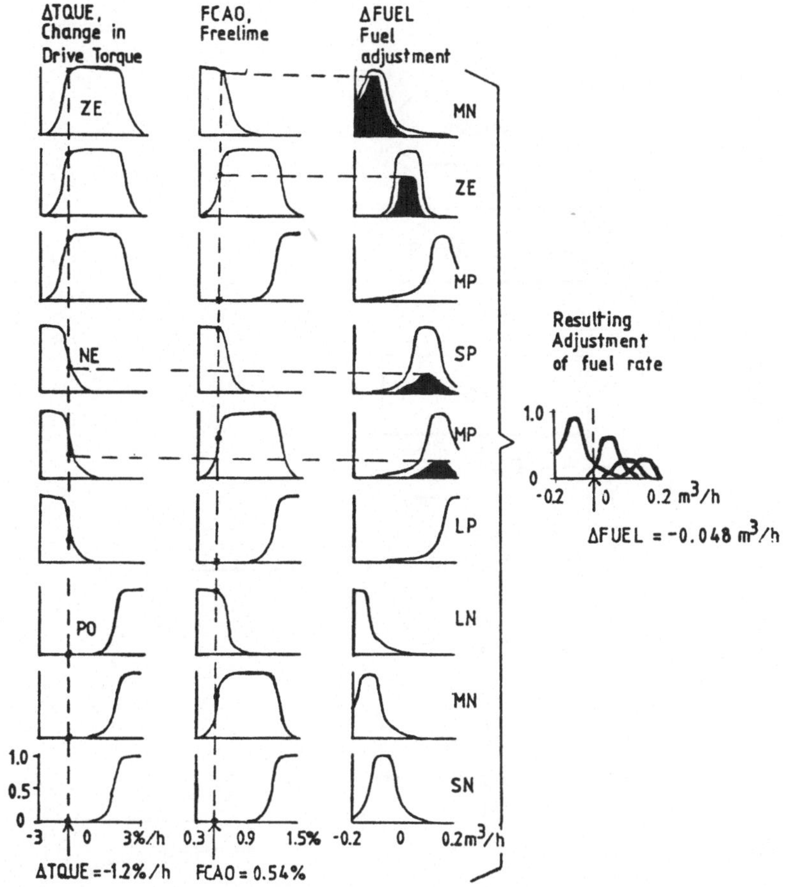


Fig. 5. Cement kiln control algorithm example.

In the third column the membership functions representing the control adjustments are weighed down according to the corresponding DOF. The dark areas represent the control contribution of the various rules. The different control contributions are then combined to form the fuzzy control action to be taken. From this a deterministic action is obtained by using the "centre of gravity" procedure, i.e. the action is given by the abscissa value that divides the area under the fuzzy action membership into two equal parts. If Fig. 5 this will be ΔFUEL = -0.048 $m^3$/hour.

## 2.2 Simplified Fuzzy Control Algorithm

Consider the control rules of the form

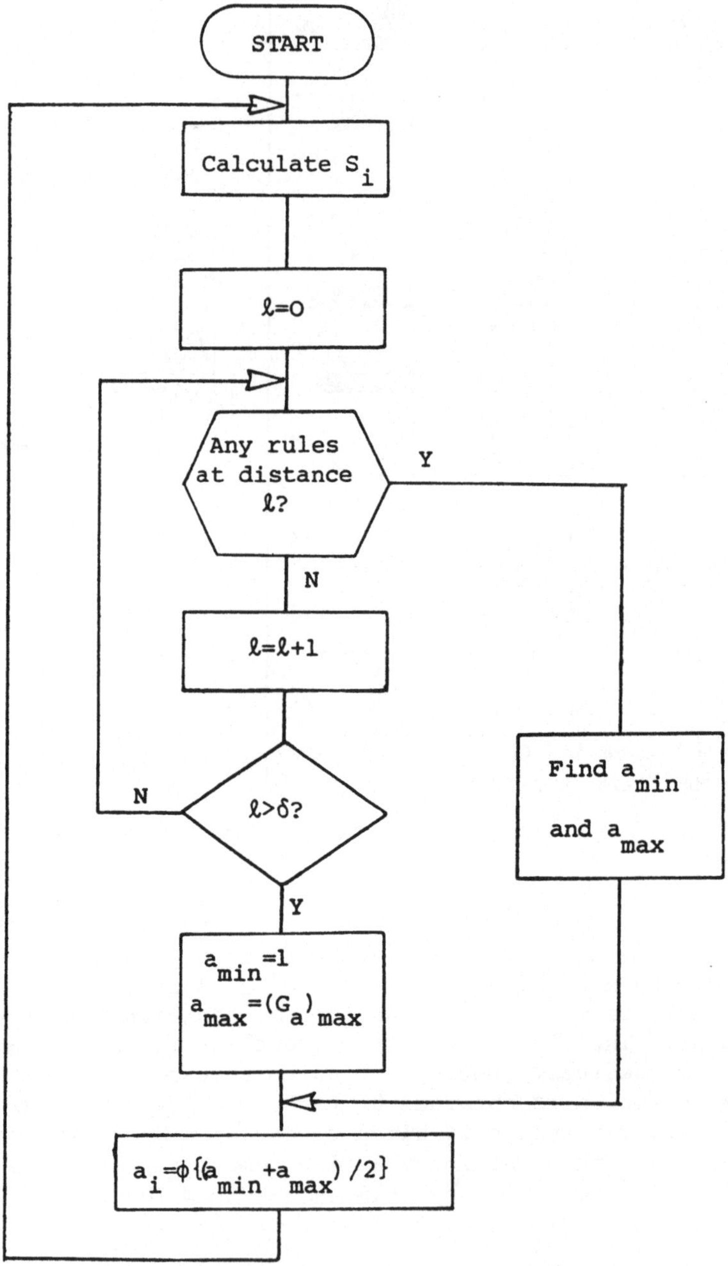


Fig. 6. Simplified controller algorithm.

$$s^j \to a^j \equiv s^{j1} \times s^{j2} \times \ldots \times s^{jp} \to a^j$$

characterized by p input elements $s^j$ and one output element $a^j$. These rules can be represented in a p dimensional array [R], where the elements $s^{j1}$, $s^{j2}$, ..., $s^{jp}$ are the array indices and the output elements $a^j$ are the array contents.

An element in [R] is undefined if there exists no rule corresponding to the element's indices. To represent the fuzzy nature of the control rules, we consider all the defined elements in [R] to correspond to maximum membership function values. Thus the effect of the rule spreads around the defined elements in [R] to a distance $\delta$ in all directions, $\delta$ being half the spread of the fuzzy sets defining the rule inputs. Having defined the control rules thus the flow chart of the control algorithm is shown in Fig. 6. Now given an input $s_{i1} \times s_{i2} \times \ldots \times s_{ip}$, it is easy to check if there exists a corresponding rule in [R]. If it does then this rule will solely affect the control action. If it does not then the rules in the immediate neighborhood (within distance $\delta$) will affect the control action. When the contributing rules are found the rules with the smallest ($a_{min}$) and the largest ($a_{max}$) outputs are selected, and the value $(a_{min} + a_{max})/2$ is calculated. Then the result is rounded off to the nearest element in the universe of discourse of the output variable. This then gives the deterministic action to be taken by the controller. The last two steps represent the "mean of maxima" procedure of obtaining deterministic control action and are part of the input interface as shown in Fig. 2.

As an example, consider a two input, single output fuzzy control algorithm. Let the universes of the inputs and output be integers in the intervals $G_1 = [1,9]$, $G_2 = [1,5]$, $G_a = [1,99]$ respectively. Also let $\delta = 2$ and let there be 7 rules $R^1$, $R^2$, ..., $R^7$ characterized by 7 elements in [R], where

$R^1$ : (5,3,50)
$R^2$ : (2,2,90)
$R^3$ : (3,2,80)
$R^4$ : (7,2,50)
$R^5$ : (7,4,21)
$R^6$ : (9,1,50)
$R^7$ : (9,5,99)

The array [R] is shown in Fig. 7. Now, let us examine the output from the control algorithm for 5 different inputs

(a) Input = (2,5): It can be seen from [R] that no rule is within

| $G_1$ \ $G_2$ | 1 | 2 | 3 | 4 | 5 |
|---|---|---|---|---|---|
| 1 | - | - | - | - | - |
| 2 | - | 90 | - | - | - |
| 3 | - | 80 | - | - | - |
| 4 | - | - | - | - | - |
| 5 | - | - | 50 | - | - |
| 6 | - | - | - | - | - |
| 7 | - | 50 | - | 21 | - |
| 8 | - | - | - | - | - |
| 9 | 50 | - | - | - | 99 |

Fig. 7. Rule matrix for simplified controller example.

distance $\ell = \delta$ from (2,5), so we use for $a_{min}$ and $a_{max}$ the minimum and maximum values in their universe ($G_a$). Thus $a = \phi\{(1+99)/2\} = 50$, where $\phi$ is a function which rounds off the result to the nearest element in $G_a$.

(b) Input = (5,2): One rule, $R^1$, is at distance $\ell = 1$ from (5,2) so the output is

$$a = \phi\{(50+50)/2\} = 50$$

(c) Input = (5,3): This input corresponds exactly to the rule $R^1$. Thus

$$a = \phi\{(50+50)/2\} = 50$$

(d) Input = (7,3): Two rules, $R^4$ and $R^5$, are at distance $\ell = 1$ from (7,3). Thus

$$a = \phi\{(50+21)/2\} = \phi\{35.5\} = 36$$

(e) Input = (9,3): Four rules, $R^4$, $R^5$, $R^6$ and $R^7$, are at distance $\ell = 2$ from (9,3). The maximum output is 99 and the minimum

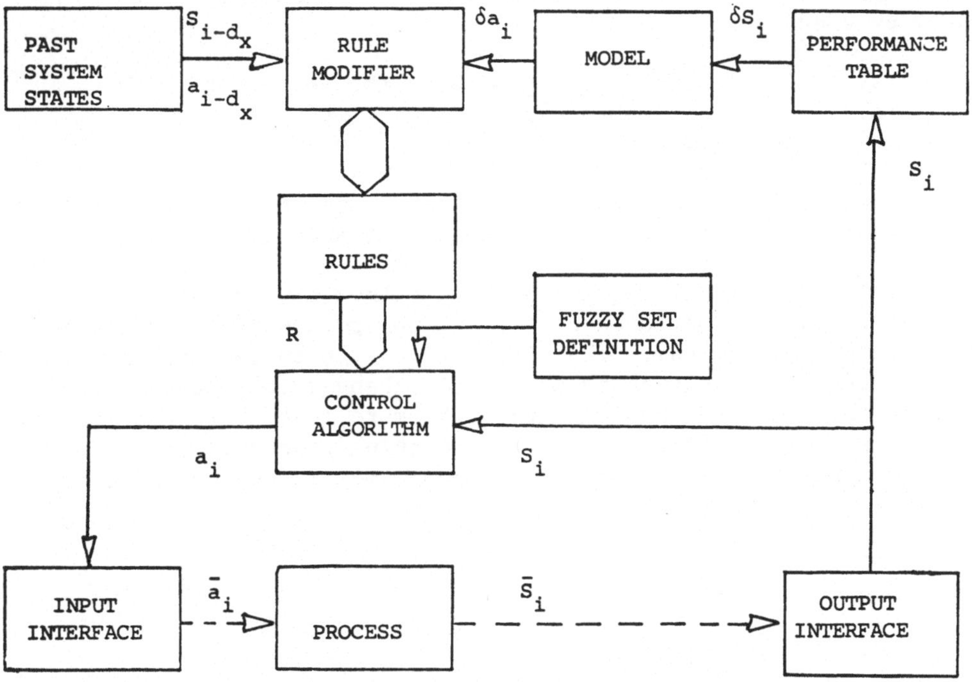


Fig. 8. Block diagram of Self-Organizing Control system.

is 21, which gives $a = \phi\{(21) + 99)/2\} = 60$.

## 3. A SELF-ORGANIZING CONTROLLER DESIGN

The self-organizing controller (SOC) consists of a simple fuzzy controller, like the ones described in 2, to which an extra hierarchical level has been added. A block diagram for SOC appears in Fig. 8. The goal which the controller must aim for is the desired closed loop response.

The performance of the controller in relation to each process output is measured by the deviation of the actual response from the desired one. This is expressed as $\delta s_i$, which is a rough indication of the magnitude of the desired corrections required at the process output. The response of an output is monitored by its error $e_i$ and change in error $c_i$. Thus the *performance table* takes the form of a decision maker which issues the process output corrections required from a knowledge of $e_i$ and $c_i$. It expresses the designer's idea of a minimum tolerable process response and is therefore not specific to the type of process being controlled. The zero entries in the table (see Fig. 9) correspond to states which require no correction and the area forms the set of desired responses. These entries ensure

(a) Sufficiently fast approach to the set point
(b) Good damping when close to set point
(c) Some tolerance around the set point.

Thus, the further away from the desired trajectories, the greatest the required output correction, one way or another.

The *model* translates the process output deviations $\delta s_i$ obtained from the performance table into process input corrections, or reinforcements, $\delta a_i$. Thus the model is an incremental model of the process that relates process input changes to process output changes and need not be accurate. The more inaccurate the model is the more time it will take for the learning procedure to produce good control rules. If [M] is the model matrix we can write

$$\delta \bar{s}_i = [M]\, \delta \bar{a}_i$$

or

$$\delta \bar{a}_i = [M^{-1}]\, \delta \bar{s}_i$$

where $\delta \bar{a}_i$, $\delta \bar{s}_i$ are the values of $\delta a_i$ and $\delta s_i$ at the process input and output. Thus by normalizing [M] we obtain

$$\delta a_i = [M_s^{-1}]\, \delta s_i \ .$$

| | | Change in error. | | | | | | | | | | | | |
|---|---|---|---|---|---|---|---|---|---|---|---|---|---|---|
| | | Towards set-point | | | | | | | Away from set-point | | | | | |
| | | −6 | −5 | −4 | −3 | −2 | −1 | 0 | +1 | +2 | +3 | +4 | +5 | +6 |
| | −6 | 0 | 0 | 0 | 0 | 0 | 0 | 6 | 6 | 6 | 6 | 6 | 6 | 6 |
| | −5 | 0 | 0 | 0 | 2 | 2 | 3 | 6 | 6 | 6 | 6 | 6 | 6 | 6 |
| Below | −4 | 0 | 0 | 0 | 2 | 4 | 5 | 6 | 6 | 6 | 6 | 6 | 6 | 6 |
| set-point | −3 | 0 | 0 | 0 | 2 | 2 | 3 | 4 | 4 | 4 | 4 | 5 | 5 | 6 |
| | −2 | 0 | 0 | 0 | 0 | 0 | 0 | 2 | 2 | 2 | 3 | 4 | 5 | 6 |
| | −1 | 0 | 0 | 0 | 0 | 0 | 0 | 1 | 1 | 1 | 2 | 3 | 4 | 5 |
| Error | −0 | 0 | 0 | 0 | 0 | 0 | 0 | 0 | 0 | 0 | 1 | 2 | 3 | 4 |
| | +0 | 0 | 0 | 0 | 0 | 0 | 0 | 0 | 0 | 0 | −1 | −2 | −3 | −4 |
| | +1 | 0 | 0 | 0 | 0 | 0 | 0 | −1 | −1 | −1 | −2 | −3 | −4 | −5 |
| | +2 | 0 | 0 | 0 | 0 | 0 | 0 | −2 | −2 | −2 | −3 | −4 | −5 | −6 |
| Above | +3 | 0 | 0 | 0 | −2 | −2 | −3 | −4 | −4 | −4 | −4 | −5 | −5 | −6 |
| set-point | +4 | 0 | 0 | 0 | −2 | −4 | −5 | −6 | −6 | −6 | −6 | −6 | −6 | −6 |
| | +5 | 0 | 0 | 0 | −2 | −2 | −3 | −6 | −6 | −6 | −6 | −6 | −6 | −6 |
| | +6 | 0 | 0 | 0 | 0 | 0 | 0 | −6 | −6 | −6 | −6 | −6 | −6 | −6 |

Fig. 9. The Performance Table.

The *rule modifier* (see Fig. 10) inserts new rules to the rules store, if necessary, and deletes any redundant rules in view of the latest rule generated.

Now let us consider the operation of SOC. First, the controller input variables error and change in error are calculated, thus producing

$$\bar{s}_i = \bar{e}_i \times \bar{c}_i$$

and then these are scaled and quantized to produce

$$s_i = e_i \times c_i \quad .$$

The performance table, $\pi$, is then locked up to find the output corrections

$$\delta s_i = \pi\{s_i\}$$

and the scaled model is used to calculate the input reinforcements

$$\delta a_i = [M_s^{-1}]\ \delta s_i \quad .$$

If $\delta a_i = 0$, no correction is required and the rules are not modified. However, if $\delta a_i \neq 0$, new rules are formed and the rule store contents modified. A new rule is formed by assuming a control action $d_x$ samples in the past was one that contributed to the present poor performance ($d_x$ is called the delay in reward and it can take more than one value). Thus, as this past state, the controller inputs and outputs would have been

$$s_{i-d_x} \rightarrow a_{i-d_x} \quad .$$

The controller output that would have been desired is $(a_{i-d_x} + \delta a_i)$ rather than $a_{i-d_x}$. Therefore

$$s_{i-d_x} \rightarrow (a_{i-d_x} + \delta a_i) \qquad d_x \geq 1, \quad x = [1, x_{max}]$$

effectively constitutes a new rule to be processed by the rule modifier. The modifier checks if the new rule antecedents are identical with those of an old rule already in the rules store. If they are then the old rule is deleted and the new rule put in its place. If they are not identical then the new rule is inserted in the rules store as an additional $(n+1)^{th}$ rule

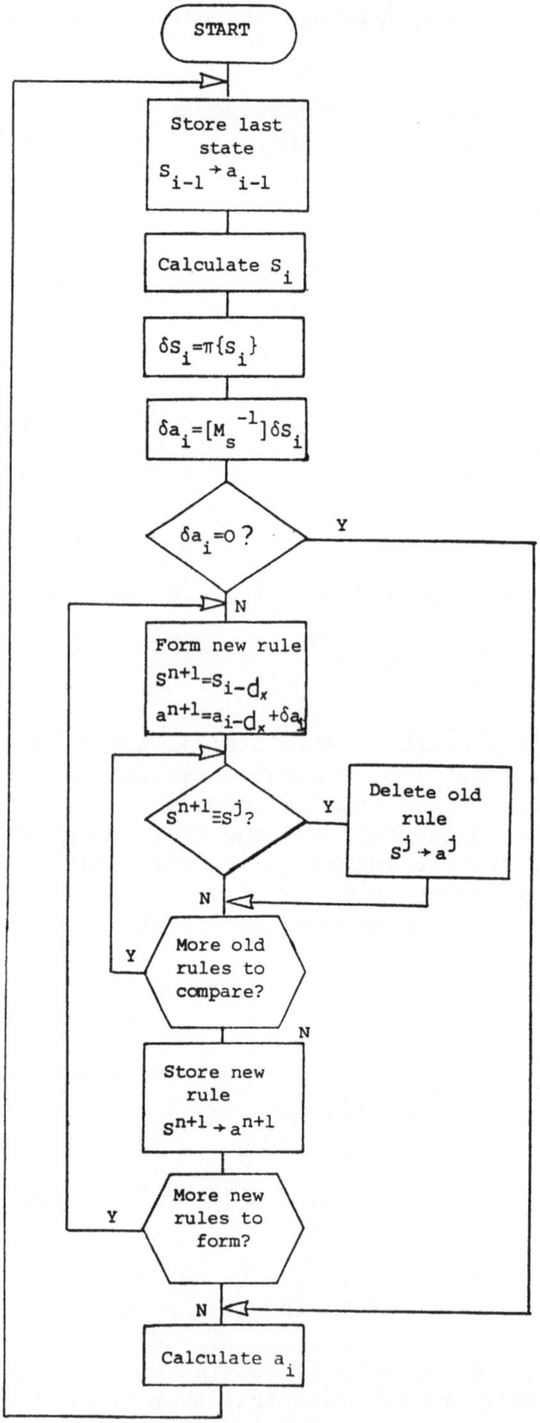


Fig. 10. Self-Organizing Controller algorithm.

$$\left.\begin{aligned} s^{n+1} &= s_{i-d_x} \\ a^{n+1} &= (a_{i-d_x} + \delta a_i) \end{aligned}\right\} s^{n+1} \rightarrow a^{n+1} .$$

Finally, the control algorithm is used to process the current rules and infer the output $a_i$ which is converted by the process input interface to $\bar{a}_i$ - the control action applied to the process.

## 4. REFERENCES

1. R. M. Tong, A Control Engineering Review of Fuzzy Systems, Automatica, 1977, vol. 13, pp. 559-569.
2. I. P. Holmblad and J. J. Østergaard, Fuzzy Logic Control: Operator Experience Applied in Automatic Process Control, FLS Review, 1981, (F. L. Smidth & Co., Vigerslev Alle 77, DK-2500 Valby, Copenhagen).
3. E. H. Mamdani and B. R. Gaines, eds., "Fuzzy Reasoning and its Applications", Academic Press (London), 1981.
4. T. J. Procyk and E. H. Mamdani, A Linguistic Self-Organizing Process Controller, Automatica, 1979, vol. 15, pp. 15-30.

# A NEW APPROACH TO DESIGN OF FUZZY CONTROLLER

M. Sugeno and T. Takagi

Department of Systems Science
Tokyo Institute of Technology
4259 Nagatsuta, Midori-ku
Yokohama 227, Japan

## 1. INTRODUCTION

We have heuristically designed fuzzy controllers so far since we have lacked a fuzzy model of a system. The authors have recently developed a method of multi-dimensional fuzzy reasoning that enables one to build a dynamic model of a system just as we do in terms of differential equations. This paper presents a new idea of designing a fuzzy controller based on a fuzzy model of a system.

It is necessary for understanding the idea to have such concepts as fuzzy reasoning based on Likasiewicz's infinite valued logic, truth qualification in fuzzy logic and multi-dimensional fuzzy reasoning, etc. These are briefly described in Appendix.

## 2. DERIVATION OF FUZZY CONTROL RULES

Though it is most important to identify a system structure and build its model, let us begin with a given fuzzy model of a system, i.e., a set of fuzzy implications such that

$$(u_n \text{ is } P,\ y_{n-1} \text{ is } P) \rightarrow y_n \text{ is } P^2 \quad (1)$$

$$(u_n \text{ is } N,\ y_{n-1} \text{ is } P) \rightarrow y_n \text{ is } P^{1/2} \quad (2)$$

$$(u_n \text{ is } P,\ y_{n-1} \text{ is } N) \rightarrow y_n \text{ is } N^{1/2} \quad (3)$$

$$(u_n \text{ is } N,\ y_{n-1} \text{ is } N) \rightarrow y_n \text{ is } N^2, \quad (4)$$

where the linguistic truth values of all implications are assumed to be u-true.

Here P and N are fuzzy variables implying "positive" and "negative", respectively. $P^2$ and $P^{1/2}$ imply "very positive" and "rather positive" or something like that. Those membership functions are:

$$P(x) = \frac{1}{2a} x + \frac{1}{2} \tag{5}$$

$$N(x) = P(-x) \tag{6}$$

$$P^2(x) = P(x)^2 \tag{7}$$

$$P^{1/2}(x) = P(x)^{1/2} \tag{8}$$

where x stands for $u_n$, $y_{n-1}$ and $y_n \in [-a, a]$.

The above fuzzy model represents a system with first order delay in discrete time. When $(u_n, y_{n-1})$ is given, the next $y_n$ is easily calculated by multi-dimensional fuzzy reasoning (see Appendix). Fuzzy implications in a fuzzy model may be called fuzzy system behaviors, where each implication corresponds to a local behavior of a system.

Now we deal with the case that control objective is to set the output at zero. Let us design fuzzy controllers with respect to $e_{n-1}$ and $e_n$ where $e_{n-1} = -y_{n-1}$ and $e_n = -y_n$. For simplicity we use $y_{n-1}$ and $y_n$ instead of $e_{n-1}$ and $e_n$.

Controller 1

Assume that $(y_{n-1}, y_n)$ is observed ("positive", "very positive") as is shown in Fig. 1. Then we have to decrease the output at n+1 period. For example $y_{n+1}$ would be desirable for the control objective to become "positive". Then the problem is what u should be like. Let

$$(u_{n+1} \text{ is } X,\ y_n \text{ is } P^2) \rightarrow y_{n+1} \text{ is } P \tag{9}$$

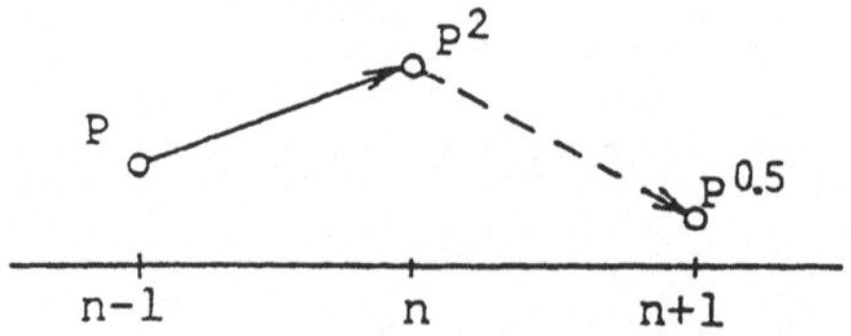


Fig. 1. Control 1 when $(y_{n-1}, y_n)$ is $(P, P^2)$

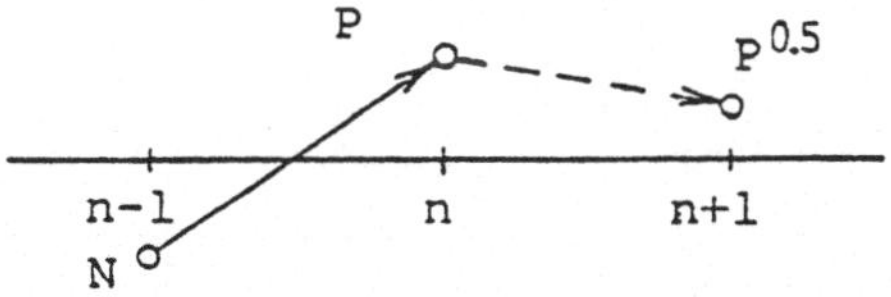


Fig. 2. Control 1 when $(y_{n-1}, y_n)$ is (N, P)

The clue to find X which decreases y from $P^2$ to P lies at the second behavior, i.e., Eq. (2) in the fuzzy model. X is easily found by comparing the variables in Eq. (2) with those in Eq. (9).

Set the semantical equations such that

$$P \simeq P^{1/2} \text{ is } \tau_1 \tag{10}$$

$$P^2 \simeq P \text{ is } \tau_2 \tag{11}$$

$$X \simeq N \text{ is } \tau_3 \tag{12}$$

where $\tau_i$ is a linguistic truth value.

Then $\tau_3$ would be approximately equal to $\tau_1$ and $\tau_2$. In this case we can find that $\tau_1 = \tau_2 =$ "very true". It follows from truth qualification that $X = N^2$ by letting $\tau_3$ = very true. Also when $(y_{n-1}, y_n)$ is (N, P) as is shown in Fig. 2, let us decrease y from P to $P^{1/2}$. The value $u_{n+1}$ is found to be N in a similar manner. Finally we obtain the following four control rules:

$$(y_{n-1} \text{ is } P,\ y_n \text{ is } P^2) \rightarrow u_n \text{ is } N^2 \tag{13}$$

$$(y_{n-1} \text{ is } P,\ y_n \text{ is } N) \rightarrow u_n \text{ is } P \tag{14}$$

$$(y_{n-1} \text{ is } N,\ y_n \text{ is } P) \rightarrow u_n \text{ is } N \tag{15}$$

$$(y_{n-1} \text{ is } N,\ y_n \text{ is } N^2) \rightarrow u_n \text{ is } P^2 , \tag{16}$$

where the truth values of the control rules are also assumed to be u-true.

## Controller 2

Let us design another controller under the idea shown in Figs. 3 and 4. The procedure is the same as in Controller 1. We obtain

$$(y_{n-1} \text{ is } P,\ y_n \text{ is } P^2) \rightarrow u_n \text{ is } N^{1/2} \tag{17}$$

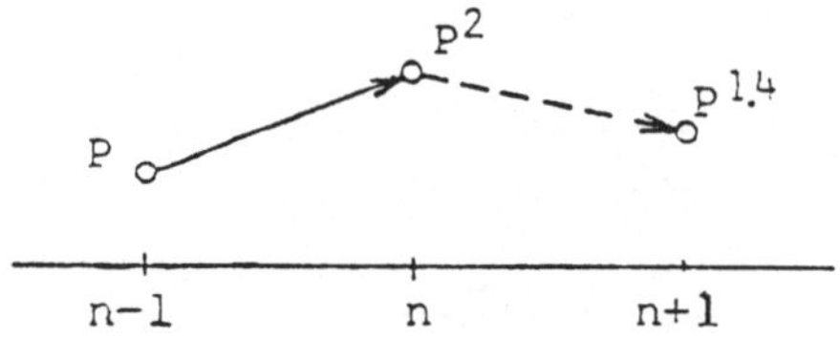


Fig. 3. Control 2 when $(y_{n-1}, y_n)$ is $(P, P^2)$

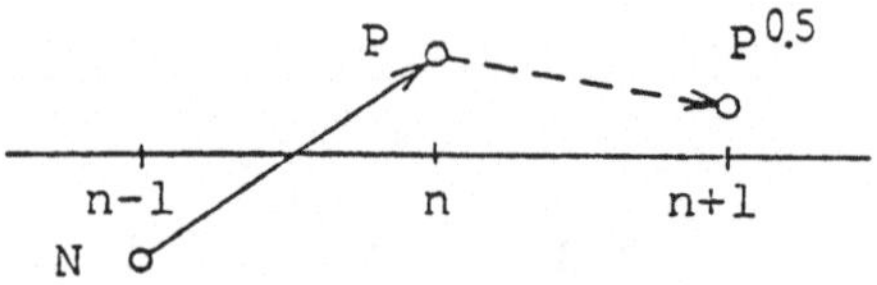


Fig. 4. Control 2 when $(y_{n-1}, y_n)$ is (N, P)

$$(y_{n-1} \text{ is } P,\ y_n \text{ is } N) \rightarrow u_n \text{ is } P \tag{18}$$

$$(y_{n-1} \text{ is } N,\ y_n \text{ is } P) \rightarrow u_n \text{ is } N \tag{19}$$

$$(y_{n-1} \text{ is } N,\ y_n \text{ is } N^2) \rightarrow u_n \text{ is } P^{1/2}\ . \tag{20}$$

Controller 3

The following control rules are also derived by referring Figs. 5 and 6.

$$(y_{n-1} \text{ is } P,\ y_n \text{ is } P^2) \rightarrow u_n \text{ is } N^{3/2} \tag{21}$$

$$(y_{n-1} \text{ is } P,\ y_n \text{ is } N) \rightarrow u_n \text{ is } P^{3/2} \tag{22}$$

$$(y_{n-1} \text{ is } N,\ y_n \text{ is } P) \rightarrow u_n \text{ is } N^{3/2} \tag{23}$$

$$(y_{n-1} \text{ is } N,\ y_n \text{ is } N^2) \rightarrow u_n \text{ is } P^{3/2} \tag{24}$$

As is easily seen, two implications, the first and the fourth, in the fuzzy model are not necessary to derive fuzzy control rules since those describe diverging behavior of the process. In general, finding X in Eq. (9) is not always easy since $\tau_1$ and $\tau_2$ may be different from each other. We omit a general method in this paper.

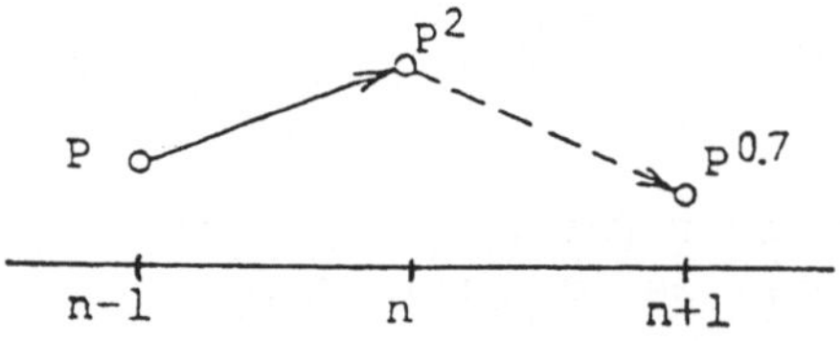


Fig. 5. Control 3 when $(y_{n-1}, y_n)$ is $(P, P^2)$

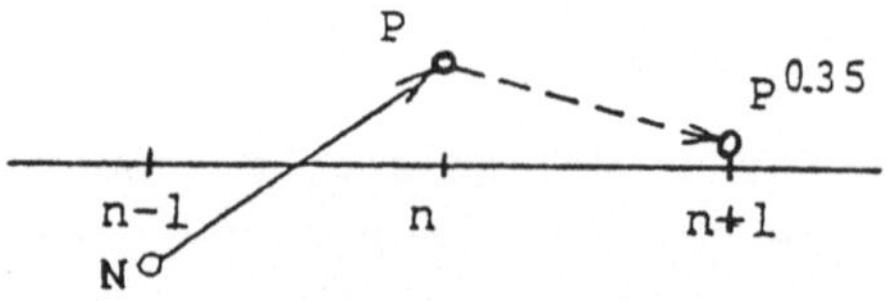


Fig. 6. Control 3 when $(y_{n-1}, y_n)$ is (N, P)

## 3. SIMULATION

Let us apply three controllers to the process. The results are illustrated in Fig. 7. As is seen in Fig. 7, the first controller is the best. Next let us examine the ability of each controller by putting pure dead time to the process. Two dead times of $n_d = 1$ and 4 are put into the process. The results are shown in Figs. 8 and 9. The third controller shows the best performance when $n_d = 1$ and the second one the best when $n_d = 4$. The results of the first and the third controller are omitted in Fig. 9 since those are something mad.

The reason why the second controller is the best when $n_d = 4$ is found by comparing Fig. 3 with Figs. 1 and 5. That is, one should not decrease y too much when y is changing from "positive" to "very positive" if the process has dead time: the excessive decrease of y causes oscillation.

## 4. CONCLUSIONS

We have presented a method to derive fuzzy control rules from a fuzzy model of a system. The results of simulation are satisfactory. Together with a method of fuzzy modeling in terms of fuzzy implications, the present method is expected to be very powerful. It gives a way to analyze theoretically a fuzzy control system and make clear such problems as control performance and stability, etc.

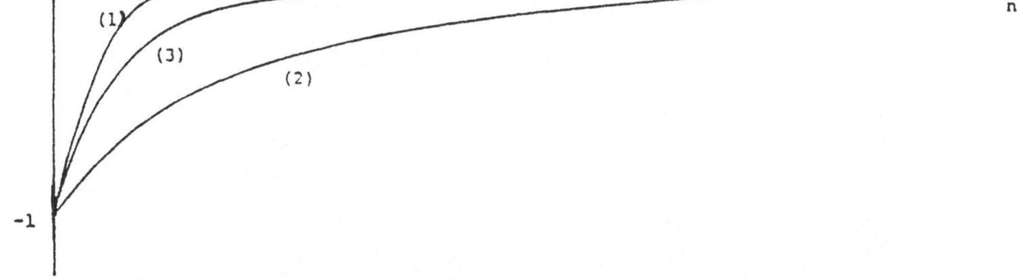

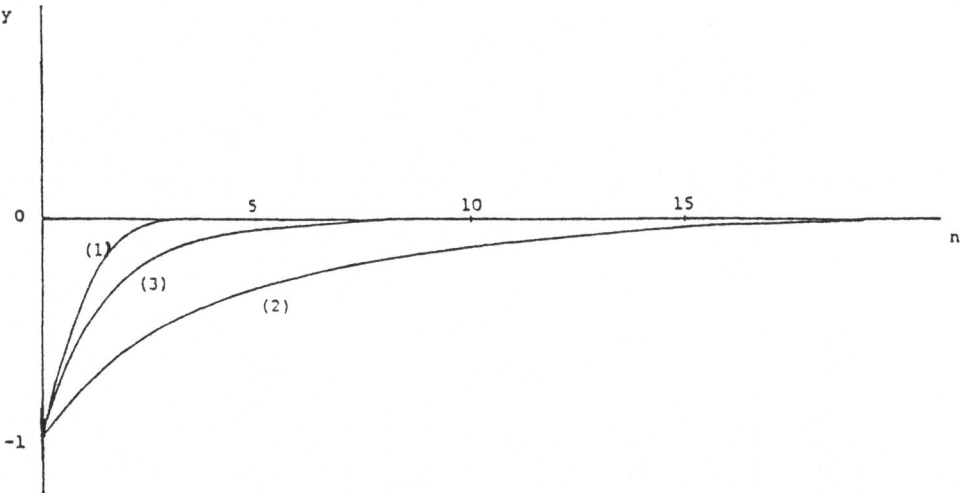

Fig. 7. Fuzzy controls of process with no dead time.

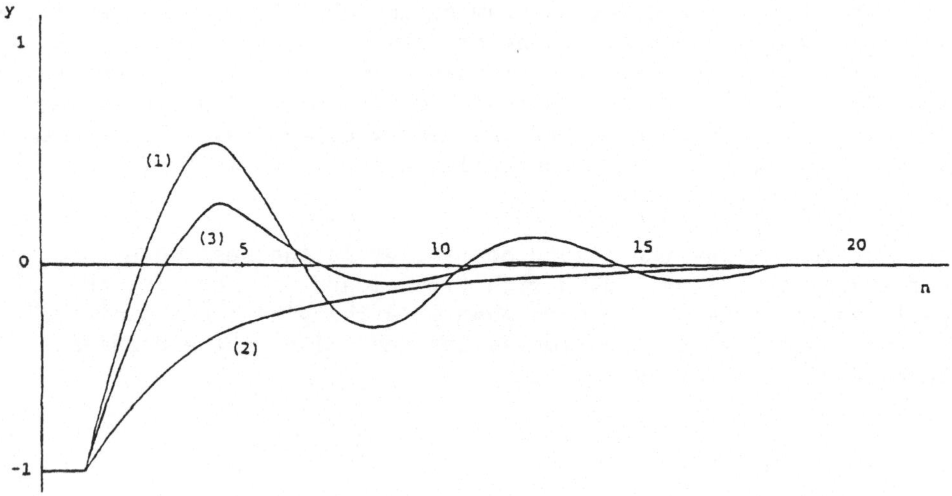


Fig. 8. Fuzzy controls of process with dead time $n_d = 1$

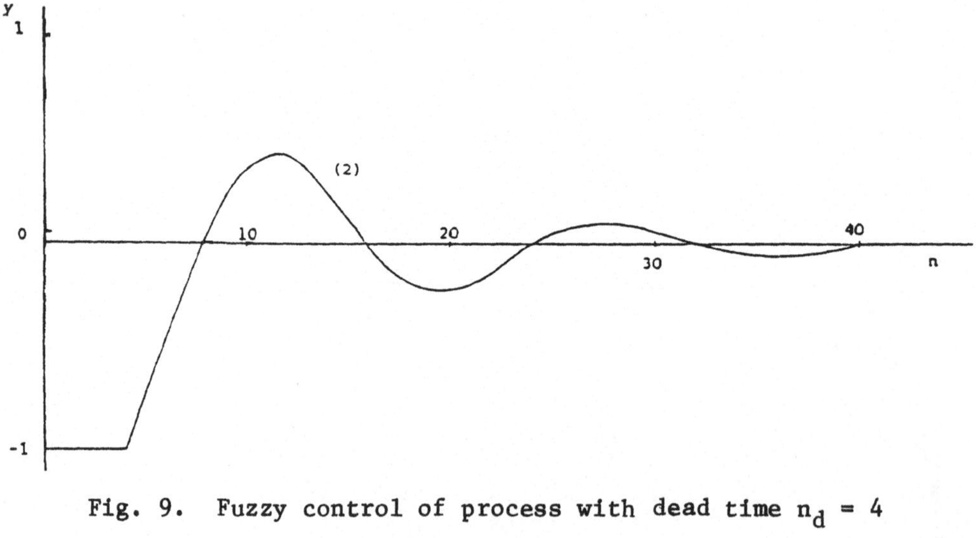


Fig. 9. Fuzzy control of process with dead time $n_d = 4$

The philosophy of derivation of fuzzy control rules is the same even if a system has nonlinearity. A nonlinear system as well as a linear system can be described by a set of fuzzy system behaviors. There is no difference between them except the number of fuzzy behaviors of a system. Fuzzy control rules depending on control object are derived by referring some, not all, of fuzzy system behaviors, where we can easily consider necessary situations for the design purpose as is shown in Figs. 1 and 2.

## REFERENCES


M. Sugeno and T. Takagi, Multi-dimensional fuzzy reasoning, submitted to International Journal of Fuzzy Sets and Systems.

L. Zadeh, Fuzzy logic and approximate reasoning, Synthese, 30, 407/428, 1975.

Y. Tsukamoto, An Approach to Fuzzy Reasoning Method, in "Advances in Fuzzy Set Theory and Applications", M. M. Gupta et al., ed., 1979.


## APPENDIX

### Fuzzified Lukasiewicz's logic and its reasoning

The truth value of implication in Lukasiewicz's infinite valued logic is expressed as

$$/A \rightarrow B/ = (1 - /A/ + /B/) \wedge 1,$$

where $/A/ \varepsilon [0, 1]$ is the truth value of a proposition A.

Let P, Q be fuzzy propositions and $\underline{P}$, $\underline{Q}$ be linguistic truth values that are fuzzy subsets of [0, 1] as usual. Then $\underline{P \rightarrow Q}$ is derived by extension principle from the above equation.

$$\underline{P \rightarrow Q} = (1 - \underline{P} + \underline{Q}) \wedge 1$$

where the operations $-$, $+$ and $\wedge$ are extended ones for fuzzy sets.

For calculation it is better to take strong $\alpha$ - cut. We have

$$\underline{P \rightarrow Q}_\alpha = (1 - \underline{P}_\alpha + \underline{Q}_\alpha) \wedge 1$$

where $\underline{P}_\alpha = \{u | h_{\underline{P}}(u) > \alpha\}$ and $h_{\underline{P}}(u)$, $u \varepsilon [0, 1]$, is the membership function of $\underline{P}$.

In many cases we can assume that $\underline{P \to Q}$ is normal, its membership function is non-decreasing in its domain [0, 1] and also $\underline{P}$ is normal and convex. Denote R = P → Q. Then it follows from the above assumptions that

$$\underline{R}_\alpha = (r(\alpha),\ 1]$$

$$\underline{P}_\alpha = (p_1(\alpha),\ p_2(\alpha)),$$

where $r(\alpha) \varepsilon [0, 1]$ is determined from $h_{\underline{R}}$ and $p_1(\alpha)$, $p_2(\alpha)$ from $h_{\underline{P}}$. Now $\underline{Q}_\alpha$ is obtained as

$$\underline{Q}_\alpha = ((p_1(\alpha) + r(\alpha) - 1) \vee 0,\ 1].$$

Then $h_{\underline{Q}}$ is easily drawn since the above expression implies that $h_{\underline{Q}}$ is a non-decreasing function.

Fuzzy reasoning based on an implication P → Q with $\underline{P \to Q}$ is carried out as follows, where fuzzy modus ponens is written

$$\frac{\text{x is P}',\ \text{x is P} \to \text{y is Q}}{\text{y is Q}'}$$

1) Given a premise P', set P' ≃ P is $\underline{P}$. By converse of truth qualification, $\underline{P}$ is found as

$$\underline{P} = h_P(P'),$$

where $h_P$ is the extended function of $h_P$ for a fuzzy set P'.

2) Calculate $\underline{Q}$ from $\underline{P}$ and $\underline{P \to Q}$ as shown above.

3) Set Q' = Q is $\underline{Q}$. Then Q' is the consequence of fuzzy reasoning. By truth qualification, Q' is obtained as

$$Q' = h_Q^{-1}(\underline{Q}),$$

where $h_Q^{-1}$ is also the extended one.

## Multi-dimensional fuzzy reasoning

Apart from computational technique, it is sufficient for the purpose to deal with two dimensional case. Let us start with four implications such that

$$(\text{x is } A_1,\ \text{y is } B_1) \to \text{z is } C_{11}$$

$$(\text{x is } A_1,\ \text{y is } B_2) \to \text{z is } C_{12}$$

$(x \text{ is } A_2,\ y \text{ is } B_1) \rightarrow z \text{ is } C_{21}$

$(x \text{ is } A_2,\ y \text{ is } B_2) \rightarrow z \text{ is } C_{22}.$

The situation of four implications is shown in Fig. A1.

Our problem is to infer "z is C" from a given premise (x is A, y is B) where A is assumed to be between $A_1$ and $A_2$ and also B between $B_1$ and $B_2$.

The outline of the algorithm is shown in Fig. A2. First let us infer the value of z at the point ($A_1$, B) from ($A_1$, $B_1$) as indicated by an arrow in Fig. A2. Modus ponens is written

$$\frac{(x \text{ is } A_1,\ y \text{ is } B),\ (x \text{ is } A_1,\ y \text{ is } B_1) \rightarrow z \text{ is } C_{11}}{z \text{ is } C_{11}'}$$

Here $C_{11}'$ is easily obtained by one dimensional reasoning since $A_1$ is fixed along the arrow. That is, set

$$(x \text{ is } A_1,\ y \text{ is } B) \simeq (x \text{ is } A_1,\ y \text{ is } B_1) \text{ is } \tau,$$

then clearly

$$y \text{ is } B \simeq y \text{ is } B_1 \text{ is } \tau$$

for the same $\tau$.

Let us next infer z at the same point ($A_1$, B) this time from ($A_1$, $B_2$). That is,

$$\frac{(x \text{ is } A_1,\ B),\ (x \text{ is } A_1,\ B_2) \rightarrow z \text{ is } C_{12}}{z \text{ is } C_{12}'}$$

Now let

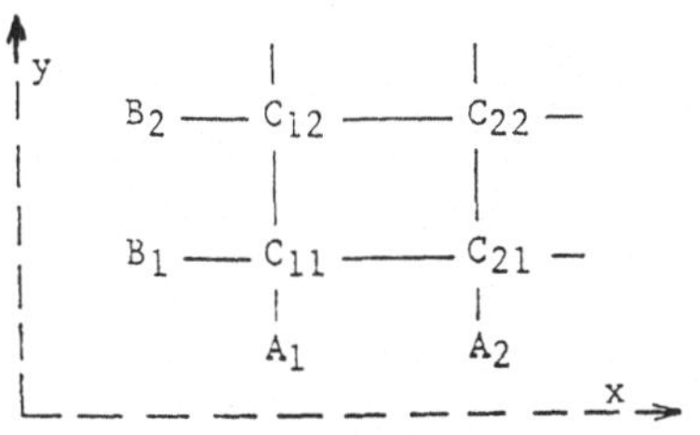


Fig. A1. Situation of four implications

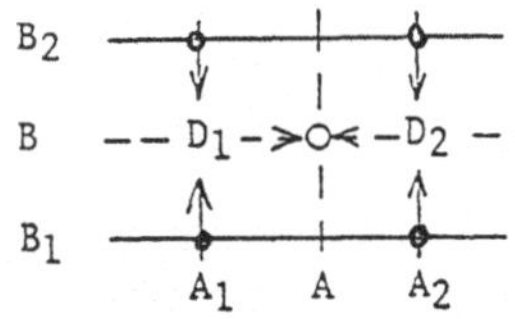


Fig. A2. Outline of algorithm

$$D_1 = \frac{1}{2}(C_{11}' + C_{12}'),$$

where scalar multiplication and addition are the extended operations. Then we obtain a new implication with respect to the point $(A_1, B)$ such that

$$(x \text{ is } A_1,\ y \text{ is } B) \rightarrow z \text{ is } D_1.$$

Here $D_1$ may be obtained in another way: $D_1 = C_{11}' \cap C_{12}'$. For the point $(A_2, B)$ we can also obtain an implication such that

$$(x \text{ is } A_2,\ y \text{ is } B) \rightarrow z \text{ is } D_2.$$

Finally if we follow the same procedure along the dotted arrows in Fig. A2, we can infer the value of z at (A, B).

In general there may be a certain number of implications available at a different situation from Fig. A1. The algorithm is easily extended to this case.

# ADVANCED RESULTS ON APPLICATIONS OF FUZZY SWITCHING FUNCTIONS TO HAZARD DETECTION

Masao Mukaidono

Faculty of Engineering
Meiji University
1-1-1 Higashi-mita Tama-ku
Kawasaki-shi 214 Japan

## 1. INTRODUCTION

Binary logic (i.e., Boolean algebra) is usually employed to analyze and design binary switching circuits. The circuits designed by binary logic behave in an expected manner in steady states but do occasionally show unexpected behaviors in transient states. This is because, although binary logic can describe all steady state behaviors, it cannot adequately describe transient behaviors changing from 0 to 1 or conversely from 1 to 0. Hazards are one example of such unexpected behaviors occurring in combinational circuits in a transient state. D. A. Huffman[1] and E. J. McClusky[2], firstly, pointed out hazards contained in switching circuits and discussed procedures for detecting and removing hazards. M. Yoeli[3] showed that the B-ternary logic could be utilized to detect static hazards in combinational switching circuits. Furthermore, the method was extended by E. B. Eichelberger[4] for application to hazards with multiple-input changes and hazards in sequential circuits. Recently, these results have been improved by M. Mukaidono[5].

A combinational switching circuit always realizes a binary switching function. A fuzzy switching function realized by the combinational switching circuit is a function obtained from the binary switching function by extending it such that every input variable takes any value of the closed interval [0,1] instead of 0 or 1. Using fuzzy switching functions, we can describe the transient behaviors of combinational switching circuits to some extent. It has been shown[6,7] that fuzzy logic and B-ternary logic satisfy the same algebraic system. Thus, fuzzy logic and B-ternary logic have the same ability to represent the transient behaviors corresponding to hazard detection of switching circuits. In fact, A. Kandel[8] applied fuzzy logic to the detection of static hazards in the case of one input variable change, and S. Hughes and A. Kandel[9] extended it to the case of many input variable changes, where the solutions were obtained by solving the logic equations.

In this paper we show a new method to detect and identify various kinds of static hazards contained in a combinational switching circuit through the canonical forms and their properties of the fuzzy switching function realized by the circuit. The method shown in the present paper detects directly and easily all static hazards contained in the circuit in the cases of one or more input changes, and this method is much simpler than the Kandel and Hughes methods.

## 2. STATIC HAZARDS IN COMBINATIONAL SWITCHING CIRCUITS

Let us consider the combinational switching circuit illustrated in Figure 1. For example, if the input $(x_1,x_2,x_3)$ takes (0,1,1) and (1,1,1) respectively in the circuit, then the outputs of the circuit are 0's in both cases. Suppose that the value of the input variable $x_1$ in Figure 1 changes from 0 to 1 while other input variables $x_2$ and $x_3$ are fixed to 1's, respectively. In this case, if the time delay of the pass A in the figure is less than it of the pass B, then a transient erroneous pulse (hazard) occurs at the output as shown in Figure 1.

<u>Definition 1</u>. A combinational switching circuit $F$ contains a static

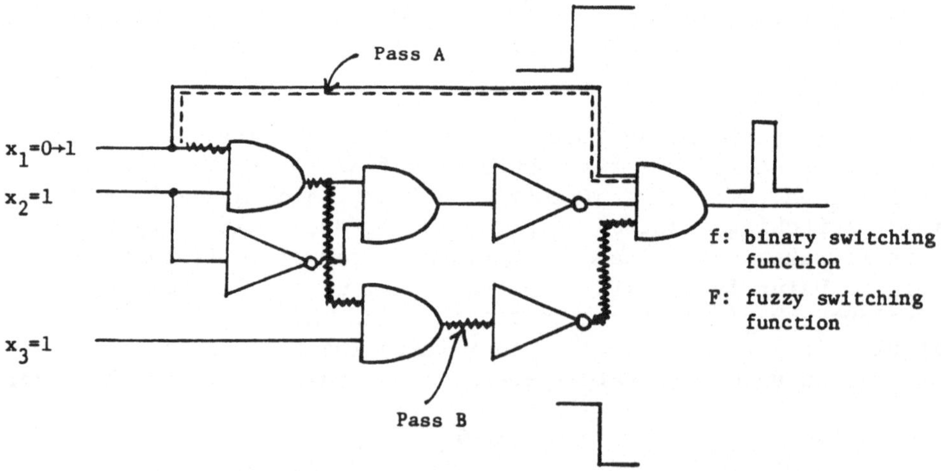


Fig. 1. An example combinational switching circuit.

hazard due to an input change (in this paper an input change may involve one or more input variables) if and only if (1) the output of $\mathcal{F}$ before the change is equal to the output after the change and (2) during the change spurious pulses may appear in the output.

Concerning the above definition, it should be noted that the statement "A circuit contains a static hazard" means that it is possible for the circuit to contain a spurious pulse due to certain combinations of time delays which exist in the wiring or logic elements and does not mean that a spurious pulse is always generated in the output. That is, Definition 1 is based on worst case assumption. This means that it is always possible to produce a spurious pulse by inserting delay elements in certain circuit locations.

A combinational switching circuit realizes always a binary switching function $f:B^n \to B$, where $B=\{0,1\}$. For example, the circuit of Figure 1 realizes the binary switching function f represented by the logic formula:

$$f = x_1\cdot\overline{(x_1\cdot x_2\cdot \bar{x}_2)}\cdot\overline{(x_1\cdot x_2\cdot x_3)} \tag{1}$$

The binary switching function represented by (1) is illustrated by Karnaugh map of Figure 2.

Let $\mathcal{F}$ denote an n-input-variable combinational switching circuit and let f denote the binary switching function realized by $\mathcal{F}$. We designate an input before change and an input after change by $A=(a_1,\ldots,a_n)$ and $B=(b_1,\ldots,b_n)$ respectively, where A and B are elements of $\{0,1\}^n = B^n$.

The hazards which exist in a combinational switching circuit are classified into two categories; one is a static hazard and the other is a dynamic hazard. In this paper, we will consider only the static hazards. A static hazard is classified into two groups;

| $x_3$ \ $x_1 x_2$ | 0 0 | 0 1 | 1 1 | 1 0 |
|---|---|---|---|---|
| 0 | 0 | 0 | 1 | 1 |
| 1 | 0 | 0 | 0 | 1 |

Fig. 2. Karnaugh map representing the binary switching function f of the example.

a 0 hazard and a 1 hazard; and, further, into another two groups, a logic hazard and a function hazard.

<u>Definition 2</u>. In a combinational switching circuit $\mathcal{F}$ a static hazard generated by an input change from a steady state A to another steady state B (henceforth written as A→B) is said to be a static 0 (1) hazard if and only if f(A)=f(B)=0 (f(a)=f(B)=1).

<u>Definition 3</u>. In a combinational switching circuit $\mathcal{F}$ a static hazard in $\mathcal{F}$ generated by an input change A→B is said to be a logic hazard if and only if all values of f are equal to each other for the inputs A and B and all the inputs which may result during the change A→B.

<u>Definition 4</u>. A combinational switching circuit $\mathcal{F}$ contains a function hazard due to an input change A→B if and only if (1) f(A)=f(B) and (2) there is an input state A' such as f(A)≠f(A') in the inputs which may result during the change A→B.

Function hazards are inherent in the binary function f realized by a circuit and do not depend on the construction of the circuit. It is known[4] that logic hazards can be removed by using all prime implicans in disjunctive form.

## 3. FUZZY SWITCHING FUNCTIONS[10,11,12]

A logic formula is a formula composed of each variable $x_i(i=1,\ldots,n)$, the constants 0 and 1 and logic operations AND(·), OR(+) and NOT(¯). If variables $x_i$ take a value in the closed interval [0,1] and the logic operations are defined as follows:

$$x_1 \cdot x_2 = \min(x_1,x_2), \quad x_1+x_2 = \max(x_1,x_2), \quad \overline{x}_1 = 1-x_1,$$

then a logic formula represents a <u>fuzzy switching function</u> or fuzzy logic function which is a mapping from $[0,1]^n$ to [0,1]. For example, a logic formula (1) in the section 2 represents a fuzzy switching function $F(x_1,x_2,x_3):[0,1]^3 \to [0,1]$, where, for example, $F(0.8,0.5,0.3)=0.8\cdot\overline{0.5}\cdot\overline{0.3}=0.5$.

Hereafter, for simplicity, we will identify a logic formula with the fuzzy switching function represented by it and omit the symbol · in the logic formula so far as there is no confusion. Then, the above example fuzzy switching function is written as

$$F = x_1\overline{(x_1x_2\overline{x}_2)}\,\overline{(x_1x_2x_3)} \tag{2}$$

In fuzzy logic, as in conventional binary logic, the commutative, associative, absorptive, distributive, idempotent and De Morgan's laws are satisfied. Yet, the fuzzy logic is characterized by the fact that the complementary laws $x \cdot \bar{x}=0$ and $x+\bar{x}=1$ do not hold.

A <u>literal</u> is a variable $x_i$ or $\bar{x}_i$, the negation of $x_i$. A <u>phrase</u> or product term is a conjunction of one or more literals and a <u>clause</u> or sum term is a disjunction of one or more literals. In the definition of the phrase or clause, it is assumed that any repeated literals are removed from it. Every fuzzy switching function can be expanded into a <u>disjunctive form</u> ( disjunction of phrases) and into a <u>conjunctive form</u> ( conjunction of clauses). However, since the complementary laws do not hold, the forms may contain some terms in which a variable and its negation exist simultaneously as a pair factor. That is, there are two kinds of phrases: one is a <u>complementary phrase</u> or contradictory phrase, which contains a factor $x_i\bar{x}_i$ for at least one variable $x_i$, the other is a <u>simple phrase</u>, which does not contain the above. Similarly, there are two kinds of clauses: one is a <u>complementary clause</u> or tautological clause, which contains a factor $x_i+\bar{x}_i$ for at least one variable $x_i$; the other is a <u>simple clause</u>, which does not. If a complementary phrase and a complementary clause contain all variables as factors, then they are called <u>minterm</u> and <u>maxterm</u>, respectively. Hereafter, we will write a disjunctive form of a fuzzy switching function F as

$$F = F_{sd}+F_{cd} \tag{3}$$

and a conjunctive form of F as

$$F = F_{sc} \cdot F_{cc} \tag{4}$$

where $F_{sd}$ ( $F_{cd}$) is a disjunction of simple phrases (complementary phrases) and $F_{sc}$ ( $F_{cc}$) is a conjunction of simple clauses ( complementary clauses). For example, the fuzzy switching function (2) can be represented by a disjunctive form

$$F = x_1x_2\bar{x}_3+x_1\bar{x}_2+x_1\bar{x}_1 \tag{5}$$

where $F_{sd}=x_1x_2\bar{x}_3+x_1\bar{x}_2$, $F_{cd}=x_1\bar{x}_1$, and by a conjunctive form

$$F = x_1(\bar{x}_1+\bar{x}_2+\bar{x}_3)(\bar{x}_1+\bar{x}_2+x_2) \tag{6}$$

where $F_{sc}=x_1(\bar{x}_1+\bar{x}_2+\bar{x}_3)$, and $F_{cc}=(\bar{x}_1+\bar{x}_2+x_2)$.

As mentioned above, every fuzzy switching function can always be represented by the disjunctive and conjunctive forms, but in general, may have several such forms.

Every complementary phrase ( clause) can be expanded into a disjunction ( conjunction) of complementary minterms ( maxterms).

It is known[10] that there are two kinds of canonical forms of a fuzzy switching function F:

the <u>canonical disjunctive form</u>

$$F = F_{sd}+F_{md} \tag{7}$$

and the <u>canonical conjunctive form</u>

$$F = F_{sc}\cdot F_{mc} \tag{8}$$

where $F_{md}$ is a disjunction of complementary minterms and $F_{mc}$ is a conjunction of complementary maxterms. These canonical forms are determined uniquely for any fuzzy switching function by ignoring the order in which phrases or clauses occur. The following two formulas (9) and (10) are the canonical disjunctive and conjunctive forms, respectively, of the fuzzy switching function F of our example (2):

$$F = x_1x_2\bar{x}_3+x_1\bar{x}_2+x_1\bar{x}_1x_2x_3 \tag{9}$$

where $F_{md}=x_1\bar{x}_1x_2x_3$, and

$$F = x_1(\bar{x}_1+\bar{x}_2+\bar{x}_3)(\bar{x}_1+\bar{x}_2+x_2+x_3) \tag{10}$$

where $F_{mc}=(\bar{x}_1+\bar{x}_2+x_2+x_3)$.

Let V and $V^n$ be a set of $\{0,1/2,1\}$ and the n-dimensional Cartesian product of it, respectively. A simple phrase ( clause) corresponds to an element of $V^n$ and vice versa as follows: a simple phrase ( clause) $\alpha$ corresponds to an element $A=(a_1,\ldots,a_n)$ if $a_i=1$ ( 0) iff $x_i$ exists in $\alpha$, $a_i=0$ ( 1) iff $\bar{x}_i$ exists in $\alpha$ and $a_i=1/2$ iff $x_i$ and $\bar{x}_i$ do not exist in $\alpha$; for example, a simple phrase $x_1\bar{x}_2$ ( clause $\bar{x}_1+x_2$) corresponds to an element (1,0,1/2) of $V^3$ and so on in the case of n=3. Furthermore, there is one-to-one correspondence between a set of complementary minterms ( maxterms) and $V^n-B^n$, where $B=\{0,1\}$; for example, a complementary minterm $x_1\bar{x}_1x_2x_3$ ( maxterm $x_1+\bar{x}_1+\bar{x}_2+\bar{x}_3$) corresponds to an element (1/2,1,1) $V^3-B^3$ and vice versa.

A partially ordered relation ¶ of $[0,1]^n$, which describes ambiguity and plays an important role in the theory of fuzzy switching functions, is defined as follows: for all $a_i$, $a_j$ of [0,1], $a_i¶a_j$ iff either $1/2\leq a_i\leq a_j$ or $1/2\geq a_i\geq a_j$. We always have $1/2¶a_i$ for all $a_i$ of [0,1]. Moreover, if $a_i=0$ and $a_j=1$, or if $a_i=1$ and $a_j=0$, then $a_i$ and $a_j$ cannot be compared with each other. $a_i¶a_j$ means that $a_i$ is more ambiguous than or equal to $a_j$. The relation ¶ is extensible to $[0,1]^n$ as follows: for $A=(a_1,\ldots,a_n)$ and $B=(b_1,\ldots,b_n)$, A¶B iff $a_i¶b_i$ for all i. The following theorem shows an important property of fuzzy switching functions:

Theorem 1.[10] Let F be a fuzzy switching function. If A¶B, then

F(A)¶F(B).

As a consequence of the theorem we have, if F(A)=1/2, then F(B)=1/2 for all B such as B¶A, and if F(A) is 0 or 1, then F(A)=F(B) for all B such as A¶B.

Definition 5. Let F be a fuzzy switching function and A be an element of $[0,1]^n$. Then A* designates the set of all elements A' of $B^n$ such that A¶A', and F(A*) designates the set of the values which F(A') takes, that is,

$$F(A^*) = \{F(A') \mid A¶A' \varepsilon B^n\}.$$

For example, if A=(1/2,1,1), then A*={(0,1,1),(1,1,1)} and F(A*)={0} for the fuzzy switching function F of our example Figure 1 or Figure 2. In any cases, the value of F(A*) is one of {0}, {1} and {0,1} for any elements A and any fuzzy switching functions F.

Lemma 1.[10] Let F be a fuzzy switching function. For all elements A of $[0,1]^n$,

(1) $F(A) = 1/2 \Leftarrow F(A^*) = \{0,1\}$,

(2) $F(A) < 1/2 \Rightarrow F(A^*) = \{0\}$,

(3) $F(A) > 1/2 \Rightarrow F(A^*) = \{1\}$.

Lemma 2.[10] Let $F_{sd}$ be a fuzzy switching function which is represented by the disjunction of simple phrases. Then

$$F(A) < 1/2 \Leftrightarrow F(A^*) = \{0\}$$

for all elements A of $[0,1]^n$.

Lemma 3.[10] Let $F_{sc}$ be a fuzzy switching function which is represented by the conjunction of simple clauses. Then

$$F(A) > 1/2 \Leftrightarrow F(A^*) = \{1\}$$

for all elements A of $[0,1]^n$.

## 4. STATIC HAZARD DETECTION BY FUZZY SWITCHING FUNCTIONS

Henceforth, let F denote the fuzzy switching function realized by a combinational switching circuit $\mathcal{F}$. An input in transient state is symbolized by an element $C=(c_1,\dots,c_n)$ ($c_i \varepsilon[0,1]$) of $[0,1]^n-B^n$. Here, $c_i \varepsilon(0,1)$, that is, $c_i \notin\{0,1\}$ means that the input variable $x_i$ is changing from 0 to 1 or from 1 to 0. If $c_i \varepsilon B$, then it means that $x_i$ does not change. The subject of this paper centers on combinational switching circuits composed of gate-type AND, OR, NOT, NAND and NOR elements and it is assumed that each such element does not contain any logic hazard.

Theorem 2. Let $\mathcal{F}$ be a combinational switching circuit and F be the fuzzy switching function realized by $\mathcal{F}$. $F(C)\varepsilon(0,1)$, that is, $F(C)\notin\{0,1\}$, for an element $C=(c_1,\dots,c_n)$ of $[0,1]^n$ iff the output of $\mathcal{F}$ may change as a result such that the input variables $x_i$ change simultaneously if $c_i \varepsilon(0,1)$ and other input variables $x_j$ are fixed to the values $c_j$ if $c_j \varepsilon\{0,1\}$.

(Proof) It is evident by the definitions of each element that the output values of AND, OR, NOT, NAND and NOR elements are in the open interval (0,1) iff the outputs of the elements may change when the variables taking the values in the open interval (0,1) change simultaneously and the input variables taking in the values of {0,1} are fixed to the values. Since a combinational switching circuit is constructed by these elements, the theorem is proved. (Q.E.D.)

In our example of Figure 1, if $(x_1,x_2,x_3)=(0.2,1,1)$, then $F(0.2,1,1)=0.2$, which means that the output of the circuit may change if $x_1$ changes and $x_2$ and $x_3$ are fixed to 1's, respectively.

If the input to the circuit changes from a steady state input $A=(a_1,\dots,a_n)\varepsilon B^n$ to another steady state input $B=(b_1,\dots,b_n)\varepsilon B^n$, then at least one variable $x_i$ should change from 0 to 1 or from 1 to 0. Therefore, we will define AUB for two steady state inputs $A,B\varepsilon B^n$ such as in Figure 3 for each component. For example, if $A=(0,1,1)$ and $B=(1,1,1)$ then $AUB=(1/2,1,1)$. If $a_i U b_i = 1/2$, then it means that the input variable $x_i$ changes in the input change from the steady state A to another steady state B or from B to A. Notice that AUB is an element of $V^n=\{0,1/2,1\}^n$ if A and B are elements of $B^n$.

| $b_i$ \ $a_i$ | 0 | 1 |
|---|---|---|
| 0 | 0 | 1/2 |
| 1 | 1/2 | 1 |

Fig. 3. $a_i U b_i$

Theorem 3. A combinational switching circuit $\mathcal{F}$ contains a static hazard caused by an input change A→B if and only if

(1) F(A) = F(B),

(2) F(AUB) = 1/2.

(Proof) Since inputs A and B are elements of $B^n$, we obtain f(A)= F(A)=F(B)=f(B). The value of the fuzzy switching function F(AUB) is one of 0, 1/2 and 1, because each component of AUB takes only 0, 1/2 or 1. Hence, the theorem is shown by Definition 1 and Theorem 2. (Q.E.D.)

If a combinational switching circuit $\mathcal{F}$ contains a static hazard for an input change A→B, then $\mathcal{F}$ also contains a static hazard for an input change B→A by Theorem 2 because of AUB=BUA. Therefore, we need not distinguish between input changes A→B and B→A with regard to static hazards. Hereafter, we write an input change as A↔B.

Corollary 1. A combinational switching circuit $\mathcal{F}$ contains a 0 ( 1) static hazard for an input change A↔B if and only if

(1) F(A) = F(B) = 0 (1),

(2) F(AUB) = 1/2.

(Proof) It is evident from Definition 2 and Theorem 3. (Q.E.D.)

Corollary 2. A combinational switching circuit $\mathcal{F}$ contains a logic hazard for an input change A↔B if and only if

(1) F(AUB) = 1/2,

(2) F((AUB)*) ≠ {0,1}.

(Proof) (AUB)* is the set consisting of all elements of $B^n$ derived by replacing each 1/2 in AUB with 0 or 1. Therefore, it expresses the set of A and B and all steady state inputs occurring during the transition A↔B. F((AUB)*)≠{0,1} means F(A)=F(B). Hence, we obtain the theorem by Definition 3 and Theorem 3. (Q.E.D.)

Corollary 3. A combinational switching circuit $\mathcal{F}$ contains a static function hazard for an input change A↔B if and only if

(1) F(A) = F(B),

(2) F((AUB)*) = {0,1}.

(Proof) Since F((AUB)*)={0,1}, (AUB)* contains elements A' and A" satisfying $F(A')=0$ and $F(A'')=1$. In consequence, the theorem is shown by $F(A)\neq F(A')$ or $F(A)\neq F(A'')$ and Definition 4. (Q.E.D.)

A logic ( function) hazard is called logic ( function) 0 ( 1) hazard if it is a static 0 ( 1) hazard.

In our example of Figure 1, for example, the input change $(0,1,1) \leftrightarrow (1,1,1)$ has a logic 0 hazard because of $F((0,1,1)\cup(1,1,1))=F(1/2,1,1)=1/2$, $F(0,1,1)=F(1,1,1)=0$ and $F((1/2,1,1)^*)=\{0\}$, and the input change $(1,1,0)\leftrightarrow(1,0,1)$ has a function 1 hazard because of $F((1,1,0)\cup(1,0,1))=F(1,1/2,1/2)=1/2$, $F(1,1,0)=F(1,0,1)=1$ and $F((1,1/2,1/2)^*)=\{F(1,0,0), F(1,0,1), F(1,1,0), F(1,1,1)\}=\{0,1\}$.

From Theorem 3 we can easily determine whether a static hazard is contained in a combinational switching circuit $\mathcal{F}$ for an input change $A\leftrightarrow B$ and we can further determine from Corollary 1 whether it is a static 0 hazard or a static 1 hazard. Although the detection of whether logic hazards or function hazards is possible by Corollary 2 and 3, such detection is difficult because we need to examine the value of F for all elements of $(A\cup B)^*$.

## 5. STATIC HAZARD DETECTION THROUGH THE CANONICAL FORMS OF FUZZY SWITCHING FUNCTIONS

For a combinational switching circuit $\mathcal{F}$, there exists a logic formula $\psi$ corresponding to $\mathcal{F}$. The logic formula $\psi$ represents a binary switching function f and a fuzzy switching function F.

<u>Theorem 4</u>. Let $\psi$ be the formula corresponding to a combinational switching circuit $\mathcal{F}$, let $\psi'$ be any formula obtained from $\psi$ by applying the laws covered by the fuzzy logic and let $\mathcal{F}'$ be the combinational switching circuit corresponding to $\psi'$. Then, for an input change $A\leftrightarrow B$, $\mathcal{F}$ contains a static hazard iff $\mathcal{F}'$ contains a static hazard.

(Proof) Even if a law covered by the fuzzy logic is applied to a logic formula, the fuzzy switching function represented by the formula is the same. Yet, since the existence of a static hazard is determined by the fuzzy switching function realized by the circuit (Theorem 3), we obtain this theorem. (Q.E.D.)

From the above theorem it is seen that every law held in the fuzzy logic is a transformation that preserves static hazards. On the other hand, every fuzzy switching function can be expanded into the disjunctive ( conjunctive) form. Therefore, it is sufficient to show the effectiveness of the methods of detection and identification of static hazards involved in the above two forms. In the first half of this section we will deal with the detection and identification of various kinds of static hazards by obtaining the disjunctive form of the fuzzy switching function realized by the circuit.

The logic formula (1), for example, corresponded to the combinational switching circuit illustrated in Figure 1. The formula (2) was expanded into the disjunctive form (5) as a fuzzy switching function by using the equalities held in fuzzy logic. In the combinational switching circuit illustrated in Figure 4, which corresponds to the disjunctive form (5), if the time delay of the pass A is less than it of the pass B, the circuit contains a 0 hazard for an input change $(0,1,1)\leftrightarrow(1,1,1)$ similarly as in the circuit of Figure 1.

Theorem 5. If the disjunctive ( conjunctive) form of the fuzzy switching function realized by a combinational circuit $F$ consists only of simple phrases ( clauses), then all static 0 ( 1) hazards in $F$ are function hazards.

(Proof) If a static 0 ( 1) hazard is contained in $F$ for an input change $A\leftrightarrow B$, then $F(A)=F(B)=0$ ( 1) and $F(A\cup B)=1/2$ are satisfied ( Corollary 1). According to Lemma 2 ( Lemma 3), if the disjunctive ( conjunctive) form of the fuzzy switching function F is composed only of simple phrases ( clauses), then $F(A\cup B)=1/2$ is equivalent to either $F((A\cup B)^*)=\{1\}$ ( $\{0\}$) or $F((A\cup B)^*)=\{0,1\}$. The former contradicts $F(A)=F(B)=0$ ( 1). Therefore, it is a function hazard ( Corollary 3). (Q.E.D.)

Theorem 6. If the disjunctive ( conjunctive) form of the fuzzy switching function realized by a combinational switching circuit $F$ is composed only of complementary phrases ( clauses), then $F$ does not contain static 1 ( 0) hazards and function hazards.

(Proof) If the disjunctive ( conjunctive) form of the fuzzy switching function F consists only of complementary phrases ( clauses), then it is evident that $F(A^*)=\{0\}$ ($\{1\}$) for all elements A of $V^n$. Therefore, Corollary 1 and 3 indicate that a static 1 ( 0) hazard and a function hazard are not contained in $F$. (Q.E.D.)

As described in section 3, the disjunctive ( conjunctive) form of a fuzzy switching function F was written as $F=F_{sd}+F_{cd}$ ( $F=F_{sc}\cdot F_{cc}$). By using $F_{sd}$ ( $F_{sc}$) and $F_{cd}$ ( $F_{cc}$), the following conclusions are derived from the above two theorems.

Theorem 7. A combinational switching circuit $F$ contains a function 0 hazard for an input change $A\leftrightarrow B$ iff

(1) $F_{sd}(A\cup B) = 1/2$,

(2) $F_{sd}(A) = F_{sd}(B) = 0$.

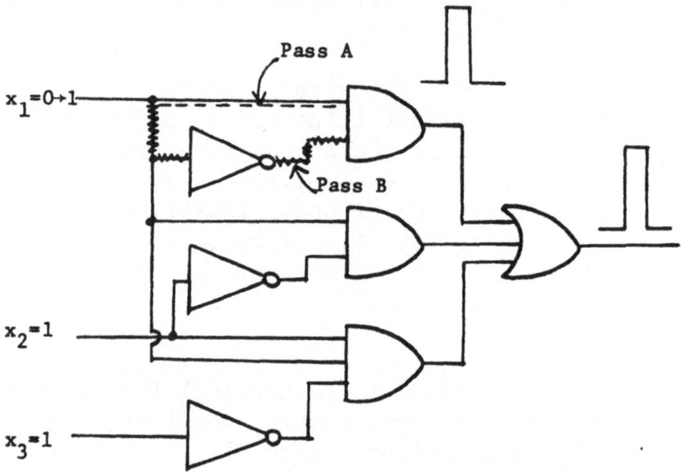


Fig. 4. The combinational switching circuit corresponding to the disjunctive form (5).

(Proof) We need not consider function hazards involved in $F_{cd}$ because $F_{cd}$ does not contain such hazards ( Theorem 6). Since $F_{sd}$ does not contain a logic 0 hazard ( Theorem 5), the theorem is derived from Corollary 1. (Q.E.D.)

<u>Theorem 8</u>. A combinational switching circuit $\mathcal{F}$ contains a logic 0 hazard for an input change $A \leftrightarrow B$ iff

(1) $F_{sd}(A \cup B) = 0$,

(2) $F_{cd}(A \cup B) = 1/2$.

(Proof) Because $F_{sd}(A \cup B)=0$ and $A \cup B \neq A, B$, we have $F_{sd}(A)=F_{sd}(B)=0$. Furthermore, since A and B are elements of $B^n$, $F_{cd}(A)=F_{cd}(B)=0$ holds. Therefore, we obtain $F(A)=F(B)=0$. On the other hand, it follows from $F=F_{sd}+F_{cd}$ that $F(A \cup B)=1/2$. Hence, a static 0 hazard is contained in $\mathcal{F}$. Since this hazard is not a function 0 hazard because $F_{sd}(A \cup B)=0$ ( Theorem 7), it is a logic 0 hazard. (Q.E.D.)

<u>Theorem 9</u>. A combinational switching circuit $\mathcal{F}$ contains a function 1 hazard for an input change $A \leftrightarrow B$ iff

(1) $F_{sc}(A \cup B) = 1/2$,

(2) $F_{sc}(A) = F_{sc}(B) = 1$. (The proof is omitted).

<u>Theorem 10</u>. A combinational switching circuit $\mathcal{F}$ contains a logic 1 hazard for an input change $A \leftrightarrow B$ iff

(1) $F_{sc}(A \cup B) = 1$,

(2) $F_{cc}(A \cup B) = 1/2$. (The proof is omitted).

We can distinguish easily between logic hazards and function hazards by Theorems 7 - 10 than Corollary 2 and 3. The detection of all static hazards contained in a given combinational switching circuit requires examination of all inputs by using Theorem 2. We can examine them fairly efficiently if we use the properties discussed up to now. The last half of this section shows a simple method to detect all logic hazards algebraically through the canonical forms of the fuzzy switching function realized by the circuit. If a combinational switching circuit $\mathcal{F}$ contains a logic hazard for an input change $A \leftrightarrow B$, then $\mathcal{F}$ contains a logic hazard for any input change $A' \leftrightarrow B'$ satisfying $A \cup B = A' \cup B'$, because the condition for existence of a logic hazard is given by Corollary 2. Therefore, a logic hazard can be represented by an element C of $V^n - B^n$ rather than by a pair of changing inputs. That is, the statement "$\mathcal{F}$ contains a logic hazard for C" means that $\mathcal{F}$ contains a logic hazard for all input changes $A \leftrightarrow B$ satisfying $A \cup B = C$.

Let $F = F_{sd} + F_{md}$ ( $F = F_{sc} \cdot F_{mc}$) be the canonical disjunctive ( conjunctive) form of the fuzzy switching function F realized by a combinational switching circuit $\mathcal{F}$, as defined in section 3. In the following, $A \cap B = \phi$ for two elements $A = (a_1, \ldots, a_n)$ and $B = (b_1, \ldots, b_n)$ means that A and B are not comparable to each other concerning with the partially ordered relation ¶, that is, there is at least one i such as $a_i = 0$ and $b_i = 1$ or $a_i = 1$ and $b_i = 0$.

<u>Theorem 11</u>. A combinational switching circuit $\mathcal{F}$ contains a logic 0 ( 1) hazard for an input change C if and only if there is a complementary minterm ( maxterm) $\beta$ in the canonical disjunctive ( conjunctive) form of the fuzzy switching function F realized by $\mathcal{F}$ satisfying $C ¶ C_\beta$ and $C \cap C_\alpha = \phi$ for every simple phrase ( clause) $\alpha$ of F, where $C_\alpha$ and $C_\beta$ are elements corresponding to $\alpha$ and $\beta$, respectively.

(Proof) $C \cap C_\alpha = \phi$ for a simple phrase $\alpha$ iff $\alpha(C) = 0$. Therefore, $F_{sd}(C) = 0$ by the assumption. On the other hand, if a complementary minterm $\beta$ exists in the canonical disjunctive form of F, then $F_{md}(C_\beta) = 1/2$, which derives $F_{md}(C) = 1/2$ because of $C ¶ C_\beta$. The above fact shows that $\mathcal{F}$ contains a logic 0 hazard for an input change C ( Theorem 8). Conversely, if $\mathcal{F}$ contains a logic 0 hazard for an input change C, $F_{sd}(C) = 0$ and $F_{md}(C) = 1/2$ ( Theorem 8). $F_{sd}(C) = 0$ is equal to $\alpha(C) = 0$ for all simple phrases $\alpha$ of $F_{sd}$. On the other hand, if $F_{md}(C) = 1/2$, then there exists a complementary minterm $\beta$ in $F_{md}$ satisfying $\beta(C) = 1/2$, which is equal to $C ¶ C_\beta$. It can be shown in the similar manner for a logic 1 hazard. (Q.E.D.)

<u>Corollary 4</u>. If there is a complementary minterm ( maxterm) $\beta$ in

the canonical disjunctive ( conjunctive) form of the fuzzy switching function F realized by $\mathcal{F}$, then $\mathcal{F}$ contains a logic 0 ( 1) hazard for the input change $C_\beta$ corresponding $\beta$.

(Proof) If there is a complementary minterm ( maxterm) $\beta$ in the canonical disjunctive ( conjunctive) form, then $C_\beta \cap C_\alpha = \phi$ for every simple phrase ( clause) $\alpha$ of F. This is because, if $C_\beta \cap C_\alpha \neq \phi$, then $\alpha \supseteq \beta$; that is, $\beta$ is omitted by $\alpha$. This is contradictory to the hypothesis that F is the canonical disjunctive ( conjunctive) form. Hence, the proof follows from Theorem 11. (Q.E.D.)

In our example of Figure 1 or Figure 4, the canonical disjunctive and conjunctive forms of the fuzzy switching function F realized by the circuit are represented by (9) and (10), respectively. The set of elements corresponding to simple phrases of $F_{sd}=x_1\bar{x}_2+x_1x_2\bar{x}_3$ is $\{(1,0,1/2),(1,1,0)\}$ and it of the elements corresponding to complementary minterm of $F_{md}=x_1\bar{x}_1x_2x_3$ is $\{(1/2,1,1)\}$. Therefore all logic 0 hazards contained in the circuit are only for an input change $(1/2,1,1)=(0,1,1)\leftrightarrow(1,1,1)$, which was already pointed out in section 2. The set of the elements corresponding to simple clauses of $F_{sc}=x_1(\bar{x}_1+\bar{x}_2+\bar{x}_3)$ is $\{(0,1/2,1/2),(1,1,1)\}$ and it of elements corresponding to complementary maxterms of $F_{mc}=(\bar{x}_1+\bar{x}_2+x_2+x_3)$ is $\{(1,1/2,0)\}$. Therefore, all logic 1 hazards contained in the circuit are only for an input change $(1,1/2,0)=(1,1,0)\leftrightarrow(1,0,0)$.

## 6. CONCLUSION

There are two kinds of static hazards in a combinational switching circuit: one is a function hazard and the other is a logic hazard. All function hazards which are possible to occur in a combinational switching circuit are essentially detectable by the binary switching function realized by the circuit, but the logic hazards are not detectable by the binary switching function. This paper has shown that all logic hazards which are possible to occur in a combinational switching circuit are detectable easily through the canonical forms of the fuzzy switching function realized by the circuit.

The results obtained in this paper give an answer to the synthesis problem in the construction of a combinational circuit containing specified logic hazards.

## REFERENCES

1. D. A. Huffman, The design and use of hazard-free switching net works, J. ACM, 4, 1, p. 47 (Jan. 1957).
2. E. J. McCluskey, Transients in combinational logic circuits, in: "Redundancy Techniques for Computing Systems", Spartan

Books, Washington, D.C., p. 9 (1962).

3. M. Yoeli and S. Rinon, Application of ternary algebra to static hazards, J. ACM, 11, 1, p. 84 (Jan. 1964).
4. E. B. Eichelberger, Hazard detection in combinational and sequential switching circuits, IBM J., 9, 2, p. 90 (Mar. 1965).
5. M. Mukaidono, The B-ternary logic and its applications to the detection of hazards in combinational switching circuits, Proceedings of the 8-th International Symposium on Multiple-valued Logic, p. 269 (1978).
6. F. P. Preparata and R. T. Yeh, Continuously valued logic, J. Computer and System Science, 6, p. 387 (1972).
7. M. Mukaidono, On some properties of a quantization in fuzzy logic, Proceedings of the 7-th ISMVL, p. 103 (1977).
8. A. Kandel, Application of fuzzy logic to the detection of static hazards in combinational switching systems, International Journal of Computer and Information Sciences, 3, 2, p. 129 (1974).
9. J. S. Hughes and A. Kandel, Applications of fuzzy algebra to hazard detection in combinational switching circuits, International Journal of Computer and Information Sciences, 6, 1, p. 71 (1977).
10. M. Mukaidono, On some properties of fuzzy logic, Trans. I.E.C.E., Japan, 58-D, 3, p. 150 (Mar. 1975); available in English in Systems·Computers·Controls, 6, 2, p. 36 (Mar.-Apr. 1975).
11. M. Mukaidono, An algebraic structure of fuzzy logic functions and their minimal and irredundant form, Trans. I.E.C.E., Japan, 58-D, 12, p. 748 (Dec. 1975); available in English in Systems·Computers·Controls, 6, 6, p. 60 (Nov.-Dec. 1975).
12. A. Kandel and S. Lee, Fuzzy Switching and Automata: Theory and Applications, Crane Russack, New York (1979).

# THE APPLICATION OF FUZZY SET THEORY TO A RISK ANALYSIS MODEL OF COMPUTER SECURITY

Wilker Shane Bruce and Abraham Kandel

Department of Mathematics and Computer Science
Florida State University
Tallahassee, Florida 32306

## 1. INTRODUCTION

The advent of the computer as a problem solving tool is one of the major advances of twentieth century society. The introduction of the computer into any organization can speed the record keeping operations of the organization, allow the storage of large amounts of data in one location and automate the manufacturing processes of the organization. While all of these computer induced changes have the ability to increase the productivity and the profitability of the organization, they also create security problems in the organization for which comprehensive protection programs have not yet been fully perfected.

In general, two different approaches have been taken in the literature to the computer security problem. The first approach to the problem is a discussion of specific protection mechanisms which can be used to combat specific threats to the system. This type of approach can be found in several volumes, some of which are [2, 4-8, 11, 14, 16-17]. The second approach to the problem is a discussion of how to choose a coherent set of mechanisms once threats have been recognized. Some standard volumes which attempt to solve the problem in this manner are [3, 13, 15].

The development of a computer security program in any organization must deal with three basic issues. These are 1) What are the specific security problems of the organization; 2) What mechanisms are available which will provide protection for the computer related assets of the organization; and 3) What degree of security is appropriate for each of the assets of the organization. A corollary issue which must be faced is the determination of the degree of protection which any specific mechanism might provide. However, the complexity of modern day computer systems often makes it impossible to completely understand both the full effect of an undesired action upon the system and the actual degree of protection which any specific mechanism might give the system. As a result of the uncertainty caused by the complexity of the system, generalized computer security models have been developed.

It is the nature of the modeling process that some information inherent in the system is lost during modeling. This is not undesirable since one of the purposes of creating a model is to generalize a system whose complexity makes it incomprehensible. However, the uncertainty which results from the loss of information in the model must be handled by the methods used in the model or the results which are obtained from the model will not be reliable.

One method which has been developed for working with systems which have lost information in the modeling process is fuzzy set theory. Since fuzzy sets do not require strict partitioning of objects into groups, it is possible to handle situations where one desires the creation of sets where different elements of the set have differing amounts of membership in the set. An example of this type of set is the set of all mechanisms which will stop a specific undesired action where the effectiveness of any specific mechanism in stopping the undesired action is not fully understood.

In the past, both probabilistic methods and fuzzy methods have been applied to different aspects of the computer security problem. However, fuzzy methods have not yet been applied to the problem of creating an entire computer security program. Probabilistic methods have been applied to this problem. The result of this application is the probabilistic risk analysis model. It will be the purpose of this paper to create a version of the risk analysis model which uses fuzzy methods. The probabilistic model which we shall modify in this paper is that of Brocks [1].

## 2. FUZZY TOOLS

Fuzzy set theory was originally developed by Zadeh [18]. The purpose of the development of fuzzy set theory was to generalize classical set theory in such a manner as to allow the possibility of partial membership in a set. In the real world, one finds that membership and nonmembership in a set is often not the crisp distinction which classical set theory would suggest. In everyday life one can find many examples of sets for which membership is not well defined. Some examples of this type of set are the set of all tall men, the set of all old women, the set of all very large trees, the set of all bald men and the set of all protection mechanisms which provide security against a certain threat. Each of these sets have some elements which can and should be considered to be partial members of the set. It is this type of set with which fuzzy set theory deals.

Definition 1. Let X be a space of points (objects) with a generic element of X being denoted by x. Thus $X = \{x\}$. A fuzzy set A in the universe X is a set of ordered pairs so that $A = \{x, \chi_A(x)\}$, $x \in X$. In each ordered pair, the $\chi_A(x)$ represents the grade of membership of element x in fuzzy set A. An object x is defined to have greater membership in fuzzy set A than an object y if and only if $\chi_A(x) \geq \chi_A(y)$. $\chi_A(x)$, which usually is defined to be in the interval [0,1], is called the membership function of set A [18].

As is the case with classical set theory, operations have also been defined for fuzzy sets. Zadeh, in his original paper on fuzzy sets, defined the equality relation and the operations of union, intersection and complement of fuzzy sets [18]. For the purpose of the examples to follow in this study, a discrete fuzzy set and its membership function values will be notated in the form $\chi(x_1)/x_1 + \ldots + \chi(x_n)/x_n$.

Often it becomes necessary to define a fuzzy set based upon more than one variable. An example of this would be the fuzzy set of how many eggs and pieces of bacon can be eaten by an adult at breakfast. Clearly there are two parameters in this fuzzy set. The concept of the fuzzy relation has been developed to handle this type of situation.

Definition 2. An n-ary fuzzy relation is a fuzzy set A in the product space $X \times X \times \ldots \times X$. The membership function of an n-ary fuzzy relation is of the form $\chi_A(x_1,\ldots,x_n)$, where $x_i \in X$, $1 \leq i \leq n$ [18].

An application of fuzzy set theory to statistics is the fuzzy expected value. This concept was originally developed by Kandel and Byatt [9]. The motivation behind its development was the need to apply the subjectivity involved in fuzzy set theory to the standard probabilistic concept of expected value. They give two important reasons to justify this. First, in a probability system events are not necessarily well defined and more often than not are members of fuzzy sets. Second, even if an event is well defined, the probability function for the events in the universe of events that are possible may not be well defined. It is common for predictions of expected value to be soft predictions.

Definition 3. A classical probability system is a triple $(\Omega', S, P)$ where $\Omega'$ is an arbitrary set which includes all possible outcomes of a situation, S is a set of events and P is a real valued function defined for each $A \subset S$ such that:

1) $0 \leq P(A) \leq 1$
2) $P(\Omega') = 1$
3) If $A_1, A_2, \ldots$ is any sequence of pairwise disjoint sets in S, then $P(U_n A_n) = \sum_n P(A_n)$ [9].

A function P which satisfies the three conditions above is called a probability measure. The third property of a probability system is known as countable additivity. It is this system which Kandel and Byatt fuzzified in defining the concept of fuzzy expected value.

Definition 4. Let B be a Borel field ($\sigma$-algebra) of subsets of a sample space $\Omega$. A set function $\mu(\cdot)$ defined on B is called a fuzzy measure if it has the following properties:

1) $\mu(\phi) = 0$
2) $\mu(\Omega) = 1$
3) If $\alpha, \beta \in B$ with $\alpha \in \beta$, then $\mu(\alpha) \leq \mu(\beta)$
4) If $\{\alpha_j \mid 1 \leq j \leq \infty\}$ is a monotone sequence, then $\lim_{j\to\infty}[\mu(\alpha_j)] = \mu[\lim_{j\to\infty}(\alpha_j)]$ [9].

It is important to note that fuzzy measures, unlike probability measures, do not have the countable additivity property.

Definition 5. $(\Omega,B,\mu)$ is a fuzzy measure space. The analog of a fuzzy measure space in probability is $(\Omega',S,P)$. The fuzzy measure of $(\Omega,B)$ is $\mu(\cdot)$ [9].

Definition 6. Let $\chi$ be the membership function of a fuzzy set A. Also let $\chi:\Omega \to [0,1]$ and $\xi_T = \{x \mid \chi(x) \geq T\}$. The function $\chi$ is called a B-measurable function if $\xi_T \in B$ for all $T \in [0,1]$ [9].

Definition 7. Let $\chi$ be a B-measurable function. The fuzzy expected value (FEV) of $\chi$ over a fuzzy set A with respect to the measure $\mu(\cdot)$ is defined to be

$$\underset{T \to [0,1]}{\text{Sup}}\{\text{Min}[T,\mu(\xi_T)]\} \qquad (2.1)$$

where $\xi_T \in A$, $\xi_T = \{x \mid \chi(x) \geq T\}$ [9].

The previous definition for the fuzzy expected value is seemingly a very pessimistic one with its maximum of all the minimums method. However, it has been shown that if one takes an optimistic minimum of all the maximums view, the fuzzy expected value of $\chi$ over set A remains the same.

Theorem 1. $\text{Sup}\{\text{Min}[T,\mu(\xi_T)]\} = \text{Inf}\{\text{Max}[T,\mu(\xi_T)]\} = \text{FEV}(\chi)$. The proof for this is in Kandel and Byatt [9].

Several interesting theoretical results have been developed about the fuzzy expected value. For the purpose of this study, only two of these results are required. For the reader who is interested in other theoretical results about the fuzzy expected value, the paper by Kandel and Byatt [9] from which the majority of these fuzzy expected value definitions are being taken covers many of the other results.

Definition 7 for the fuzzy expected value can be extended from the interval [0,1] to any real interval [a,b] by extending T, $\mu$ and $\xi_T$ to the interval [a,b] under the same transformation which $\chi$ undergoes.

Definition 8. The fuzzy expected value of $\chi$ over a set A with respect to the measure $\mu(\cdot)$ when the membership function lies in the interval [a,b] is

$$\sup_{T^* \to [a,b]}\{\mathrm{Min}[T^*, \mu^*(\xi_{T^*})]\} \tag{2.2}$$

where T, $\mu$ and $\xi_T$ become $T^*$, $\mu^*$ and $\xi_{T^*}$ respectively by undergoing the same transformation from the interval [0,1] as the function $\chi$ [9].

As a result of this definition, it can be shown that changing the interval in which the fuzzy expected value is defined does nothing more than scale the fuzzy expected value to the new interval. For example, it can be shown that $\mathrm{FEV}(a\chi + b) = b + a \cdot \mathrm{FEV}(\chi)$. This proof divides into separate cases for $a \geq 0$ and $a < 0$. The proof of the first case is from [10]. The proof for the second case has not yet been published, but was obtained in a private communication with Margolis [12].

Theorem 2. $\mathrm{FEV}(a\chi + b) = b + a \cdot \mathrm{FEV}(\chi)$.

Proof: Case 1. Let $a \geq 0$ and b be constants and $\chi:\Omega \to [0,1]$. Then

$$\mathrm{FEV}(a\chi + b) = \sup_{T^* \to [b,a+b]}\{\mathrm{Min}[T^*, \mu^*(\xi_{T^*})]\}$$

where

$$T^* = aT + b,$$

$$\mu^* = a\mu(\xi_{T^*}) + b$$

and

$$\begin{aligned}\xi_{T*} &= \{x \mid a\chi(x) + b > T*\}\\ &= \{x \mid a\chi(x) + b \geq aT + b\}\\ &= \{x \mid a\chi(x) \geq aT\}\\ &= \{x \mid \chi(x) \geq T\}\\ &= \xi_T.\end{aligned}$$

Thus

$$\begin{aligned}FEV(a\chi + b) &= \underset{T \to [0,1]}{\text{Sup}}\{\text{Min}[aT + b, a\mu(\xi_T) + b]\}\\ &= b + \underset{T \to [0,1]}{\text{Sup}}\{\text{Min}[aT, a\mu(\xi_T)]\}\\ &= b + a \cdot \underset{T \to [0,1]}{\text{Sup}}\{\text{Min}[T, \mu(\xi_T)]\}\\ &= b + a \cdot FEV(\chi).\end{aligned}$$

Case 2. Let $a < 0$ and $b$ be constants and $\chi:\Omega \to [0,1]$. Then

$$FEV(a\chi + b) = \underset{T^* \to [a+b,b]}{\text{Sup}}\{\text{Min}[T^*, \mu^*(\xi_{T^*})]\}$$

where

$$T^* = aT + b,$$

$$\mu^* = a\mu(\xi_{T^*}) + b$$

and

$$\begin{aligned}\xi_{T*} &= \{x \mid a\chi(x) + b \leq T*\}\\ &= \{x \mid a\chi(x) + b \leq aT + b\}\\ &= \{x \mid a\chi(x) \leq aT\}\\ &= \{x \mid \chi(x) \geq T\}\\ &= \xi_T.\end{aligned}$$

Thus

$$\begin{aligned}
\mathrm{FEV}(a\chi + b) &= \sup_{T \to [0,1]}\{\mathrm{Min}[aT + b,\ a\mu(\xi_T) + b]\} \\
&= b + \sup_{T \to [0,1]}\{\mathrm{Min}[aT,\ a\mu(\xi_T)]\} \\
&= b + a \cdot \inf_{T \to [0,1]}\{\mathrm{Max}[T,\ \mu(\xi_T)]\} \\
&= b + a \cdot \mathrm{FEV}(\chi).
\end{aligned}$$

The other theoretical result which must be presented for the fuzzy expected value is an alternative method of computing the fuzzy expected value. At this point it is important to state that the function $\chi$ which is used in computing the fuzzy expected value is also called the compatibility function.

Definition 9. Assume there exists a finite set of data points in which there are n + 1 distinct levels of compatibility such that $0 \leq a_1 \leq a_2 \ldots \leq a_{n+1} \leq 1$. This implies that there exist n distinct levels of fuzzy measure $\mu(\xi_T)$, excluding 0 and 1. The median of the set of the 2n + 1 numbers obtained by combining the n + 1 levels of compatibility and the n levels of fuzzy measure and sorting them in increasing order of magnitude is the fuzzy expected value. The proof of this is in Kandel and Byatt [9].

## 3. POSSIBILITY THEORY

In dealing with computer security considerations, it is often the case that the data which is being used is neither exact nor lends itself to exact analysis. This so called soft data can be inexact in several ways. First, it may not be possible to determine whether or not a piece of information which will enable an individual to overcome a system's security measures is available to that individual. Second, even if it is possible to make an exact verification of whether this information is available, it may not be within the ability of the system to obtain this data within a reasonable cost. Often probability theory has been used to handle soft data in the security structures. However, probability theory has the inherent difficulty that there is often a difference between what is probable and what is possible. In order to provide the maximum amount of system security, it would seem that one would wish to protect against the possible as well as the probable. A system which takes into account this difference is possibility theory [19].

The basic concept of possibility theory is the possibility distribution. A possibility distribution arises from another closely related concept, the fuzzy restriction.

Definition 10. Let X be a variable which takes values in a universe of discourse U with the generic element of U being denoted by u and X = u signifying that X is assigned the value u, u ε U. Let F be a fuzzy subset of U which is characterized by the membership function $\chi_F$. F is a fuzzy restriction on X if F acts as an elastic restraint upon the values which the variable X can be assigned. Thus an assignment to X takes the form $X = u:\chi_F(u)$ where $\chi_F(u)$ represents the degree to which the restraint placed upon X by F is satisfied when the value u is assigned to X [19].

Definition 11. Let F be a fuzzy set with a universe U. Let X be a variable which takes values in universe U. If F acts upon X as a fuzzy restriction, then the assignment of F as a fuzzy restriction upon X associates with X a possibility distribution, $\chi_X$, where $\pi_X = \chi_F(u_1)/u_1 + \ldots + \chi_F(u_n)/u_n$ for all u ε U [19].

As an example, allow the fuzzy set SMALL INTEGERS to be 1/1 + 1/2 + 1/3 + 0.8/4 + 0.5/5 + 0.2/6 + 0/7 where the universe U is defined to be the positive integers. When we assign a restriction to the variable X by making the statement X is a small integer, we associate X a possibility distribution which states what values it is possible for X to have and with what degree of ease X may be assigned each value. Thus if X is a small integer, then the possibility distribution for X is 1/1 + 1/2 + 1/3 + 0.8/4 + 0.5/5 + 0.2/6 + 0/7 + ... . This possibility distribution can be interpreted to say that it is totally impossible for X to be assigned a value greater than 6, totally possible that X can be assigned the values 1, 2 and 3, and somewhat possible that X can be assigned the values 3, 4, 5 and 6.

The negation of a restriction on a variable is similar to the concept of a complement of a fuzzy set. If we define a fuzzy set F, this type of restriction takes the form of a statement X is not F. Thus the possibility distribution of X is the membership function of the fuzzy set NOT F which is in turn the complement of the fuzzy set F [19].

It is not uncommon in dealing with every day situations to encounter restrictions placed upon variables based upon either the conjunction or disjunction of both related and unrelated restrictions. In terms of developing possibility distributions for variables or ordered n-tuples of variables, Zadeh has shown that there are six obvious ways in which these conjunctions and disjunctions can occur [19].

First, two or more unrelated restrictions upon a single variable can be placed in conjunction. The method for this is to take the intersection of the fuzzy sets being used as restrictions upon the variables.

Second, two or more unrelated restrictions can be placed in disjunction upon a variable. In this case, the method is to take the union of the fuzzy sets being placed upon the variable as restrictions.

Third, n unrelated restrictions upon n variables can be joined in conjunction to create a possibility distribution for the n-tuples belonging to the cartesian product of the n universes. There are two steps to the method in this case. First, each of the fuzzy sets being used as restrictions upon the variables are projected into the set of n-tuples obtained by taking the cartesian product of the universes of the variables. Second, the projected fuzzy sets are joined in intersection to create a single fuzzy set.

Fourth, n unrelated restrictions upon n variables can be joined in disjunction to create a possibility distribution for the n-tuples belonging to the cartesian product of the n universes. This case is similar to the third. The first step is to project each of the fuzzy sets into the set of n-tuples belonging to the cartesian product of the universes. Then, the projected fuzzy sets are joined in union to create a single fuzzy set.

Fifth, n related restrictions upon n variables can be joined in conjunction to create a possibility distribution for the n-tuples belonging to the cartesian product of the universes. Finally, n related restrictions upon n variables can be joined in disjunction to create a possibility distribution for the n-tuples belonging to the cartesian product of the universes. In both of these cases, the method is to create a fuzzy relation between the variables to compute compatibility values for each n-tuple of variable values. The fuzzy set created by the application of this fuzzy relation over all the possible n-tuples in the cartesian product of the universes is the new restriction.

It is important to note that the concept of a possibility distribution and the concept of a probability distribution are not equivalent. The fundamental distinction between the two concepts is that while some events have high possibility, they do not necessarily also have high probability. In other words, an event may be totally possible but highly improbable.

Definition 12. The consistency of probability and possibility distributions can be stated as follows:
1) An event which is impossible is bound to be improbable.
2) A high level of possibility does not quarantee a high level of probability.
3) The lessening of the possibility of an event tends to also lessen its probability, but not vice versa [19].

The final portion of this section shall concern itself with the major problem faced in this paper. That problem is how can one make a projection of the event which a population feels is most likely to occur when the only information available is a set of possibility distributions created by the subjective evaluation of this propulation. This problem can be broken into two parts. First, given a single possibility distribution, what method should one use to determine what is the most probable event in that distribution? Second, given that each of the possibility distributions for each member of the population have been analyzed to determine their most probable events, how can we combine these subjective evaluations of probability to choose a single event which is typical of the feelings of the group as a whole.

The problem of how to determine a typical event from a single possibility distribution is closely related to the probability-possibility consistency principle which was discussed earlier. As a general rule, when a situation occurs in which one desires to determine the probability of a set of events based only upon the possibility distribution of those events three guiding rules occur. First, an event which has no possibility or extremely low possibility can not have high probability. Second, when several events have the same possibility and nothing else is known about those events, it must be assumed that those events have the same probability. Finally, because the nature of the concept of a possibility value implies that if one possibility value is greater than another possibility value there is something inherent in the nature of the second event to make that event more difficult to occur than the first, an event with a higher possibility value than another event must be assigned a probability value greater than or equal to that of the other event.

As an example of this problem, let us consider the possibility distribution $0/1 + 1/2 + 0.5/3 + 0/4 + 0/5$. There are three approaches which can be taken to question what is the most probable event in this universe.

The first approach which can be taken is based upon the concept of ease of occurrence. This approach argues that only those events which have the highest possibility will have the highest probability. Thus, this approach would argue that the only events which can be typical of a distribution like the example are the events with the highest possibility. The typical event of the example if this approach is used is the number 2 with a probability of 1.0

The second approach which can be taken for this type of distribution argues that since the maximum amount of probability which an event can have is directly related to the amount of possibility that event has, the probability value for that event should be scaled according to the amount of possibility which the event has. A possible interpretation of this principle would be the assignment of a probability value to an event based upon the computation the possibility value of that event divided by the sum of all the possibility values in the distribution. This method would have the advantage of giving all the values which have some possibility some probability based upon their relative level of possibility with the event having the highest possibility receiving the highest probability. If this method is applied to the sample distribution, the typical values are the numbers 2 and 3 with the number 2 receiving a probability value of 0.66 and the number 3 receiving a probability value of 0.34.

A final approach which can be taken in this situation is the approach which says that only the events which have either the highest possibility or possibility values within a close distance of the highest possibility can have the highest probability and therefore only those events should be considered to be typical values of the distribution. Thus, in this approach one would set a threshold upon which those events with possibility values above the threshold would all be considered typical values of the distribution with each event being assigned an equivalent amount of probability.

The use of any one of these approaches in specific situations will depend upon the unique characteristics of that situation and the subjective understanding which the person making the determination of the typical value of the distribution has of the concept of possibility. Each method has some advantages and disadvantages. It is this author's personal bias that the first approach to this problem is the best. Therefore, this will be the approach used in the remainder of this study.

The second problem which must be solved in the search for a method to find a typical value given only the possibility distributions of a population is how to combine all the typical events derived from each single possibility distribution. Because the determination of the events regarded as typical is based upon the evaluation of a fuzzy or subjective distribution of the possibility of each single event for every person in the population being questioned, it would seem that the fuzzy expected value would be a proper tool.

The use of the fuzzy expected value in this situation proceeds as follows: First, each single distribution is analyzed to find its typical event(s). Then, once that event is found a counter containing the number of occurrences of that event as a typical event in the population is incremented. If there is more than one typical event in a single distribution, the amount by which the counter for each typical event is incremented is the probability value which has been determined for that event in that specific distribution. After all the single distributions have been evaluated in this fashion, a monotonic numeric series of values are assigned to the events in the universe if the events are not numeric. The next step is to compute the fuzzy expected value of the events using the numeric values assigned to the events as the compatibility values and the weights of occurrence as data for computing the measure ($\xi_T$). The fuzzy expected value in this case will be computed in the interval defined by the low and high numeric values assigned to the events. If the events are numeric events the fuzzy expected value gives a typical event which is produced by that polulation of possibility distributions. If the events are not numeric, the fuzzy expected value must be converted back to the original set of events. In either case, if the fuzzy expected value is a real number and the events are defined only in terms of integrer numbers, the fuzzy expected value can be rounded to the closest integer to predict a typical event.

Now that the principle theoretical tool of this study, the algorithm to find a single typical value from a population of possibility distributions, has been developed, the remaining step of this study is to apply this tool to the concept of computer security.

## 4. BROCKS' PROBABILISTIC RISK ANALYSIS MODEL

One generalized model which has been used with some success in the preparation of a program for designing and modifying security programs for a computing system is the risk analysis model. Martin [13], Pritchard [15], and Farr et al. [3] have all written quite extensive volumes discussing the use of this model in designing security systems. It is this type of model, specifically the model presented in a paper by Brocks [1], which we shall fuzzify in this study.

The risk analysis model is based upon two important presuppositions. First, absolute protection of a computing system is unobtainable. Second, the amount of money which any organization is able to spend in pursuit of protection is limited [13]. It is necessary as a result of these presuppositions to develop a method for determining which set of security measures will maintain the highest level of cost effectiveness in the security system.

<u>Definition 13</u>. A threat is defined to be any action or event whose occurrence would adversely effect the computing system [2]. Another term for threat is hazard.

<u>Definition 14</u>. The vulnerability of a system to a certain threat is defined to be the cost which the organization would incur

if that threat takes place [2]. Other terms for vulnerability are loss expectancy and cost.

Definition 15. The risk of a system for a certain threat is defined to be the vulnerability of the system to that threat multiplied by the probability of the occurrence of that threat within a given period of time [2]. Another term for risk is exposure.

The risk analysis model as presented by Brocks lists four separate stages in the development of a security program for a computing system. First, one must identify the threats to which the system is exposed. Second, one must define the vulnerability of the system to each of these threats and determine the probability of these events occurring. Third, one must select the most suitable protection mechanisms for each threat. Finally, one must implement and monitor the protection mechanisms [1].

The first step of the model, identification of the threats to the system, involves attempting to list every possible situation which would cause the compromise of the system via loss of availability, loss of integrity or loss of confidentiality. This step has been studied quite thoroughly. Even though every organization will have its own unique set of threats, there are many threats which are common to all systems. Lists of these common threats have been compiled in a number of volumes, such as [3-8, 11, 13-17].

The second step in Brocks' model, the assessment of the amount of risk created by each threat, is where the actual risk analysis of the system begins. Brocks divides this step in three parts. First, management is asked to determine an amount which is to be considered the lowest "crippling loss" for the organization. Then, an index value between 1 and 100 is given to each threat. The specific index value computed for any threat is defined to be 100 times the vulnerability of the system to the threat divided by the organization's crippling loss [1]. An example of this would be a situation where a corporation's vulnerability to a certain threat is $300,000 and the corporation's crippling loss is $1,000,000. This threat would receive a vulnerability index of 30. Finally, an estimate is made as to the probability of occurrence of each threat. This estimate is to be a subjective evaluation of the likelihood of the threat's occurrence made on the basis of the frequency of the system's exposure, the relative hostility of the environment and past experience [1].

The third step in Brocks' model, the choice of suitable protection mechanisms, is similar to the second. The management of the organization in this step sets a minimum "crippling cost" to the organization for providing protection against each threat. Each of the possible protection mechanisms for a threat is given a cost index by the formula 100 times the cost of the mechanisms divided by the crippling cost [1].

The protection mechanisms are evaluated by the comparison of the mechanism cost index with the "expected value of loss" index which is obtained by multiplying the vulnerability index by the probability index. Through the comparison of the indices, management can decide which protection mechanisms are cost effective in terms of the organization's security requirements and which are not [1].

The fourth step of Brocks' model, implementation and monitoring of the protection mechanisms, simply stresses the obvious fact that the situation in which security for a computing system is being provided is always changing. As a result of this, there must be an organized and on going effort to maintain the optimal level of security possible based upon the changing cost restraints and security requirements of the system.

## 5. THE FUZZY RISK ANALYSIS MODEL

There are several changes which could be made to improve Brocks' risk analysis model. First, Brocks admits that many of the probabilities required by his model are soft probability estimates. In these situations, the fuzzy expected value could be used in a group decision making environment to give a better estimate of the soft expected value. A second weakness is that there is no method given in Brocks' model for handling the varying amounts of protection which different mechanisms provide for the same threat. A final weakness is that there is no definitive method for management to use in making a decision as to what set of security measures will provide the maximum amount of protection for the system as a whole. If one creates a fuzzy set MEASURES WHICH PROVIDE FULL PROTECTION AGAINST THE THREAT for each threat, then an application of the typical value of a population of possibility distributions algorithm which was developed in the previous chapter provides a method for choosing the best set of protection mechanisms in a group decision making environment. These techniques will now be applied to the risk analysis model to give a fuzzified version of the model.

The first step in the fuzzy risk analysis model is the identification of threats to the system. It is obvious that any and all models of security must have some knowledge of what threats they are attempting to deter. Thus, the first step of the new model will consist of the determination of a set of n threats T with a generic element of T being denoted by the symbol $T_i$, $1 \leq i \leq n$.

The second step in the model is the assessment of the risk which each threat induces. This step can be broker 'nto four distinct parts.

The first part of the risk assessment step is the determination of the vulnerability of the system to the specific threat being considered. This can be done in two ways. First, if empirical evidence of the cost of recovering from the threat is available, it should be used. An example of this would be the estimation of the vulnerability of the computing center to fire by the use of the present replacement cost of all of the equipment in the computing center. The second technique can be used if no empirical evidence of vulnerability is available. This technique is to take soft estimates of the vulnerability to the threat from the members of the group creating the security program. The estimates will fall between some upper and lower bound. A typical estimate of vulnerability can be computed by finding the fuzzy expected value of the estimates in the interval between the lower bound and the upper bound. The set of vulnerability estimates is called V with a generic element of V being denoted by the symbol $V_i$.

The second part of the risk assessment step is the estimation of a probability of occurrence for each threat $T_i$. The set of probability estimates is called P with a generic element of P being denoted by $P_i$. Brocks states that the probability estimates which are used in his model are for the most part subjective evaluations. In the fuzzy model, there are two ways for estimating a member of P. First, if empirical evidence is available, use it to get a hard probability estimate. Second, if a subjective estimate of probability is required, have each member of the computer security group assign a membership value for the threat in the fuzzy set EVENTS WHICH WILL OCCUR DURING THE SPECIFIED TIME PERIOD. The typical value of probability for that threat will be the fuzzy expected value of each of the membership values for the threat as given by the group.

The third part of the risk assessment step is the computation of the organization's risk for each specific threat. The set of risks is called R with a generic element of R being denoted by $R_i$. The risk $R_i$ of the organization to a threat $T_i$ is computed by the formula $R_i = V_i \times P_i$. $R_i$ represents the amount which the organization could reasonably expect to lose if no protection is provided against threat $T_i$.

The final part of the risk assessment step is the computation of a set of priority percentages $PP_i$ corresponding to the set of threats T. The priority percentages are found according to the formula

$$PP_i = R_i \quad \sum_{j=1}^{n} R_j . \tag{5.1}$$

The priority percentages represent the relative amount of damage which can be expected from each threat when compared to the damage which can be expected from all of the threats.

The third step in the fuzzy risk analysis model is the determination of a set of most effective protection mechanisms for the set of threats confronting the system. This step can be divided into five distinct parts.

The first part of the mechanism selection step is the setting of the maximum amount which the organization can spend to provide security against all of the threats. This amount is called TOTAL.

The second part of the mechanism selection step is the creation of a list of possible protection mechanisms for each threat. This list should be created by consulting with staff members who have knowledge of the system and by studying the literature for suggested mechanisms. The list of mechanisms created for any threat T is called $M_i$. An element of $M_i$ shall be denoted by $M_{ij}$.

The set $M_i$ should include as single elements each mechanism by itself, mechanisms which can be used in combination with each other and the null mechanism. A possible set of mechanisms for the protection of a terminal room from unauthorized entry might be 1) Posting a guard at the door; 2) Providing new locks for the door; 3) Using a closed circuit television system to record access to the terminal room; 4) Posting a guard and providing new locks for the door; 5) Posting a guard and using closed circuit television; 6) Providing new locks and using closed circuit television; 7) Posting a guard, using closed circuit television and providing new locks for the door; and 8) The null mechanism.

The third part of the mechanism selection step is to have each person in the security group estimate a compatibility value, $\chi(M_{ij})$, which represents the compatibility of mechanism $M_{ij}$ to the security needs of the organization against threat $T_i$. These compatibility values are to be assigned in the interval [0,1] with 1 representing the fact that mechansim $M_{ij}$ will fully protect the organization against threat $T_i$, 0 representing the fact that mechanism $M_{ij}$ provides no protection against threat $T_i$, and values between 0 and 1 representing partial protection against the threat. Each set of mechanisms $M_i$ along with its corresponding set of compatibility values are a fuzzy set MECHANISMS WHICH PROVIDE FULL PROTECTION AGAINST THE THREAT $T_i$. For each mechanism $M_{ij}$, each of member of the security group should also estimate a cost $C_{ij}$ for that mechanism. This estimate can either be hard or soft depending upon the availability of empirical evidence as to the cost. If we allow a variable $X_i$ to take as its universe the mechanisms in set $M_i$, the assignment $X_i$ is restricted by the fuzzy set MECHANISMS

WHICH PROVIDE FULL PROTECTION AGAINST THE THREAT $T_i$ creates a possibility distribution for $X_i$. This possibility distribution gives the possibility that each mechanism in $M_i$ will fulfil the security requirements for the threat $T_i$.

The fourth part of the mechanism selection step is the conjunction of the possibility distributions for each $X_i$ to give a possibility distribution for the n-tuples of mechanisms created by taking the cartesian product of the sets $M_i$. This can be classified as the conjunction of several related restrictions upon n variables. The restrictions are related for three reasons. First, the amount of security which each mechanism provides to the system as a whole is different. Second, the amount of security which the system as a whole receives is the sum of the security which each of the mechanisms provide. Finally there is a threshold cost which the n-tuple of mechanisms to be chosen can not exceed.

For each member of the computer security group, there should be n possibility distributions corresponding to the n threats to the system. Each threat $T_i$ should have a priority percentage $PP_i$ which was computed earlier. Each mechanism $M_{ij}$ in the set of possibility distributions should have a cost $C_{ij}$ and a possibility value $Poss(M_{ij})$ which was created by the assignment of the fuzzy set MECHANISMS WHICH PROVIDE FULL PROTECTION AGAINST THE THREAT $T_i$ as a restriction upon the variable $X_i$. The fuzzy relation which we will use in this model to give a possibility value for each ordered n-tuple of mechanisms which can be combined to provide security against the n threats is the formula

$$Poss(M_{1j},\ldots,M_{nj}) = \begin{cases} 0 \text{ if } \sum_{i=1}^{n} C_{ij} > \text{Total} \\ \\ \sum_{i=1}^{n} PP_i \times Poss(M_{ij}) \text{ if otherwise} \end{cases} \tag{5.2}$$

The application of this fuzzy relation to every n-tuple in the cartesian product of the sets $M_i$ creates a possibility distribution representing the possibility that those n-tuples will meet the system's protection requirements against the set of threats while not exceeding the total allowable expenditure for security. At this point in the model there should be one possibility distribution for each person in the computer security group.

The final part of the mechanism selection step is the application of the typical value of a population of possibility distributions algorithm created previously to the possibility distributions of the computer security group. The result of this algorithm will be the set of mechanisms which the group as a whole has chosen as both providing a maximum amount of protection while also staying within the organization's security budget.

The final step in the fuzzy risk analysis model is the implementation and monitoring of the security mechanisms chosen. This step is the same as in Brocks' model. The only addition which this model makes is the requirement that the security group meet at set intervals of time to reevaluate the protection mechanisms selected.

## 6. AN EXAMPLE OF THE USE OF THE FUZZY MODEL

As the final subject of this chapter, let us consider the following application of the fuzzy risk analysis model to the design of a security program in a hypothetical situation. A synopsis of the entire model is given in Table 1 to make reference to the steps in the model easier. It should be noted that the mechanism cost estimates used in this example were chosen arbitrarily and may not reflect the actual costs of the corresponding mechanisms in the real world.

An organization has decided to add a new terminal room in its plant for the use of staff members in two confidential project groups. This terminal room contains ten terminals which are to be used by staff members working on the first project and six terminals

Table 1. Steps in the Fuzzy Risk Analysis Model

1. Identify set of threats T
2. Assess risk of system to each threat $T_i$
   a. Estimate vulnerability $V_i$
   b. Estimate probability of occurrence $P_i$
   c. Compute risk $R_i$
   d. Compute priority percentage $PP_i$
3. Select appropriate set of security mechanisms
   a. Determine maximum amount to be spent on security
   b. Identify possible protection mechanisms $M_{ij}$
   c. Estimate compatibility $\chi(M_{ij})$ and cost $C_{ij}$
   d. Take conjunction of possibility distributions
   e. Apply typical value of a population of possibility distributions algorithm to choose n-tuple of mechanisms
4. Implement and Monitor Mechanisms Chosen

which are to be used by the staff of the second project. The organization desires to provide security for the room, but must do so within the maximum cost of \$600 per year.

The first step in the fuzzy model is the determination of threats to the terminal room. The security group of three members lists three elements in set T. These are 1) A terminal may be stolen or damaged intentionally; 2) A fire may destroy the terminal room; and 3) A staff member of one of the projects may attempt to use a terminal which is to be used only by staff members of the other project.

The second step of the fuzzy model is the assessment of risk. The members of the security group estimate $V_1$ and $V_2$ on the basis of empirical data, which in this case is the replacement cost of the terminals. Thus $V_1$ = \$1000 and $V_2$ = \$16,000. No empirical data is available on the vulnerability of the organization to threat $T_3$. To obtain a vulnerability value, each member of the group estimates the vulnerability. The estimates are \$600, \$520 and \$650. The group estimate for $V_3$ is the fuzzy expected value of the estimates in the interval [520,650]. Thus $V_3$ = \$600.

The second part of the risk assessment step is the estimation of the probability of occurrence for each threat $T_i$. In this case, no empirical evidence is available for any of the threats. As a result of this, the members of the security group each estimate what he or she believes the probability of each threat's occurrence might be. The estimates for $P_1$ are 0.05, 0.06 and 0.02. The estimates for $P_2$ are 0.01, 0.02 and 0.04. The estimates for $P_3$ are 0.90, 0.70 and 0.70. To find a typical estimate of the probability of the occurrence of each threat, the fuzzy expected value of each set of estimates in the interval [0,1] is computed. The estimates of the security group are 0.06 for $P_1$, 0.04 for $P_2$ and 0.70 for $P_3$.

The third part of the risk assessment step in the model is the computation of the risk for each threat. Using the formula $R_i = V_i \times P_i$, the risks are computed to be $R_1$ = \$60, $R_2$ = \$720 and $R_3$ = \$420.

The final part of the risk assessment step is the computation of the priority percentages. For this example $PP_1$ is computed to be 0.05, $PP_2$ is computed to be 0.60 and $PP_3$ is computed to be 0.35.

The first part of the mechanism selection step is the creation of lists of possible protection mechanisms for each threat. After studying the literature, the security group decides upon the protection mechanisms given in Table 2.

For the rest of this example the mechanisms described in Table 2 shall be specified by the combination of the mechanism set number

Table 2. Threats T and Proposed Counter Measures $M_i$

$T_1$ Stolen Terminal
- $M_{11}$ Bolting the terminals to the tables
- $M_{12}$ Installing a new lock on terminal room door
- $M_{13}$ Bolting the terminals to the tables and installing a new lock on terminal room door
- $M_{14}$ The null mechanism

$T_2$ Fire in Terminal Room
- $M_{21}$ Installing a sprinkler system
- $M_{22}$ Installing fire extinguishers
- $M_{23}$ The null mechanism

$T_3$ Unauthorized Use of Terminal
- $M_{31}$ Using badge system to activate terminals
- $M_{32}$ Using passwords to limit access to terminals
- $M_{33}$ The null mechanism

i (where this is not clear from the context) and the mechanism number j.

The second part of the mechanism selection step is the assignment of grades of membership in the set MECHANISMS WHICH PROVIDE FULL PROTECTION AGAINST THE THREAT for each mechanism in $M_i$. The compatibility values as estimated by each member of the security group are given in Table 3.

Along with defining the fuzzy sets, each member of the security

Table 3. Estimated Compatibility of Mechanism to Threat

First Member of Security Group

$$\chi(M_1) = .75/1 + .5/2 + .8/3 + 0/4$$
$$\chi(M_2) = .8/1 + .4/2 + 0/3$$
$$\chi(M_3) = .4/1 + .3/2 + 0/3$$

Second Member of Security Group

$$\chi(M_1) = .6/1 + .6/2 + .8/3 + 0/4$$
$$\chi(M_2) = .8/1 + .2/2 + 0/3$$
$$\chi(M_3) = .5/1 + .4/2 + 0/3$$

Third Member of Security Group

$$\chi(M_1) = .4/1 + .4/2 + .7/3 + 0/4$$
$$\chi(M_2) = .7/1 + .5/2 + 0/3$$
$$\chi(M_3) = .6/1 + .7/2 + 0/3$$

Table 4. Estimated Cost of Each Mechanism

First Member of Security Group

$C_1$ = \$25/1 + \$50/2 + \$75/3 + \$0/4
$C_2$ = \$6000/1 + \$200/2 + \$0/3
$C_3$ = \$150/1 + \$300/2 + \$0/3

Second Member of Security Group

$C_1$ = \$30/1 + \$40/2 + \$75/3 + \$0/4
$C_2$ = \$6000/1 + \$200/2 + \$0/3
$C_3$ = \$150/1 + \$400/2 + \$0/3

Third Member of Security Group

$C_1$ = \$25/1 + \$55/2 + \$70/3 + \$0/4
$C_2$ = \$6000/1 + \$200/2 + \$0/3
$C_3$ = \$150/1 + \$160/2 + \$0/3

group also estimates the cost of each of the protection mechanisms. Three of these estimates are based upon hard data which was obtained from manufacturers. These are the sprinkler system, whose cost is $6000, the fire extinguishers, whose cost is $200, and the badge system, whose cost is $150. (Remember that these cost figures are arbitrarily chosen for this example and may not reflect the actual cost of such mechanisms.) The cost estimates of the group are given in Table 4.

After the estimation of the grades of membership in the sets and the estimation of the costs of the mechanisms, the remaining steps in selecting the proper security mechanisms are algorithmic in nature. The security group uses the fuzzy relation defined earlier in this chapter to take the conjunction of the three related possibility distributions corresponding to the three threats. This gives a single possibility distribution corresponding to the protection provided to the system by the n-tuples of mechanisms for each person in the security group. The n-tuples and their corresponding possibility values as estimated by each member of the security group are given in Tables 5, 6 and 7.

The final part of the mechanism selection step is the application of the typical value of a population of possibility distributions algorithm to the three distributions given in the preceeding paragraphs. In this case, the typical value of the distribution for security group members one and two is the 3-tuple (3,2,1). The typical value for group member three is the 3-tuple (3,2,2). Thus the counter for 3-tuple (3,2,1) is set at 2 and the counter for 3-tuple (3,2,2) is set to 1. Counters for all the other 3-tuples are set to 0. Since the 3-tuples of mechanisms are not intrinsically numeric events, it is necessary that each 3-tuple be arbitrarily

Table 5. Possibility Distribution for First Member of the Group

| N-Tuple | Possibility | N-Tuple | Possibility |
|---|---|---|---|
| (1,1,1) | 0.0000 | (3,1,1) | 0.0000 |
| (1,1,2) | 0.0000 | (3,1,2) | 0.0000 |
| (1,1,3) | 0.0000 | (3,1,3) | 0.0000 |
| (1,2,1) | 0.4175 | (3,2,1) | 0.4200 |
| (1,2,2) | 0.3825 | (3,2,2) | 0.3850 |
| (1,2,3) | 0.2775 | (3,2,3) | 0.2800 |
| (1,3,1) | 0.1775 | (3,3,1) | 0.1800 |
| (1,3,2) | 0.1425 | (3,3,2) | 0.1450 |
| (1,3,3) | 0.0375 | (3,3,3) | 0.0400 |
| (2,1,1) | 0.0000 | (4,1,1) | 0.0000 |
| (2,1,2) | 0.0000 | (4,1,2) | 0.0000 |
| (2,1,3) | 0.0000 | (4,1,3) | 0.0000 |
| (2,2,1) | 0.4050 | (4,2,1) | 0.3800 |
| (2,2,2) | 0.3700 | (4,2,2) | 0.3450 |
| (2,2,3) | 0.2650 | (4,2,3) | 0.2400 |
| (2,3,1) | 0.1650 | (4,3,1) | 0.1400 |
| (2,3,2) | 0.1300 | (4,3,2) | 0.1050 |
| (2,3,3) | 0.0250 | (4,3,3) | 0.0000 |

Table 6. Possibility Distribution for Second Member of the Group

| N-Tuple | Possibility | N-Tuple | Possibility |
|---|---|---|---|
| (1,1,1) | 0.0000 | (3,1,1) | 0.0000 |
| (1,1,2) | 0.0000 | (3,1,2) | 0.0000 |
| (1,1,3) | 0.0000 | (3,1,3) | 0.0000 |
| (1,2,1) | 0.3250 | (3,2,1) | 0.3350 |
| (1,2,2) | 0.0000 | (3,2,2) | 0.0000 |
| (1,2,3) | 0.1500 | (3,2,3) | 0.1600 |
| (1,3,1) | 0.2050 | (3,3,1) | 0.2150 |
| (1,3,2) | 0.1700 | (3,3,2) | 0.1800 |
| (1,3,3) | 0.0300 | (3,3,3) | 0.0400 |
| (2,1,1) | 0.0000 | (4,1,1) | 0.0000 |
| (2,1,2) | 0.0000 | (4,1,2) | 0.0000 |
| (2,1,3) | 0.0000 | (4,1,3) | 0.0000 |
| (2,2,1) | 0.3250 | (4,2,1) | 0.2950 |
| (2,2,2) | 0.0000 | (4,2,2) | 0.2600 |
| (2,2,3) | 0.1500 | (4,2,3) | 0.1200 |
| (2,3,1) | 0.2050 | (4,3,1) | 0.1750 |
| (2,3,2) | 0.1500 | (4,3,2) | 0.1400 |
| (2,3,3) | 0.0300 | (4,3,3) | 0.0000 |

Table 7. Possibility Distribution for Third Member of the Group

| N-Tuple | Possibility | N-Tuple | Possibility |
|---|---|---|---|
| (1,1,1) | 0.0000 | (3,1,1) | 0.0000 |
| (1,1,2) | 0.0000 | (3,1,2) | 0.0000 |
| (1,1,3) | 0.0000 | (3,1,3) | 0.0000 |
| (1,2,1) | 0.5300 | (3,2,1) | 0.5450 |
| (1,2,2) | 0.5650 | (3,2,2) | 0.5800 |
| (1,2,3) | 0.3200 | (3,2,3) | 0.3350 |
| (1,3,1) | 0.2300 | (3,3,1) | 0.2450 |
| (1,3,2) | 0.2650 | (3,3,2) | 0.2800 |
| (1,3,3) | 0.0200 | (3,3,3) | 0.0350 |
| (2,1,1) | 0.0000 | (4,1,1) | 0.0000 |
| (2,1,2) | 0.0000 | (4,1,2) | 0.0000 |
| (2,1,3) | 0.0000 | (4,1,3) | 0.0000 |
| (2,2,1) | 0.5300 | (4,2,1) | 0.5100 |
| (2,2,2) | 0.5650 | (4,2,2) | 0.5450 |
| (2,2,3) | 0.3200 | (4,2,3) | 0.3000 |
| (2,3,1) | 0.2300 | (4,3,1) | 0.2100 |
| (2,3,2) | 0.2650 | (4,3,2) | 0.2450 |
| (2,3,3) | 0.0200 | (4,3,3) | 0.0000 |

Table 8. Fuzzy Expected Value Information

| 3-Tuple | (X,Y,Z) | $\mu^*(\xi_{T*})$ | 3-Tuple | (X,Y,Z) | $\mu^*(\xi_{T*})$ |
|---|---|---|---|---|---|
| (1,1,1) | 36 | 1.00 | (3,1,1) | 18 | 1.00 |
| (1,1,2) | 35 | 1.00 | (3,1,2) | 17 | 1.00 |
| (1,1,3) | 34 | 1.00 | (3,1,3) | 16 | 1.00 |
| (1,2,1) | 33 | 1.00 | (3,2,1) | 15 | 24.33 |
| (1,2,2) | 32 | 1.00 | (3,2,2) | 14 | 36.00 |
| (1,2,3) | 31 | 1.00 | (3,2,3) | 13 | 36.00 |
| (1,3,1) | 30 | 1.00 | (3,3,1) | 12 | 36.00 |
| (1,3,2) | 29 | 1.00 | (3,3,2) | 11 | 36.00 |
| (1,3,3) | 28 | 1.00 | (3,3,3) | 10 | 36.00 |
| (2,1,1) | 27 | 1.00 | (4,1,1) | 9 | 36.00 |
| (2,1,2) | 26 | 1.00 | (4,1,2) | 8 | 36.00 |
| (2,1,3) | 25 | 1.00 | (4,1,3) | 7 | 36.00 |
| (2,2,1) | 24 | 1.00 | (4,2,1) | 6 | 36.00 |
| (2,2,2) | 23 | 1.00 | (4,2,2) | 5 | 36.00 |
| (2,2,3) | 22 | 1.00 | (4,2,3) | 4 | 36.00 |
| (2,3,1) | 21 | 1.00 | (4,3,1) | 3 | 36.00 |
| (2,3,2) | 20 | 1.00 | (4,3,2) | 2 | 36.00 |
| (2,3,3) | 19 | 1.00 | (4,3,3) | 1 | 36.00 |

assigned a compatibility value which in this case will be an integer between 1 and 36. The arbitrary assignment in this case will be in reverse lexigraphic order. After the scaled values for $\mu_*(\xi_{T*})$ are computed from the counters, the fuzzy expected value of the set of 3-tuples is computed in the interval [1,36]. The compatibility value and $\mu_*(\xi^*)$ for each 3-tuple is given in Table 8.

The fuzzy expected value of this set of 3-tuples is 15. Thus, the set of mechanisms which in the opinion of the security group gives the best protection to the terminal room while at the same time meeting the security budget is the set represented by the 3-tuple (3,2,1).

The final step in the fuzzy risk analysis model is the implementation and monitoring of the measures. To provide protection for the terminal room, the organization takes the following steps: 1) The terminals are bolted to the tables; 2) New locks are placed on the terminal room doors; 3) Fire extinguishers are placed in the terminal room; and 4) A badge system is installed to limit access to the use of the terminals. As a final step, the security group agrees to review the success of the protection mechanism chosen every three months.

Although this example includes only a small number of threats and protection mechanisms, the fuzzy model can be used with the same degree of effectiveness in situations which have more complex sets of threats and mechanisms.

## 7. CONCLUSIONS

In this study we have presented a risk analysis model of computer security which uses fuzzy set theory. This model has three distinct advantages over the probabilistic risk analysis model as presented by Brocks [1]. First, there exists in the fuzzy model a method for selecting a set of protection mechanisms for the system. Brocks' model, after providing those making the selection with the basic tools of risk analysis, does not give any solid algorithm for choosing the protection mechanism. Second, the use of the fuzzy expected value in a group decision making environment to determine soft probability gives a much more central and thus reliable estimate of probability to use in the risk calculations. Finally, Brocks' model does not address the question of the relationship between the amount of protection provided by a mechanism and the amount of protection needed to make the system totally secure against a threat. The use of fuzzy sets to handle this problem makes it possible to add to the model what is without a doubt vital information in the system.

The validity of the fuzzy model presented has not been demonstrated by experimental or observational evidence. Comparing the effectiveness of the fuzzy model against the effectiveness of other models remains a problem for future study. Original experimentation in this area should probably consist of in-depth comparisons of the effectiveness of the set of protection mechanisms suggested by the fuzzy model and the set of protection mechanisms suggested by the probabilistic risk analysis model in identical situations. Due to the complexity of these models, these first experiments should probably be limited to systems whose security requirements are fairly obvious. For the application of the fuzzy model to complex systems, the development of a computer program which takes the conjunction of the possibility distributions would be extremely helpful. Since this process is algorithmic in nature, such a program would probably be fairly easy to write.

There are two areas it may be possible to improve on the fuzzy risk analysis model. The first area is the fuzzy relation (5.2) which is used in the model to take several possibility distributions for single mechanisms and combine them into a single possibility distribution representing the effectiveness of n-tuples of mechanisms. It may be the case that a more effective function for this fuzzy relation could be found. The second area in which improvement may be possible is in the application of the typical value of a population of possibility distributions algorithm. Because the n-tuples of protection mechanisms are not numeric events, compatibility values for the n-tuples are arbitrarily assigned in the computation of the fuzzy expected value. It is possible that some method could be found to assign compatibility values which would be based upon some intrinsic quality of the n-tuples of protection mechanisms. These two areas remain open for further study.

The use of the fuzzy expected value and the typical value of a population of possibility distributions algorithm in the fuzzy risk analysis model is just one possible application of these methods to decision making. These methods can and should be applied to situations where decisions (particularly group decisions) must be made in an uncertain environment. Some possible areas of application of these methods are stock selection, urban planning, business decision making and computer pattern recognition.

## 8. REFERENCES

1. B. J. Brocks, "Security Problems in EDP - Assessing the Risks", Proceedings of the 1974 Eurocomp Conference, pp. 765-780 (1974).
2. D. Fannin, Guidelines for Establishing a Computer Security Program, Hudson, Massachusetts: Computer Security Institute (1979).

3. M. A. L. Farr, B. Chadwick and K. K. Wong, Security for Computer Systems, Manchester, England: National Computing Center (1972).
4. G. Green and R. C. Farber, Introduction to Security, Los Angeles: Security World Publishing Co., Inc. (1975).
5. P. Hamilton, Computer Security, London: Associated Business Programmes, Ltd. (1972).
6. C. F. Hemphill and J. M. Hemphill, Security Procedures for Computer Systems, Homewood, Illinois: Dow Jones-Irwin, Inc. (1973).
7. L. J. Hoffman, Security and Privacy in Computer Systems, Los Angeles: Melville Publishing Co. (1973).
8. D. K. Hsiao, D. S. Kerr and S. E. Madnick, Computer Security, New York: Academic Press (1979).
9. A. Kandel and W. J. Byatt, "Fuzzy Sets, Fuzzy Algebra, and Fuzzy Statistics," Proceedings of the IEEE, 66, pp. 1619-1939 (1978).
10. A. Kandel, "Fuzzy Statistics and System Security," Proceedings of the 1980 International Conference: Security through Science and Engineering, West Berlin (1980).
11. L. I. Krauss, SAFE: Security Audit and Field Evaluation for Computer Facilities and Information Systems, New York: AMACOM (1972).
12. I. Margolis, Private communication received on February 23, 1981.
13. J. Martin, Security, Accuracy, and Privacy in Computer Systems, Englewood Cliffs, New Jersey: Prentice-Hall, Inc. (1973).
14. D. B. Parker, Crime by Computer, New York: Scribner, Inc. (1976).
15. J. A. T. Pritchard, Risk Management in Action, Manchester, England: NCC Publications (1978).
16. B. J. Walker and I. F. Blake, Computer Security and Protection Structures, Stroudsburg, Pennsylvania: Dowden, Hutchingon & Ross, Inc. (1977).
17. S. Wooldridge, D. R. Corder and C. R. Johnson, Security Standards for Data Processing, New York: John Wiley & Sons (1973).
18. L. A. Zadeh, "Fuzzy Sets," Information and Control, 8, pp. 338-353 (1965).
19. L. A. Zadeh, "Fuzzy Sets as a Basis for a Theory of Possibility," Fuzzy Sets and Systems, 1, pp. 3-28 (1978).

# FUZZY MODELS OF HUMAN PROBLEM SOLVING

William B. Rouse

Center for Man-Machine Systems Research
School of Industrial and Systems Engineering
Georgia Institute of Technology
Atlanta, Georgia 30332

## INTRODUCTION

As the routine operations of many technical systems become increasingly automated, the roles of the human operators in these systems are becoming more and more oriented towards problem solving. The human operator is required to engage in problem solving when the automation encounters a situation for which it was not designed or, when the automation fails due to hardware and/or software problems.

Many researchers and practitioners are concerned about the human's abilities to perform satisfactorily in the role of problem solver[1]. A wide variety of training and aiding methods have been proposed and, to an extent, evaluated. However, only limited progress has been made. This is due in part to the lack of a widely accepted, operationally useful model of human problem solving. Such a model could be utilized to guide the design, development, and evaluation of training methods and aiding schemes.

The modeling of human problem solving presents considerable challenge because it involves the study of a fairly high level of human behavior[2]. Further, if one considers robust domains of application such as power plants, ships, and aircraft, the man-machine system of interest is certainly complex. Such complexity may require the qualitatively different approach to modeling possible with fuzzy set theory[3].

A variety of successful applications of fuzzy set theory to modeling man-machine interaction have emerged in recent years.

The control of industrial processes[4,5,6], ships[7], and automobiles[8] has been studied. Most of these efforts have been concerned with fuzzy descriptions of the encoding of displayed signals and control actions rather than the processes by which particular sequences of inputs produce particular sequences of outputs. In problem solving, however, one must be concerned with the problem solving process if any significant insight is to be gained.

In this paper, fuzzy models of human problem solving will be discussed. This will begin with consideration of the particular aspects of problem solving that might be considered fuzzy. Alternative problem solving algorithms will then be discussed. Consideration of the issues involved with measurement and evaluation will follow. Finally, a general discussion will be devoted to typical results obtained with fuzzy models and a comparison of alternative fuzzy formulations.

## FUZZY ASPECTS OF PROBLEM SOLVING

Most problem solving tasks involve a universal set of all possible solutions. For technical systems, this set may include all components, assemblies, subsystems, etc. that could possibly be faulty. For medical diagnosis, the universal set could include all possible diseases. In these two problem solving domains, one is interested in the membership of each component or disease in the fuzzy set of possible solutions.

However, the fuzzy set of possible solutions is actually only the end product of the process of problem solving. If one is concerned with this process, then the universal set might involve all paths to the solution, all problem solving rules, or all possible information seeking actions. A process-oriented model would utilize these sets to derive a solution set. The way in which this is accomplished depends on the overall point of view or approach adopted.

### Pattern Recognition Approach

The pattern recognition approach views the human problem solver as directly transforming a multi-dimensional sample of problem attributes into membership values for the fuzzy set of possible solutions. For technical systems, this may involve the transformation of pressures, temperatures, levels, flows, etc. For medical diagnosis, the transformation may include patient past history, present symptoms, signs observed upon physical examination, and results of clinical and diagnostic tests[9,10].

The aspect of the pattern recognition approach that distinguishes it from other approaches is the direct mapping of measurements to membership in the fuzzy set of possible solutions. No assumptions regarding intervening processes are necessary. The crucial assumption is that the human has a repertoire of stored patterns that is sufficient for producing acceptable solutions to the problems that will be encountered. Such a model will have difficulty with novel problems. However, of course, humans sometimes have difficulty with novel problems.

### Structural Approach

This approach views the human problem solver as using the structure of the problem to infer membership in the fuzzy set of possible solutions. For technical systems, this involves using the functional relationships within the system to define the possibility of various components causing the observed symptoms. While medical diagnosis can conceivably be viewed in this way, it typically is not because of the nature of the training of physicians.

The structural approach is distinguished from the pattern recognition approach by the fact that it explicitly accounts for the use of information that is somewhat internal to a problem. Thus, in one application of this approach to technical systems, not only was the fuzzy set of possible solutions defined, but also the fuzzy set of impossible solutions and fuzzy set of components involved in feedback loops were defined[11,12]. Of course, one might question whether or not the human is able or willing to utilize information about problem structure. As will later be discussed, the answer to this question depends on the novelty of the problem solving situation.

### Rule-Based Approach

The rule-based approach involves the use of rules that evoke actions that lead the problem solver towards a solution. It is not necessary to define the set of possible solutions. Thus, one can avoid the perhaps tenuous behavioral assumption that the human has knowledge of this set.

Not all rules produce observable actions. Some rules evoke strategies or tactics. Others provide internal updates of the problem solver's state of knowledge and, as a result, may serve to set up the preconditions for other rules[13,14].

Rules may also differ in terms of the types of information which they utilize. Some rules may look for familiar patterns in the symptoms (symptomatic or S-rules) while others may utilize problem structure (topographic or T-rules)[14,15]. In this way, the rule-based approach includes aspects of both the pattern recognition and structural approaches. However, the rule-based approach is more than just a hybrid; it is unique in that it avoids explicitly dealing with a solution set in the same way as the other approaches.

As a result, the fuzzy set of possible solutions is not a particularly useful construct within the rule-based approach. Instead, fuzzy sets of rules are needed. For example, one formulation utilizes fuzzy sets of recalled, applicable, useful, and simple rules to derive a fuzzy set of choosable rules[14,15].

### Crisp Sets

Not all sets of interest to the problem solver are fuzzy. For many problems, some of the steps in the solution process may be quite obvious (e.g., if the system is about to destroy itself, then turn if off). As a result, some fuzzy sets may have only two-valued membership functions (i.e., be non-fuzzy or crisp). Of course, this does not present any conceptual difficulties since crisp sets are special cases of fuzzy sets. However, it is useful to note that computational advantages can be obtained if fuzzy formulations are only used for those aspects of the problem solving that are in fact fuzzy[11,12,14].

## PROBLEM SOLVING ALGORITHMS

The various approaches discussed in the previous section provide different perspectives of problem solving. As a result, the algorithms by which the different fuzzy sets are manipulated vary among these approaches. In this section, these algorithms will be discussed and contrasted.

### Pattern Recognition Approach

For this approach, the goal of the algorithm is to find the member of the repertoire of stored patterns that is "closest" to the current pattern of information, according to some weighted distance criterion. Depending on the closeness of the match, the algorithm may either gather more information (guided by the match) or choose a solution. Gathering of more information results in updating the pattern of current information and a reiteration of the pattern matching process[9,10].

### Structural Approach

This approach utilizes the network of relationships among problem elements[11,12]. Given the symptoms of the problem and the network of relationships, the fuzzy set of possible solutions is defined as those problem elements that reach (i.e., have a path to) all symptoms. This set is fuzzy in that the human may not have precise knowledge of the existence or lack of a path from each element to each symptom.

The symptoms are the indications of an abnormal situation. Beyond these indications, however, there are often many normal indications which imply that at least a portion of the technical system or the patient's physiological system is functioning normally. If a problem element reaches (i.e., has a path to) any of these normal indications, this element cannot be the source of the abnormality. Thus, given the normal indications and the network of relationships among problem elements, the fuzzy set of impossible solutions can be defined.

The intersection of the possible set and the complement of the impossible set defines the fuzzy set of problem elements to be considered. Information is then gathered concerning one of these elements. If the element is not found to be the source of the abnormal situation, the information obtained about the element as well as the network of relationships are useful to update the membership values of all problem elements in the possible and impossible sets[11,12].

## Rule-Based Approach

This approach requires that one identify rules and preferences or relationships among rules. One approach to identifying rules is by using the opinions of experts who have studied the problem solving transcripts of subjects. The subjective nature of these opinions can be partially compensated for by objectively evaluating the results of problem solving simulations using the rules identified by the experts[13,14].

While simply asking humans to verbalize their rules can present difficulties, an interesting indirect variant of this method is to have humans construct functional block diagrams and specify the symptoms that would result from each failure. Using this information, the rules can then be directly derived[16]. Since this set of rules mainly reflects the human's understanding of the system in question, additional problem solving rules may have to be gathered from solution transcripts.

Preferences or relationships among rules can be represented in several ways. One approach is to assume a rank ordering with the most preferred, applicable rule being chosen. Algorithms for identifying rank orderings of rules are available[13], but have not been extended to fuzzy orderings of fuzzy rules.

An alternative approach is to define each rule in terms of a set of attributes that influence its use. For example, to be choosable, a rule must be recalled, applicable, useful, and simple[14,15]. Parameterized membership functions that define the level of each attribute in any particular situation can be used to produce choices of rules. The parameters of the membership functions can be varied until choices of model and humans are similar. In this way, the model comes to have preferences among rules, but these preferences are not the result of an a priori rank ordering.

## MEASUREMENT AND EVALUATION

Once those aspects of the problem solving situation that will be viewed as fuzzy have been defined and the problem solving algorithm derived, measurement and evaluation must be considered. In this section, a variety of approaches to these issues will be discussed.

### Measurement of Membership

The most fundamental measurement problem involves determining membership values for the elements of the various fuzzy sets of interest. While one of the benefits of fuzzy formulations is that membership can be expressed linguistically rather than numerically, the desire for an operational model may require that linguistic statements be quantified.

One approach to this problem is to ask the humans involved to provide numerical estimates of membership values. Unfortunately, humans have great difficulty in being consistent when making estimates of this nature. Indirect methods that avoid some of this difficulty have been proposed and utilized[10,17].

However, the difficulties associated with measurement of membership values are much greater than simply the issue of how to ask the questions. For many realistic problem solving domains, there are several fuzzy sets and many problem elements. As a result, it may be logistically impossible to measure all of the membership values needed.

### Inferring Membership

Rather than trying to measure membership values, one can attempt to infer them. This can be accomplished using parameterized membership functions that express membership as a function of one or more independent variables. If one can determine the appropriate independent variables, which is by no means guaranteed, then parameter adjustment methods can be used to find the parameter settings that result in the model's behavior most closely matching human behavior[11,12,14].

Of course, care must be taken in interpreting the membership values inferred in this way. All that one can really claim is that these membership values are consistent with the observed human behavior. One does not know that these are the "true" values. However, this limitation usually does not cause difficulty, unless the purpose of the modeling effort is to determine membership values rather than describe problem solving behavior.

### Evaluation

To compare the behavior of a model to the behavior of humans, one needs a measure or measures which indicate the degree to which the comparison is favorable or not. This is particularly true for the parameter estimation approach to inferring membership that was just discussed. The choice of measure depends on the goal of the modeling effort and may range from very fine-grained comparisons of behaviors to comparisons of only overall performance measures.

For tasks involving the control of dynamic systems, which usually are not problem solving tasks, the deviations of key state variables are typical measures of interest[4,5,6,7,8]. For diagnostic tasks, the goal may be to have the model produce the same diagnoses as humans[9,10]. These measures reflect the product of performing the task. As such, they provide only indirect insights into the process of performing the task.

Measures of the process of problem solving provide a more rigorous test of a model than possible with product measures. This is due to the much more fine-grained nature of process measures. For example, one might specify that the model not only should produce the same problem solution as humans but should also take the same path to the solution (i.e., take the same actions in the same order). A slightly less ambitious criterion would be to require similar paths in that the model utilizes the same problem elements, performs the same operations, or evokes the same problem solving rules, but does not necessarily take the exact same actions in the same order[13,14]. This simplified approach to process measures can provide considerably more insight than possible with product measures while, at the same time, avoiding a criterion that may be impossible to satisfy.

## DISCUSSION

The previous sections have dealt with alternative formulations of problem solving as a fuzzy process, derivations of problem solving algorithms, and approaches to measurement and evaluation. In this section, typical results obtained with fuzzy models of human problem solving will be reviewed and alternative fuzzy formulations compared.

## Typical Results

There are two main ways in which fuzzy models of human problem solving have been useful. One way involves using the model as a basis for automating a task that had been performed by humans. The process and ship control work noted earlier are good examples of successful use of fuzzy models in this way[4,5,6,7]. In these cases, automatic controllers based on fuzzy models of human controllers performed better than automatic controllers based on traditional system theoretic approaches.

The second important way in which fuzzy models of human problem solving have been used is to gain insights into human behavior. The medical diagnosis model of Esogbue and Elder fits into this category[9,10]. Their model compared more favorably with human diagnostic decisions than had been possible with the crisp pattern recognition models advocated by others.

The fault diagnosis model of Rouse[11,12] was used to identify factors constraining the performance of maintenance personnel in troubleshooting tasks. The results obtained using the model led to the design and evaluation of training methods and aiding schemes to overcome these limitations. The work of Hunt in this area has had similar implications[14].

## Comparison of Approaches

The pattern recognition, structural, and rule-based approaches provide alternative perspectives of human problem solving. Each approach has an appropriate domain of application. For tasks where the human problem solver cannot or need not utilize problem structure, pattern recognition or rule-based formulations are suitable. On the other hand, for tasks where structure must be used (e.g., an unfamiliar situation for which there are not stored patterns), structural or rule-based formulations are appropriate.

It appears that the rule-based approach has the widest applicability. This is due to the fact that such a model can incorporate both pattern recognition rules and structural rules (e.g., the aforementioned S-rules and T-rules). However, as noted in earlier discussions, the rule-based approach is founded on different behavioral assumptions than the pattern recognition and structural approaches. In fact, the differences in the behavioral assumptions underlying the three approaches should be a strong factor in choosing among the approaches.

# CONCLUSIONS

This paper has reviewed and contrasted alternative approaches to modeling human problem solving as a fuzzy process. This topic is fairly new, particularly in terms of realistic applications. However, it is already clear that some of the features of fuzzy set theory provide an attractive means for representing the complexity of human problem solving. At this point in time, practical issues surrounding underlying behavioral assumptions, measurement, and evaluation need to be addressed if fuzzy models of human problem solving are to realize their potential.

## ACKNOWLEDGEMENT

This research was supported by the Army Research Institute for the Behavioral and Social Sciences under Contract No. MDA 903-79-C-0421.

## REFERENCES

1. J. Rasmussen and W. B. Rouse, eds., "Human Detection and Diagnosis of System Failures," Plenum Press, New York (1981).
2. W. B. Rouse, "Systems Engineering Models of Human-Machine Interaction," North-Holland, New York (1980).
3. L. A. Zadeh, Outline of a New Approach to the Analysis of Complex Systems and Decision Processes, IEEE Transactions on Systems, Man , and Cybernetics, vol. SMC-3, no. 1, pp. 28-44 (1973).
4. W. J. M. Kickert and H. R. van Nauta Lemke, Application of a Fuzzy Controller in a Warm Water Plant, Automatica, vol. 12, pp. 301-308 (1976).
5. P. J. King and E. H. Mamdani, The Application of Fuzzy Control Systems to Industrial Processes, Automatica, vol. 13, pp. 235-242 (1977).
6. D. Willaeys and N. Malvache, Contribution of the Fuzzy Sets Theory to Man-Machine System, in: "Advances in Fuzzy Set Theory and Applications," M. M. Gupta, R. K. Ragade, and R. R. Yager, eds., pp. 481-499, North-Holland, New York (1979).
7. J. Van Amerongen, H. R. van Nauta Lemke, and J. C. T. van der Veen, An Autopilot for Ships Designed with Fuzzy Sets, in: "Digital Computer Applications to Process Control," H. R. van Nauta Lemke and H. B. Verbruggen, eds., North-Holland, New York (1977).
8. U. Kramer and G. Rohr, Psycho-Mathematical Model of Vehicular Guidance Based on Fuzzy Automata Theory, Proceedings of the First Annual European Conference on Decision Making and Manual Control, Delft, The Netherlands, May (1981).
9. A. O. Esogbue and R. C. Elder, Fuzzy Sets and the Modeling of Physician Decision Processes, Part I: The Initial Interview Information Gathering Session, Fuzzy Sets and Systems, vol. 2, pp. 279-291 (1979).

10. A. O. Esogbue and R. C. Elder, Fuzzy Sets and the Modeling of Physician Decision Processes, Part II: Fuzzy Diagnosis Decision Models, Fuzzy Sets and Systems, vol. 3, pp. 1-9 (1980).
11. W. B. Rouse, A Model of Human Decision Making in a Fault Diagnosis Task, IEEE Transactions on Systems, Man, and Cybernetics, vol. SMC-8, no. 5, pp. 357-361 (1978).
12. W. B. Rouse, A Model of Human Decision Making in Fault Diagnosis Tasks that Include Feedback and Redundancy, IEEE Transactions on Systems, Man, and Cybernetics, vol. SMC-9, no. 4, pp. 237-241 (1979).
13. W. B. Rouse, S. H. Rouse and S. J. Pellegrino, A Rule-Based Model of Human Problem Solving Performance in Fault Diagnosis Tasks, IEEE Transactions on Systems, Man, and Cybernetics, vol. SMC-10, no. 7, pp. 366-376 (1980).
14. R. M. Hunt, Human Pattern Recognition and Information Seeking in Fault Diagnosis Tasks, Ph.D. Thesis, University of Illinois at Urbana-Champaign (1981).
15. W. B. Rouse and R. M. Hunt, A Fuzzy Rule-Based Model of Human Problem Solving in Fault Diagnosis Tasks, Proceedings of the Eighth Triennial World Congress of the International Federation of Automatic Control, Kyoto, Japan, August (1981).
16. R. M. Hunt, A New Approach to Heuristic Model Creation, Coordinated Science Laboratory, University of Illinois at Urbana-Champaign, December (1980).
17. M. Kochen, Application of Fuzzy Sets in Psychology, in: Fuzzy Sets and Their Applications to Cognitive and Decision Processes, L. A. Zadeh, K. S. Fu, K. Tanaka, and M. Shimura, eds., Academic Press, New York (1975).

# A CONCEPT OF A FUZZY IDEAL FOR MULTICRITERIA CONFLICT RESOLUTION*

Yee Leung

Department of Geography
The Chinese University of Hong Kong
Shatin, Hong Kong

## I. INTRODUCTION

Fuzzy sets theory [19], since its inception, has been exerting propagating influence on the methodological development in a variety of disciplines. Its impact on decision analysis is a typical example. Based on the conceptualization of a fuzzy decision-making environment, new theories and methods have been constructed and applied to decision-making processes in general (see for example [2, 5, 10, 21]), and multicriteria decision-making processes (see for example [6, 8, 17]) in particular. In spite of the proliferation of fuzzy sets approach in this field, its plausible application to conflict resolution involving multiple criteria, a special class of decision-making problems, though important, has only been subjected to limited scrutinization recently [12, 16, 22, 23]).

Over the years, an "ideal solution" has been employed as a fundamental concept in the mathematical methods of conflict resolution involving multiple criteria. As Zeleny [23] documented, the concept of an "ideal" was possibly introduced by Geoffrion [9] under the name of a "perfect solution". Similar idea also appeared in the work of Radizikowski [13] and Jüttler [11]. Saska [15] and

*This paper was completed while the author was a Visiting Senior Fellow of The Center for Metropolitan Planning and Research and a Visiting Associate Professor of the Department of Geography and Environmental Engineering at the Johns Hopkins University in the spring semester, 1982.

Dinkelbach [7] made the concept operational in a multiprogramming framework. A formalization, under the name of "utopia point" was proposed by Yu [18] to resolve conflict involving multiple decision makers.

Conventionally, an ideal solution is a solution having the highest achievement scores in all individual criteria characterizing an alternative and is in general infeasible. Nevertheless, it serves as a reference point to obtain a compromise solution whenever conflict occurs. A major characteristic of the procedure is that the ideal and the compromise solutions are point-value with exactitude. Though it is a useful approach to resolve multicriteria conflicts, it falls short in describing real-world conflict resolution processes. Due to inexact information and our cognitive and decision-making capabilities, evaluations of criteria and alternatives are in general fuzzy. Instead of numerical scores, verbal judgements, such as linguistic propositions, are more common and perhaps more flexible. Under this situation, the cognized "ideal" becomes fuzzy. Zeleny [23] has suggested that the ideal is possibly a region damarcated by fuzzy intervals of the individual criteria. Since the ideal is fuzzy, the associated compromise solutions, I believe, should be fuzzy and take on a form of a fuzzy region also. The inexactness of the ideal and compromise solutions in turn lead to new perspectives on the resolution of conflicts.

The purpose of this paper is to provide an analytical framework of multicriteria conflict resolution in a fuzzy environment. Section II deals with the linguistic and numeric characterizations of a fuzzy ideal and the derivation of the corresponding compromise solutions. Procedures which enable decision makers to zero in on the most compatible element of a fuzzy ideal are examined in Section III. The paper then concludes by evaluating the merits and problems of the proposed conflict resolution framework.

## II. FUZZY IDEAL AND FUZZY COMPROMISE SOLUTIONS

Let $X = \{X_1, \ldots, X_i, \ldots, X_n\}$ be a set of initial feasible alternatives. Let $Y = \{Y_1, \ldots, Y_j, \ldots, Y_m\}$ be a set of criteria, usually conflicting, by which the performance of an alternative is evaluated. Our decision-making process is concerned with the selection of the most desirable alternative so that the performances of the individual alternatives are satisfied as much as possible.

Let

$$\begin{bmatrix} p_1^1 & \cdots & p_j^1 & \cdots & p_m^1 \\ \vdots & & \vdots & & \vdots \\ p_1^i & \cdots & p_j^i & \cdots & p_m^i \\ \vdots & & \vdots & & \vdots \\ p_1^n & \cdots & p_j^n & \cdots & p_m^n \end{bmatrix} \tag{1}$$

be the performance matrix whose element $p_j^i$ is a verbal evaluation of the performance of an alternative $X_i$ with respect to criterion $Y_j$. For convenience of presentation, the superscript instead of the subscript i is employed to denote the alternatives. The general format of the evaluation $p_j^i$ is a linguistic proposition

$$p_j^i : Y_j \text{ is } F_j^i , \tag{2}$$

which may be stated as

"With reference to alternative $X_i$, the criterion $Y_j$ is $F_j^i$, a fuzzy subset".

For example, "With reference to alternative $X_i$, net return ($Y_j$) is *around 2 billion dollars* ($F_j^i$)".

Thus, our evaluation rules involve linguistic terms whose meanings are crucial in the selection process. To permit a formal analysis, a mathematical translation of the linquistic propositions is pertinent. Based on a theory of possibility [20], the linguistic proposition in eq. 2 induces a possibility distribution defined by the possibility assignment equation

$$Y_j \text{ is } F_j^i \rightarrow \Pi_{Y_j}^i = F_j^i , \tag{3}$$

such that

$$\text{Poss } (Y_j = y_j) = \pi_{Y_j}^i (y_j) = \mu_{F_j^i}(y_j). \tag{4}$$

Therefore, the matrix of linguistic propositions in eq. 1 can be transformed into a matrix of possibility distribution fucntions

$$\begin{bmatrix} \pi^1_{Y_1} & \cdots & \pi^1_{Y_j} & \cdots & \pi^1_{Y_m} \\ \vdots & & \vdots & & \vdots \\ \pi^i_{Y_1} & \cdots & \pi^i_{Y_j} & \cdots & \pi^i_{Y_m} \\ \vdots & & \vdots & & \vdots \\ \pi^n_{Y_1} & \cdots & \pi^n_{Y_j} & \cdots & \pi^n_{Y_m} \end{bmatrix} \qquad (5)$$

whose element $\pi^i_{Y_j}$ takes on a form defined by eq. 4. Such a distribution in turn imposes a fuzzy interval on the values of $Y_j$ under propositon $p^i_j$. Hence, the numerical counterpart of the matrix in eq. 5 is the following matrix

$$\begin{bmatrix} [\underline{y}^1_1, \overline{y}^1_1] & \cdots & [\underline{y}^1_j, \overline{y}^1_j] & \cdots & [\underline{y}^1_m, \overline{y}^1_m] \\ \vdots & & \vdots & & \vdots \\ [\underline{y}^i_1, \overline{y}^i_1] & \cdots & [\underline{y}^i_j, \overline{y}^i_j] & \cdots & [\underline{y}^i_m, \overline{y}^i_m] \\ \vdots & & \vdots & & \vdots \\ [\underline{y}^n_1, \overline{y}^n_1] & \cdots & [\underline{y}^n_j, \overline{y}^n_j] & \cdots & [\underline{y}^n_m, \overline{y}^n_m] \end{bmatrix} \qquad (6)$$

Its element $[\underline{y}^i_j, \overline{y}^i_j]$ is a fuzzy interval associated with the possibility distribution $\Pi^i_{Y_j}$ and a minimal level of satisfaction, $\alpha$, and is obtained as

$$[\underline{y}^i_j, \overline{y}^i_j] = \{y_j \mid \pi^i_{Y_j}(y_j) \geqq \alpha\}, \quad \alpha \in (0, 1]. \qquad (7)$$

The values $\underline{y}^i_j$ and $\overline{y}^i_j$ are the minimum and maximum values of the interval. (See Fig. 1).

With reference to all individual alternatives, the ideal

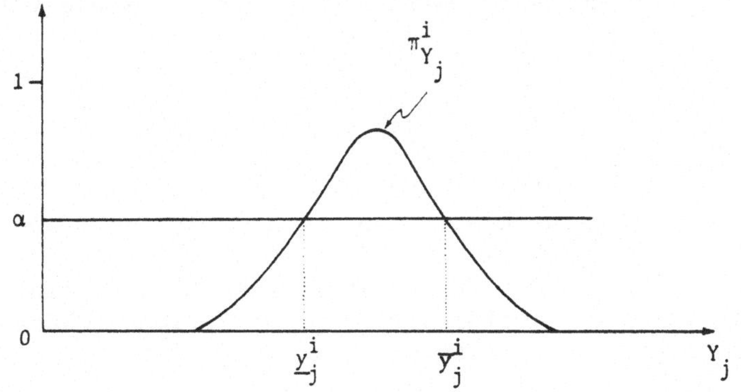


Fig. 1. Fuzzy interval imposed by a possibility distribution function and a level of satisfaction.

performance of an alternative under criterion $Y_j$ may be expressed as a composite linguistic proposition

$$P_j : Y_j \text{ is } F_j \ , \tag{8}$$

where, the term $F_j$ is a fuzzy subset defined as

$$F_j = F_j^1 \text{ and } \dots \text{ and } F_j^i \text{ and } \dots \text{ and } F_j^n \ . \tag{9}$$

That is, the verbal specification of the ideal performance of an alternative under $F_j$ is

"With reference to all alternatives, the criterion $Y_j$ is $F_j$".

Again, the composite linguistic proposition in eq. 8 can be translated into a composite possibility distribution such that

$$\pi_{Y_j}(y_j) = \mu_{F_j}(y_j), \tag{10}$$

with

$$\mu_{F_j} = \mu_{F_j^1 \cap \dots \cap F_j^i \cap \dots \cap F_j^n} \tag{11}$$

and

$$\pi_{Y_j}(y_j) = \mu_{F_j^1}(y_j) \wedge \dots \wedge \mu_{F_j^i}(y_j) \wedge \dots \wedge \mu_{F_j^n}(y_j). \tag{12}$$

Since $\pi_{Y_j}$ serves as a restriction on the possible values of $y_j$ in $Y_j$, a fuzzy interval may be obtained as

$$[\underline{y}_j^*, \overline{y}_j^*] = \{y_j | \pi_{Y_j}(y_j) \geqq \alpha\}, \quad \alpha \in (0, 1], \tag{13}$$

where, $\underline{y}_j^*$ and $\overline{y}_j^*$ are the minimum and maximum values respectively.

The description in eq. 8 and 10 represents the ideal performance of an alternative with respect to criterion $Y_j$ because for $y_j \in Y_j$, the following relationship

$$\pi_{Y_j}(y_j) \leqq \pi^i_{Y_j}(y_j) \Leftrightarrow F_j \subset F^i_j \;, \quad \text{for all } i, \tag{14}$$

holds. That is, if an alternative satisfies the condition imposed by proposition $P_j$, eq. 8, it simultaneously satisfies the conditions imposed by propositions $P^i_j$, $i = 1, \ldots, n$, eq. 2. Since the fuzzy interval $[\underline{y}_j^*, \overline{y}_j^*]$ is in fact the intersection of the individual fuzzy intervals $[\underline{y}^i_j, \overline{y}^i_j]$ with

$$\underline{y}_j^* = \min_i \underline{y}^i_j \;, \text{ and}$$

$$\overline{y}_j^* = \min_i \overline{y}^i_j \;, \tag{15}$$

then, for any $y_j$ in $[\underline{y}_j^*, \overline{y}_j^*]$, it is also an element of $[\underline{y}^i_j, \overline{y}^i_j]$. Consequently, $[\underline{y}_j^*, \overline{y}_j^*]$ can be treated as a fuzzy interval specifying a set of ideal values for $Y_j$.

*Example* Let "net return" ($Y_j$) be a criterion by which the performances of four alternatives $X_1$, $X_2$, $X_3$, and $X_4$ are evaluated. Let $P^1_j$ : net return is *between 5 and 11 million dollars*; $P^2_j$ : net return is *around 9 million dollars*; $P^3_j$ : net return is *a little more than 6 million dollars*, and $P^4_j$ : net return is *not less than 7 million dollars* be the linguistic propositions with the correponding possibility distribution functions depicted in Fig. 2a, b, c, and d respectively.

Then, the composite linguistic proposition is

$P_j$ : net return is *between 5 and 11 million dollars* and *around 9 million dollars* and *a little more than 6 million dollars* and *not less that 7 million dollars*,

and, the composite possibility distribution function and the

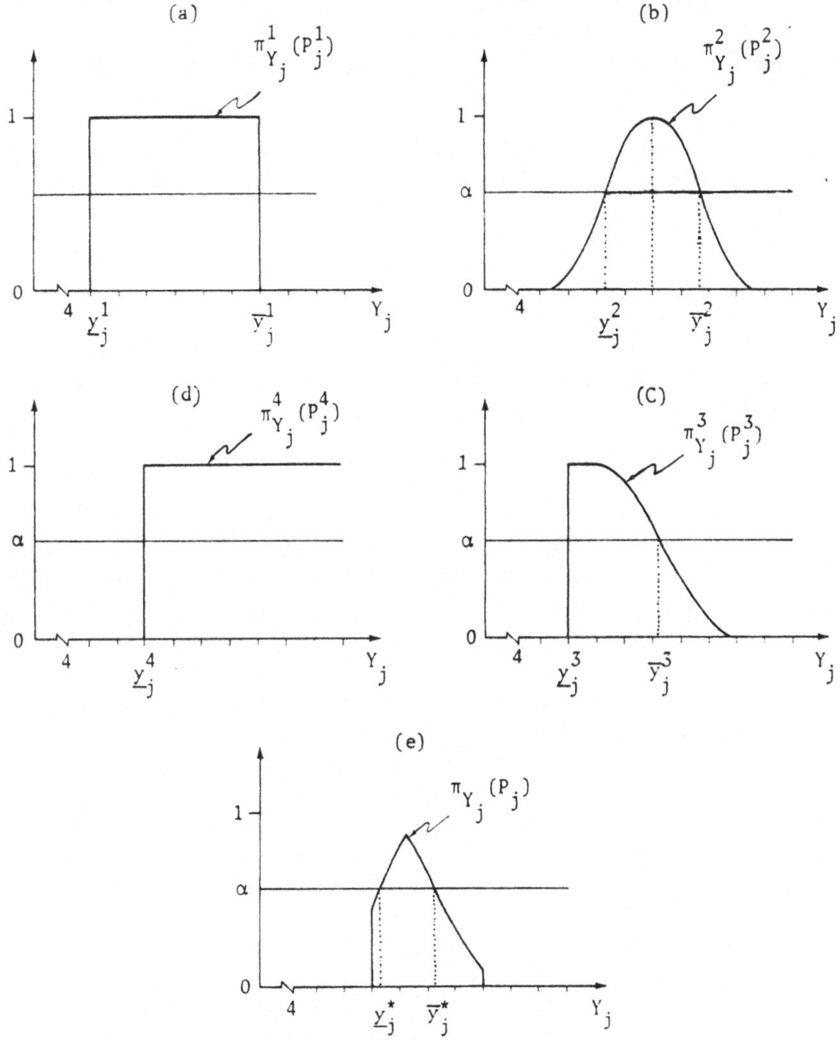


Fig. 2. Possibility distribution functions and the associated fuzzy intervals.

associated fuzzy interval are depicted in Fig. 2e.

Therefore, in determining the ideal value of a criterion, the present analytical framework differs from the conventional approach in two major aspects. First, in place of a score function which requires numeric assessment, linguistic proposition is employed to evaluate the performance of an alternative. Second, instead of a scalar, the ideal performance of an alternative with respect to a criterion is designated by a composite linquistic proposition with a translated composite possibility distribution, or a fuzzy interval indicating a set of ideal values. As a result, the ideal performance of an alternative can be linguistically and numerically characterized.

Taking all criteria into consideration, the ideal alternative, $X^*$, can be considered as a fuzzy ideal defined by an m-tuple of linguistic propositions

$$X^* = (Y_1 \text{ is } F_1, \ldots, Y_j \text{ is } F_j, \ldots, Y_m \text{ is } F_m), \tag{16}$$

whose element, $Y_j$ is $F_j$, is a composite linguistic proposition of criterion $Y_j$, eq. 8. Equivalently, through eq. 13, it may be defined as a fuzzy region characterized by an m-cell

$$X^* = \{(y_1^*, \ldots, y_j^*, \ldots, y_m^*) | \underline{y}_j^* \leqq y_j^* \leqq \overline{y}_j^*, \ j = 1,\ldots,m\}. \tag{17}$$

Since each of its coordinate $y_j$ is an element of the fuzzy interval $[\underline{y}_j^*, \overline{y}_j^*]$, $j = 1, \ldots, m$, then every element in $X^*$ may be considered as an ideal solution. For every m-cell is compact the fuzzy ideal has a minimum $(\underline{y}_1^*,\ldots,\underline{y}_j^*,\ldots,\underline{y}_m^*)$ and a maximum $(\overline{y}_1^*,\ldots,\overline{y}_j^*,\ldots,\overline{y}_m^*)$.

In general, such a fuzzy ideal is impossible to ascertain, and conflict arises as to which alternative to select so that the chosen alternative has the minimum deviation from the fuzzy ideal. A natural solution is that of a compromise solution. Conventionally, a compromise solution set for a point-value ideal is derived from minimizing a distance function [23]. The derived compromise solutions are point-value also.

Since the ideal defined here is fuzzy, the compromise solution set should be fuzzy also. In fact, if we take the maximum of the fuzzy ideal as an ideal solution, the corresponding compromise solution set can be determined through the following distance minimization procedure.

Let

$$\left[ \sum_{j=1}^{m} (1 - \mu_{\textit{close to } \overline{y}_j^*} (y_j^i)^p \right]^{1/p}, \ 1 \leqq p \leqq \infty, \tag{18}$$

be a family of distance functions defining the distance between an alternative $X_i$ and the maximum of the fuzzy ideal, with $\mu_{\textit{close to } \overline{y}_j}(y_j^i)$ indicating the proximity of $y_j^i$ to $\overline{y}_j^*$ (see [12] for discussion). For a specific value of the parameter p, the compromise solution, denoted as $\overline{x}^c = (\overline{y}_1^c, \ldots, \overline{y}_j^c, \ldots, \overline{y}_m^c)$, is the alternative which minimizes the distance function in eq. 18. That is,

$$\left[ \sum_{j=1}^{m} (1 - \mu_{close\ to\ \mathbf{y}_j^*}(\bar{y}_j^c))^p \right]^{1/p} =$$

$$\min_{X_i \varepsilon X} \left[ \sum_{j=1}^{m} (1 - \mu_{close\ to\ \bar{y}_j^*}(y_j^i))^p \right]^{1/p}, \quad 1 \leqq p \leqq \infty . \tag{19}$$

For p = 1, eq. 18 becomes

$$\sum_{j=1}^{m} (1 - \mu_{close\ to\ \bar{y}_j^*}(y_j^i)). \tag{20}$$

The sum of the individual distances to the maximum of the fuzzy ideal is then minimized in eq. 19.

For p = 2, eq. 18 becomes

$$\left[ \sum_{j=1}^{m} (1 - \quad {close\ to\ \bar{y}_j^*}(y_j^i))^2 \right]^{1/2}, \tag{21}$$

and we are minimizing the physical distance in eq. 19.

For $p = \infty$, eq. 18 becomes

$$\max_{i} \left\{ 1 - \mu_{close\ to\ \bar{y}_j^*}(y_j^i) \right\}. \tag{22}$$

Consequently, the maximal distance a feasible alternative may have to the maximum of the fuzzy ideal is minimized in eq. 19.

Thus, the compromise solution with respect to the parameter p in the distance function designates a special form of conflict resolution between the feasible alternatives and the maximum of the fuzzy ideal. The compromise solution set with respect to $(\bar{y}_1^*, \ldots, \bar{y}_j^*, \ldots, \bar{y}_m^*)$ then consists of all point-value compromise solutions for all p, $1 \leqq p \leqq \infty$.

By the same token, taking the minimum $(\underline{y}_1^*, \ldots, \underline{y}_j^*, \ldots, \underline{y}_m^*)$ of the fuzzy ideal as an ideal solution, another set of compromise solutions, with its element denoted as $\underline{X}^c = (\underline{y}_1^c, \ldots, \underline{y}_j^c, \ldots, \underline{y}_m^c)$, can be derived through the distance minimization procedure accordingly.

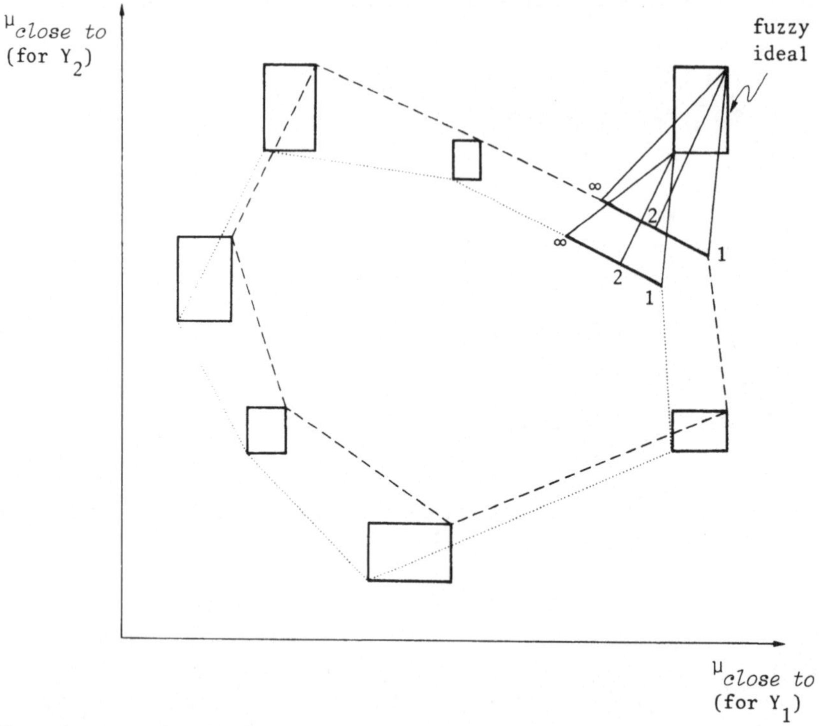


Fig. 3. Compromise solutions with respect to specific elements of a fuzzy ideal.

_ _ _ feasible region when the upper bound of the fuzzy ideal is employed

..... feasible region when the lower bound of the fuzzy ideal is employed

To illustrate the above theoretical results, a didactic example of a two-criteria conflict resolution is depicted in Fig. 3.

Therefore, for each element in the fuzzy ideal, a set of compromise solutions can be obtained by the method discussed. Instead of a set of point-value compromise solutions, decision makers are now dealing with a fuzzy region consisting of compromise solution sets. Specifically, for a specific value of the parameter p, in place of a point-value, we now have a set of point-values the compromise solution may possibly take.

Not being able to locate a point-value solution may be perturbing to some decision analysts or decision makers. However, having fuzzily demarcated ideal and compromise solutions seems to be more powerful in resolving multicriteria conflicts. On the theoretical basis, if our information is inexact and our cognitive and evaluation processes are fuzzy, it is only logical that the ideal is vaguely identified and the corresponding compromise solutions are fuzzy. Pragmatically, a single point-value solution tends to lead to deadlocks in the resolution process because it imposes too rigid a condition for a compromise. On the contrary, having a fuzzy region as an ideal, we automatically make more room for mutual consensus. If the compromise solution with respect to an element of the fuzzy ideal fails to reach a consensus, a new compromise solution can always be proposed through the selection of another element of the fuzzy ideal as the ideal solution. Thus, a higher degree of flexibility in resolving conflict can be achieved. Nevertheless, such an approach can be deceiving. Within the fuzzy ideal, some elements may be more compatible as an ideal solution and may lead to a more compatible compromise solution than others. However, to which element a decision maker should select as an ideal solution may not be too straightforward. For example, in selecting the most compatible element, the maximum of the fuzzy ideal seems to be a natural candidate. However, the selection is justifiable only if the heights of the composite possibility distribution functions characterizing the fuzzy ideal are at $\bar{y}_j^*$, for all j. Nevertheless, the most compatible element may not be an extreme point, e.g., maximum or minimum, of the fuzzy ideal in some situations. For instance, in relation to the linguistic proposition, "labor force should be about 2,000," the value 2,000 is the most compatible value. The compatibilities of the values on both sides of 2,000 decrease monotonically.

Thus, if sufficient information is available, a procedure which permits decision makers to zero in on the most compatible element of a fuzzy ideal may be necessary. In the following section, two procedures are proposed to cater to such a purpose.

## III. ZERO-IN PROCEDURE FOR A FUZZY IDEAL

A zero-in procedure is essentially a defuzzification mechanism through which a point-value is derived by focusing on a specific point of the fuzzy ideal. The compromise solution thus derived is point-value also. The focusing procedure can be mainly carried out in two different places in the conflict resolution process. First, for each criterion, we may want to zero in on a value of the fuzzy interval associated with its composite possibility distribution function. Alternatively, the zero-in procedure may be applied to the possibility distribution functions characterizing the individual criteria. In what follows, two such procedures are respectively discussed.

*Procedure 1.* Section II shows that the ideal performance of an alternative with respect to a criterion can be characterized by a composite linguistic proposition, eq. 8, which in turn can be translated into a composite possibility distribution function, eq. 10. A fuzzy interval specifying a set of ideal values is then derived in eq. 13.

To have a scalar as an ideal value, it is necessary to select within the fuzzy interval a representative value. A natural procedure is to zero in on the value of the fuzzy interval which has the highest compatibility to the composite linguistic proposition of the criterion. Since the composite possibility distribution function, eq. 10, is a translation of the composite linguistic proposition, eq. 8, then, for criterion $Y_j$, $j = 1, \ldots, m$, the most compatible element in the fuzzy interval $[\underline{y}_j^*, \bar{y}_j^*]$ should be the one, denoted as $y_j^*$, which satisfies the following condition

$$\pi_{Y_j}(y_j^*) = \sup \min \left[\pi^1_{Y_j}(y_j), \ldots, \pi^i_{Y_j}(y_j), \ldots, \pi^n_{Y_j}(y_j)\right]. \quad (23)$$

Therefore, instead of an interval, the zero-in procedure focuses on a single value as an ideal value. (See Fig. 4).

Consequently, the fuzzy ideal, originally a fuzzy region, eq. 17, can be transformed into a point-value ideal expressed as

$$X^* = (y_1^*, \ldots, y_j^*, \ldots, y_m^*), \quad (24)$$

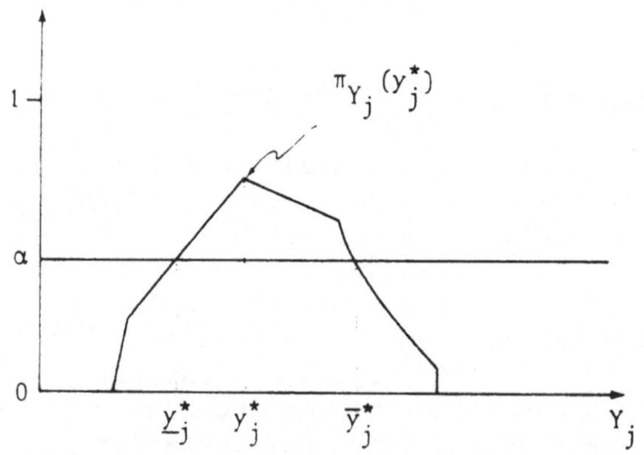


Fig. 4. Zero-in value of a fuzzy interval imposed by a composite possibility distribution function.

where $y_j^*$, $j = 1, \ldots, m$, is a zero-in value of criterion $y_j$ obtained via eq. 23. The compromise solution, denoted as $X^c = (y_1^c, \ldots, y_j^c, \ldots, y_m^c)$, with respect to a specific value of the parameter p is the one which minimizes

$$\left[ \sum_{j=1}^{m} (1 - \mu_{\textit{close to } y_j^*}(y_j^i))^p \right]^{1/p}, \quad 1 \leq p \leq \infty . \tag{25}$$

*Procedure 2.* Instead of applying a zero-in procedure to the composite linguistic distribution function, decision makers may choose the individual possibility distribution functions, eq. 4, as starters and zero in on the most compatible elements of the fuzzy intervals $[\underline{y}_j^i, \overline{y}_j^i]$, $i = 1, \ldots, n$, and $j = 1, \ldots, m$, of the matrix in eq. 6.

That is, with reference to alternative $X_i$, $i = 1, \ldots, n$, the value $y_j^{i*}$ is first selected such that

$$\pi_{Y_j}^i(y_j^{i*}) = \sup \pi_{Y_j}^i(y_j^i) . \tag{26}$$

Then, for criterion $Y_j$, $j = 1, \ldots, m$, the ideal value is obtained as

$$y_j^* = \max_i y_j^{i*} \tag{27}$$

Again, the fuzzy ideal becomes a point-value ideal similar to the one depicted in eq. 24, and the fuzzy compromise solutions are derived as point-value solutions through eq. 25.

Figure 5 depicts the compromise solution set with reference to a point-value ideal obtained through the above procedure.

Remark. Though procedures 1 and 2 attempt to force a point-value solution, there are no guarantees that such a solution exists under any circumstances. Depending on the possibility distribution function, the zero-in value $y_j^*$ can still be a fuzzy interval (see Fig. 6a, b). Unless an arbitrary value is chosen as the ideal value, point-value fuzzy ideal and compromise solutions cannot be obtained. Should a zero-in procedure be required, the employment of procedure 1 or 2 is on the decision-makers' discretion. In general, procedure 2 may lead to an ideal through which compromise solutions are difficult or impossible to obtain. For example, with respect to criterion $y_j$, alternative 1 may specify "$Y_j$ is *around* 100" ($P_1$), and alternative 2 may specify "$Y_j$ is *much greater than* 100" ($P_2$). (See Fig. 7). Employing eq. 26 and 27, the ideal value

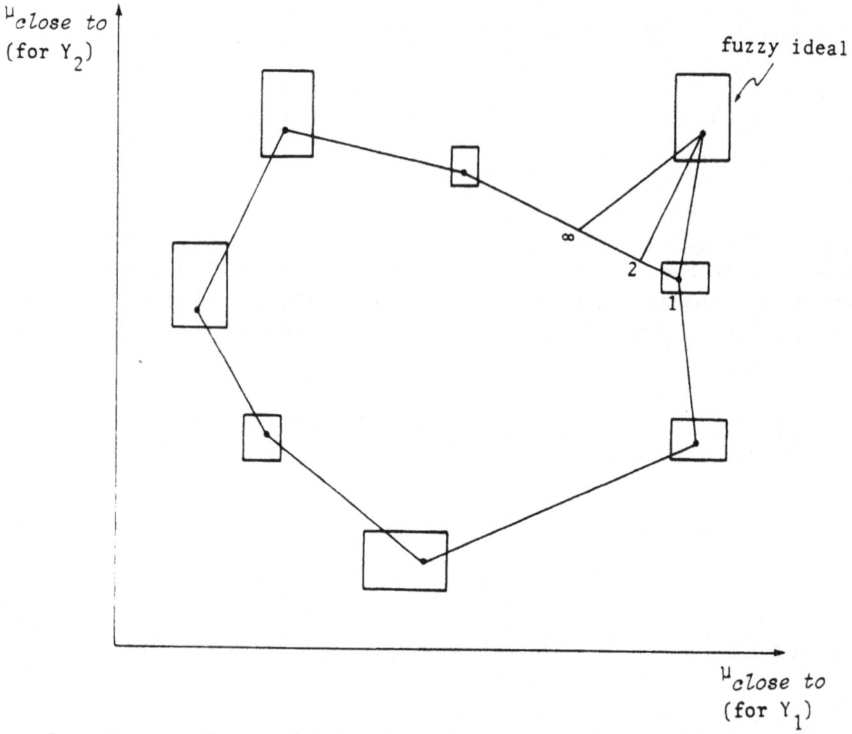


Fig. 5. Compromise solutions with respect to a zero-in point-value fuzzy ideal.

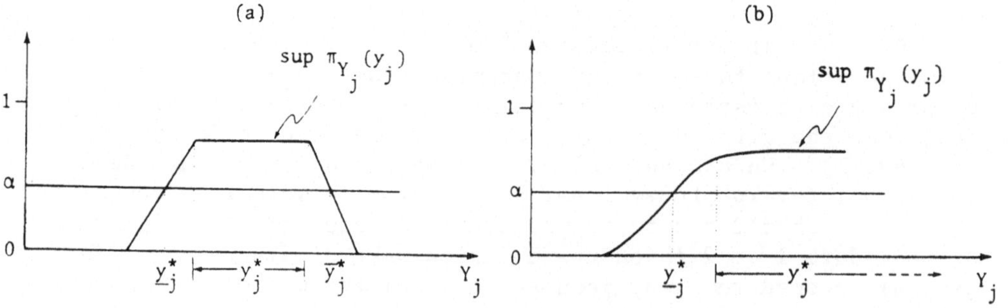


Fig. 6. Fuzzy intervals as zero-in values.

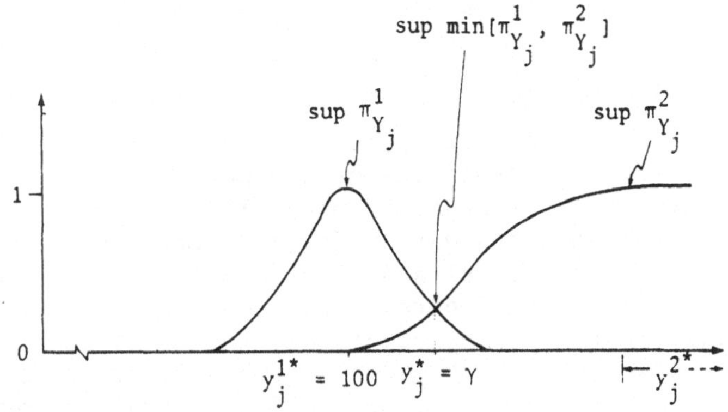


Fig. 7. The determination of the zero-in value via procedure 1 and 2.

may be determined as $y_j^* = \max(100, \infty) = \infty$. Thus, the condition specified by $P_1$ is completely violated. On the other hand, if procedure 2 is employed, the ideal value $\gamma$ is compatible to both $P_1$ and $P_2$ to a certain degree.

Once the most compatible element of the fuzzy ideal is determined, its associated compromise solution can be regarded as the most compatible alternative with respect to the fuzzy ideal. Should conflicts arise which cannot be resolved and a new compromise solution be required, we can zero in on the next most compatible element of the fuzzy ideal and derive the next most compatible compromise solution. Thus, employing the fuzzy ideal as a basis, with respect to a specific value of the parameter p, decision analysts can provide a set of compromise solutions ranging from the most compatible to the least compatible. Conflict resolution is then made more flexible and powerful.

## IV. CONCLUSION

A concept of a fuzzy ideal has been proposed to resolve multicriteria conflicts in a fuzzy environment. Under the present analytical framework, linguistic evaluation rules are employed to define a fuzzy ideal, and its numerical counterpart, a fuzzy region, is derived through the associated possibility distributions. Based on the fuzzy ideal, a collection of compromise solution sets can be derived. To determine the most compatible element within the fuzzy ideal, two zero-in procedures are proposed. The present analytical framework appears to be more flexible and powerful in resolving multicriteria conflicts on a static basis.

This study, however, has not proposed a distance minimization method by which the exact structure of the fuzzy region comprising all compromise solution sets can be determined. Such a generalization, should it be pertinent in practical conflict resolution, may be necessary in further research.

Since conflict resolution processes are ordinarily interactive and recursive, the conceptualization presented in this paper may not be sufficient in providing a dynamic resolution of conflicts. However, it serves as a foundation for the formalization of such an approach. In conventional analysis, displacement of an ideal has been considered as an important procedure to resolve conflicts in an interative manner (see for example [1, 3, 4, 12, 14, 23]). Based on the notion of a fuzzy ideal discussed here, a theory of a displaced fuzzy ideal in resolving conflicts involving multiple criteria and multiple decision-making units can be developed. This is a subject of analysis in a manuscript under preparation.

## V. ACKNOWLEDGEMENTS

The author would like to thank Mr. A. Cheung for discussions of the paper.

## VI. REFERENCES

1. J. P. Aubin and B. Näslund, An exterior branching algorithm, Working paper 72-74, European Institute for Advanced Studies in Management, Brussels (1972).
2. R. Bellman and L. A. Zadeh, Decision making in a fuzzy environment, Management Sci., 17B(4)(1970) 141-164.
3. R. Benayoun and J. Tergny, Critères multiples en programmation mathematique: une solutions dans la cas linéaire, R.A.I.R.O., 3(1969) 31-56.
4. R. Benayoun, J. de Montgolfier, J. Tergny, and O. Larichev, Linear programming with multiple objective functions: STEP method (STEM), Math. Programming, 1(1971) 366-375.
5. R. M. Capocelli and A. de Luca, Fuzzy sets and decision theory, Information and Control, 23(1973) 446-473.
6. Y. Y. M. Cheng and B. McInnis, An algorithm for multiple attribute, multiple alternative decision problems based on fuzzy sets with application to medical diagnosis, IEEE Trans. SMC, SMC-10(1980) 645-650.
7. W. Dinkelbach, Über einen lösungsansatz zum vektormaximumproblem, in: "Unternehmensforschung Heute," M. Beckman, ed., Springer-Verlag, Berlin (1971) 1-13.
8. J. Efstahiou and V. Rajkovic, Multiattribute decision making using a fuzzy heuristic approach, IEEE Trans. SMC, SMC-9 (1979) 326-333.

9. A. M. Geoffrion, A parametric programming solution to the vector maximum problem, with applications to decisions under uncertainty, Tech. Report No. 11, Operations Research Program, Standford University, CA(1965) 2.
10. M. M. Gupta, G. N. Saridis, and B. R. Gaines, eds., "Fuzzy Automata and Decision Processes," North-Holland, New York (1977).
11. J. Jüttler, Lineinaia model s Neskolkimi celovimi funkciami, Ekonomika i Matematicheskie Metody, 3(1967) 397-406.
12. Y. Leung, A value-based approach to conflict resolution involving multiple objectives and multiple decision-making units, in: "International and Regional Conflicts: Analytic Approaches," W. Isard and Y. Nagao, eds., Ballinger, Cambridge (1983).
13. W. Radzikowski, Die berücksichtigung mehrerer zielfunktionen bei aufgahen der linearen optimierung, Wirtschaftswissenschaft, 5(1967) 797-806.
14. B. Roy, Interactions et Compromis: la procédure du point de mine, Cahiers Belges de Recherche Opérationelle (1975).
15. J. Saska, Lineární multiprogramování, Ekonomicko-matematický Obzor, 4(1968) 359-373.
16. E. Takeda and T. Nishida, Multiple criteria decision problems with fuzzy domination structures, Fuzzy Sets and Syst., 3(1980) 123-136.
17. R. M. Tong and P. P. Bonissone, A linguistic approach to decision making with fuzzy sets, IEEE Trans. SMC, SMC-10 (1980) 716-722.
18. P. L. Yu, A class of solutions for group decision problems, Management Sci., 19(1973) 936-946.
19. L. A. Zadeh, Fuzzy sets, Information and Control, 8(1965) 338-353.
20. L. A. Zadeh, Fuzzy sets as a basis for a theory of possibility, Fuzzy Sets and Syst., 1(1978) 3-28.
21. L. A. Zadeh, K. S. Fu, K. Tanaka, and M. Shimura, eds., "Fuzzy Sets and their Applications to Cognitive and Decision Processes," Academic Press, New York (1975).
22. M. Zeleny, A concept of compromise solutions and the method of the displaced ideal, Computers and Operations Research, 1(1974) 479-496.
23. M. Zeleny, The theory of the displaced ideal, in: "Multiple Criteria Decision Making," M. Zeleny, ed., Springer-Verlag, Berlin (1976) 153-206.

APPENDIX

## FUZZY SET RESEARCH IN PEOPLE'S REPUBLIC OF CHINA

This Appendix is intended to supplement the information already provided in this volume by Professor Arnold Kaufmann.

While giving a seminar on fuzzy* sets in Beijing in the summer of 1974, Professor Arnold Kaufmann of France found only a small group of scientists attracted to the theory of fuzzy sets, but this group was very active and highly qualified. Returning in the spring of 1980 to give a series of lectures at Huazhong University of Science and Technology, several hundred high level researchers from all over the provinces of China came to hear his lectures on fuzzy set theory. Since the first visit of Professor Kaufmann in 1974, there has been a long pause of five years. However, numerous fuzzy set researchers such as M. Sugeno, K. S. Fu, Elie Sanchez, Paul P. Wang, etc. have visited the People's Republic of China and have lectured at various universities and research institutes.

A theory is a set of interrelated structures, definitions, basic axioms, propositions, and theorems. It is accurate to say that researchers from the People's Republic of China have contributed a great deal toward the perfection of the theory of fuzzy set. In fact, they, along with the United States of America, France, etc., are already among the leaders so far as the basic theory is concerned. In the meantime, many application papers (some of them quite unique) are beginning to appear in the public domain. While visiting this huge country, one soon realizes their enthusiasm toward fuzzy set theory to be commendable and their research programs to be quite effective. Their FUZZY MATHEMATICS journal (published by Huazhong University of Science and Technology in Wuhan) has drawn worldwide attention. Not only have the courses offered in various universities become a part of regular curriculi, but also many

---

*The task of translation of the word "Fuzzy" into Chinese language turns out to be a "fuzzy translation". The set of Chinese translations suggested up to now reads as follows: 弗晰，不分明，模糊，迷糊，不清楚，不明界限. As a result, the word "fuzzy" is the official new word in the Chinese dictionary.

research laboratories or centers have been established with the primary objective of conducting research on the theory and applications of fuzzy set theory.

It is rather difficult to list here all the institutions with active research on fuzzy sets, especially since this writer has visited only a limited number of them. For the sake of promoting the international exchange of ideas, we have prepared the following partial listing of principal investigators' addresses and their areas of interest.

(In making use of the following, "The People's Republic of China" should be added to all the addresses.)

Cai Mao-Hua
Inst. of Computing Technology
Academia Sinica
Beijing
Fuzzy logic

Cao Hong-xing
Academy of Meteorological Sci.
Central Meteorological Bureau
Beijing
Meteorological Statistics

Cao Zhi-giang
Institute of Automation
Academia Sinica
Beijing
Application of the fuzzy set theory to psychology and control systems

Cau Feng-xin
Climate Institute
Central Meteorology Bureau
Beijing
Application of the fuzzy set theory to meteorology

Chang Wei
Institute of Mathematics
Academy of Sci. of Henan Prov.
Zhenzhou
Applications of fuzzy mathematics

Chen Guo-xun
Department of Mathematics
Zhengzhou University
Zhengzhou
Fuzzy logic

Cheng Hua-cheng
Multi-logic & Fuzzy System Res. Sec.
Shanghai Inst. of Railway Technology
Shanghai
Fuzzy pattern recognition, Fuzzy filter

Chen Nianyi
Institute of Metallurgy
Academia Sinica
Shanghai
Fuzzy pattern recognition

Chen Tu-yun
Liaoning Teachers Institute
Dalian
Foundations

Chen Tin, Vice President
Huazhong Inst. of Technology
Wuhan, Hubei
Fuzzy control

Chen Xian-bo
Hangzhou Institute of Commerce
Hangzhou
Applications of fuzzy set theory to economical decision

Chen Chan-xian
Meteorological Institute
Ningxia Autonomous Region
Meteorological Bureau
Yinchuan
Meteorology

Chen Guo-fan
Academy of Meteorological Sci.
Central Meteorological Bureau
Beijing
Meteorological statistics

Chen Wei-yu
Systems Engineering Res. Inst.
Nanjing Inst. of Technology
Nanjing
Identification of fuzzy systems

Chu Shang-Yong
Dept. of Applied Mathematics
Shanghai Jiao Tong University
Shanghai
Fuzzy tree grammar and fuzzy forrest grammar

Chen Xiao-Jun
Xiang Traffic University
Xiang, Shensi
Fuzzy integral

Chen Yong-yi
Beijing Meteorological College
Bai Shi Qiao Lu No. 46
Beijing
Foundations, Cybernetics

Deng Ju-Rong
Huazhong Inst. of Technology
Wuhan, Hubei
Fuzzy control

Feng Jin-cheng
Department of Forestry
Nanjing Technological College
of Forest Products
Nanjing
Fuzzy pattern recognition and its applications

Gao Su-hua
Academy of Meteorological Sci.
Central Meteorological Bureau
Beijing
Agricultural meteorology

Geng Chun-Run
Department of Mathematics
Nanjing University
Nanjing
Fuzzy matrix

Gu Qi-jun
Department of Mathematics
Nanjing University
Nanjing
Information processing, Application of fuzzy mathematics

Guo Rung-Jiang
Institute of Automation
Academia Sinica
Beijing
Application of the fuzzy set theory to medical diagnosis

He San-yu
Institute of Automation
Academia Sinica
P.O. Box 2728
Beijing
Fuzzy control theory, System theory

Hern Zhong-Xung
Northern Traffic University
Beijing
The application of the fuzzy set theory to computer science

Hu Zhao-guang
Electrical Power Res. Institute
Qinghe, Beijing
Foundation of fuzzy set theory, Decision making analysis

Hu Shu-li
Department of Mathematics
Sichuan University
Chengdu, Sichuan
Fuzzy topology

Huang Jin-li
Department of Mathematics
Shanghai Teacher's College
Shanghai
Fuzzy control, Fuzzy integral

Huang Zhende
Huazhong Institute of Technology
Wuhan, Hubei
Fuzzy mathematics, Fuzzy pattern recognition

Jiang Jiquang
Department of Mathematics
Sichuan University
Cheng du, Sichian
Fuzzy topology

Kong You-kun
Academy of Meteorological Sci.
Central Meteorological Bureau
Beijing
Meteorological statistics

Li Bi-Xang
Huangshi Normal College
Wuhan, Hubei
Fuzzy relational equation

Li Zong-fu
Department of Mathematics
Sichuan University
Chengdu, Sichuan
Foundation of fuzzy set theor

Li Tai-Hang
Inst. of Computing Technology
Academia Sinica
Shanghai
Application of the fuzzy set theory to solve artificial intelligence problems

Li Xiang
Gui Zhow University, Gui Zhow
Fuzzy Logic

Li Zai-Fu
Wuhan Medical College
Wuhan, Hubei
Application of the fuzzy set theory to medicine

Liang Ji-hua
Department of Mathematics
Sichuan University
Chengdu, Sichuan
Fuzzy topology

Liao Qun
Electrical Power Res. Institute
Qinghe, Beijing
Fuzzy system, Fuzzy logic, Fuzzy linguastic model

Liao Zu-wei
Chinese Central Television Univ.
Beijing
Fuzzy matrix, Fuzzy relation equations, Fuzzy systems

Lin Yan-Wu
Department of Mathematics
Sichuan University
Chengdu, Sichuan
Fuzzy logic

Liu Dsosu
Department of Mathematics
Sichuan University
Chengdu, Sichuan
Fuzzy integrals and fuzzy probability theory

Liu Jun-jie
Department of Mathematics
Shanghai Teacher's College
Shanghai
Fuzzy control, Fuzzy system

Liu Lai-Fu
Department of Mathematics
Beijing Normal University
Beijing
Application of the fuzzy set theory to biology

Liu Xu-Huang
Jinlin University
Chang Chun, Jinlin
Application of the fuzzy set theory to computer science

Liu Ying-Ming
Department of Mathematics
Sichuan University
Chengdu, Sichuan
Fuzzy topology and fuzzy category

Liu Ying-sheng
Mathematics Teaching & Res. Sec.
Nanjing Institute of Posts and Telecommunications
Nanjing
Fuzzy mathematics

Liu Xu-hua
Department of Computer Science
Jilin University
Changchun City
Fuzzy logic

Liu Zhong-Kai
Agriculture Research Institute
Beijing
The application of the fuzzy set theory to agriculture

Lou Shih-Bo
Shanghai Railway Institute
Shanghai
Fuzzy controller, Fuzzy synthetic evaluation

Lu Cong-zhong
Shanxi Provincial Meteorological Bureau
Xian
Agricultural meteorology

Ma Hong
Department of Mathematics
Sichuan University
Chengdu, Sichuan
Fuzzy integral, measure and probability

Ma Mou-chao
Institute of Psychology
Academia Sinica
Beijing
Application of fuzzy set theory in psychology

Ou Yang-Mien
Department of Mathematics
Wuhan University
Wuhan, Hubei
Fuzzy mathematics

Pan Xiu-Hai
Department of Mathematics
Nanjing University
Nanjing
Fuzzy structures

Peng Yu-wei
Department of Mathematics
Sichuan University
Chengdu, Sichuan
Fuzzy Topology

Pu Bau-Ming
Department of Mathematics
Sichuan University
Chengdu, Sichuan
Fuzzy topology

Qi Zheng-Kai
Harbin Institute of Technology
Harbin
Fuzzy group

Ren Show-Ju
Department of Automation
Qinhua University
Beijing
Fuzzy control

Ro Cheng-Zhong
Department of Mathematics
Beijing Normal University
Beijing
Fuzzy relational equation

Shao Shou-Yee
Department of Automation
Shanghai Public Utilities
Research Center
Shanghai
Fuzzy filter, Fuzzy pattern recognition

Shen Rui-Ming
Electric Power Sci. Res. Inst.
Qinghe, Beijing
Fuzzy linear equation

Sheng Jia-rong
Changfeng Country Meteorological Station
Changfeng, Anhui
Weather forecasting

Shu Yongchang
Huazhong Institute of Technology
Wuhan, Hubei
Fuzzy pattern recognition

Song Da-he
North China Communication Univ.
Beijing
Fuzzy control, Fuzzy decision making

Sun Rong-guang
Department of Mathematics
Henan Normal University
Kaifeng, Henan
Fuzzy measure and integral

Tan Ying-cai
Southwestern National Institute
Fuzzy topology

Tiang Bao-zhi
Dept. of Control Engineering
Shanghai Inst. of Technology
Shanghai
Fuzzy control

Tong Zhengxiang
Shanghai Railway Institute
Shanghai
Multi-logic and fuzzy system

Wang Guo-Jun
Shangxi Normal College
Shangxi
Fuzzy topology

Wang Hong-fei
Institute of Mathematics
Academy of Sci. of Henan Prov.
Zhengzhou
Fuzzy relation equations

Wang Ming-Yi
Electric Power Sci. Res. Inst.
Qinghe, Beijing
Fuzzy category

Wang Pei-zhuang
Department of Mathematics
Beijing Normal University
Beijing
Fuzzy statistics, measure theory, falling random subsets, fuzzy controller

Wang Show Dao
Institute of Chemistry
Academia Sinica
Beijing
Application of the fuzzy set theory to chemistry

Wang Xue-sheng
Department of Mathematics
Xinxiang Teacher's College
Xinxiang, Henan
Fuzzy measure and integral

Wang Yizhi
Huazhong Institute of Technology
Wuhan, Hubei
Fuzzy pattern recognition

Wang Zheng-yuan
Department of Mathematics
Hebei University
Baoding, Hebei
Fuzzy measure, Fuzzy integral

Wang Zi-Xiao
Jinlin Normal University
Chang Chun, Jin Lin
Fuzzy measure and integral

Wu Cong-Xi
Harbin Institute of Technology
Harbin
Fuzzy linear algebra

Wu Da-xin
Huazhong Institute of Technology
Wuhan
Applied mathematics

Wu Quio-Feng
Department of Automation
Qinhua University
Beijing
Fuzzy controller

Wu Wang-ming
Department of Mathematics
Shanghai Teacher's College
Shanghai
Fuzzy control, Fuzzy group,
Fuzzy graph

Wu Xue-Mow
Wuhan 123 Box
Hubei
Pensystem logic and fuzziness

Wui Gong-Yi
Computing Center
Academia Sinica
Beijing
Fuzzy calculating mathematics

Xiang Ke-zong
Meteorological Institute
Shanxi Provincial Meteorological
Bureau
Taiyuan
Meteorology

Yan Jia-jia
Zhengzhou Inst. of Technology
Zhengzhou
Fuzzy relation equations

Yang Cai-Liang
Patent Office of China
Beijing
Fuzzy relation equations, Fuzzy
matrix, Control theory

Yang Ting
Institute of Automation
Academia Sinica
P.O. Box 2728
Beijing
Fuzzy control, System theory

Ying Xieng-ren
Institute of Automation
Academia Sinica
P.O. Box 2728
Beijing
Fuzzy controller

You Zhao-yong
Department of Mathematics
Xi'an Jiaotong University
Si'an
Fuzzy algorithms

Yu Kang-yuan
Beijing Meteorological College
Baishiqiaolu No. 46
Beijing
Application to climate

Yuan Meng
Electric Power Sci. Res. Inst.
Qinghe, Beijing
Fuzzy mathematics, Fuzzy functions,
Fuzzy analysis

Zhang Jinwen
Inst. of Computing Technology
Academia Sinica
Beijing
Fuzzy set structure, Axiomatic system of set theory, System logic
Application of system logic to artificial intelligence

Zhang Ming-Yi
Scientific Univ. of Guizhow
Guizhow
Pensystems and fuzziness

Zhang Nanlun
Department of Automation
Wuhan Inst. of Building Materials
Wuhan, Hubei
Theoretical basis of fuzzy set

Zheng Ren-sheng
Department of Mathematics
Shanghai Teacher's College
Shanghai
Fuzzy control, Fuzzy system

Zhang Weng-qian
Academy of Meteorological Sci.
Central Meteorological Bureau
Beijing
Meteorological statistics

Zhang Wen-xiu
Department of Mathematics
Xi'an Jiaotong University
Xi'an
Possibility theory

Zhao Bang-jie
Meteorology Inst. of Chengdu
Chengdu
Application of fuzzy set theory

Zhao Qin-ping
603 Dept. of Computer Science
Beijing Inst. of Aeronautics & Astronautics
Beijing
Fuzzy logic, Fuzzy automatic and semantics

Zhao Lu-Huai
Xiang Traffic University
Xiang
Fuzzy integral

Zheng Dao-peng
Department of Mathematics
Shanghai Teacher's College
Shanghai
Fuzzy measure and integral

Zheng Dau-Ming
Shanghai Normal College
Shanghai
Fuzzy integral

Zheng Wei-Min
Department of Automation
Qinhua University
Beijing
Fuzzy control

Zong Rong-xiang
Baotou Meteorological Bureau
Baotou, Nei Monggol
Weather forecasting

Zhou Haoxuen
Department of Mathematics
Sichuan University
Chengdu, Sichuan
Fuzzy topology

Zhou Kai-qi
Dalian Sea Transportations Inst.
Dalian
Foundations

Zhou Shi-wu
Industry School of Chengdu
Chengdu
Fuzzy topology

Yang Ke-Xuang
Electric Power Sci. Res. Inst.
Qinghe
Fuzzy reasoning

## AUTHOR INDEX

SUBJECT INDEX